T0189122

DNS of Wall-Bounded Turbulent Flows

Tapan K. Sengupta · Swagata Bhaumik

DNS of Wall-Bounded Turbulent Flows

A First Principle Approach

 Springer

Tapan K. Sengupta
Department of Aerospace Engineering
Indian Institute of Technology Kanpur
Kanpur, Uttar Pradesh
India

Swagata Bhaumik
Department of Mechanical Engineering
Indian Institute of Technology Jammu
Jammu, Jammu & Kashmir
India

ISBN 978-981-13-4315-5   ISBN 978-981-13-0038-7   (eBook)
https://doi.org/10.1007/978-981-13-0038-7

Printed on acid-free paper

This Springer imprint is published by the registered company Springer Nature Singapore Pte Ltd. part of Springer Nature
The registered company address is: 152 Beach Road, #21-01/04 Gateway East, Singapore 189721, Singapore

Preface

There are many books and monographs on instability and transition, including one by one of the authors of this monograph. Despite these books, this monograph is written with a specific purpose of presenting a detailed account of what constitutes direct numerical simulation (DNS) of flow from receptivity stage to fully developed turbulent flow over flat plate, while simulating the classical transition experiments for zero pressure gradient boundary layer. This has been made possible with the work of the authors' colleagues and students, proposing new theoretical and computational results, which establish a deterministic route to turbulence, as in the classical experimental efforts [3, 5, 7].

The Navier–Stokes equation governs fluid flows and Reynolds explained that the *exact* solution of the Navier–Stokes equation, even when it exists, is unable to maintain its stability with respect to omnipresent small disturbances in the flow. An equilibrium solution ensures satisfaction of conservation principles of mass, momentum, and energy balance, yet the flow in the famous pipe flow experiment of Reynolds disintegrated into sinuous motion, eventually leading to turbulent flow. This onset has been attributed to the instability of flows ever since. However, even today the transition to turbulence in pipe flow is not completely understood.

Another canonical problem which attracted the attention of researchers in the beginning is the flow over a flat plate. In studying instability of this flow field, Rayleigh developed his stability equation, but it was not amenable to study convective disturbances, which is the signature of transition for zero pressure gradient boundary layer flow. By the beginning of twentieth century, it became evident that the instability problem is intractable without the inclusion of the effects of second derivative of velocity profile and viscous actions. These latter effects have been termed as resistive instability, which is in contradiction to the assumption made by early pioneers, who mistakenly considered the action of viscous forces to attenuate disturbances and was considered not central to the study of instability.

This was the reason for Orr and Sommerfeld to independently come up with the famous Orr–Sommerfeld equation, which was investigated by Tietjens [14] and Heisenberg [4]. However, it was the definitive attempt by Tollmien that paved the understanding the concept of critical Reynolds number [15]. Following this lead, Schlichting also studied the instability of zero pressure gradient boundary layer, while making some implicit assumption which connected the temporal and spatial growth of disturbances, which has since been addressed by invoking the concept of group velocity [8]. One of the central themes of Tollmien and Schlichting's work was to show the presence of growing waves in limited part of the flow, and this is now called the Tollmien–Schlichting (TS) waves. This eigenvalue analysis is performed by neglecting the growth of boundary layer. We would like to mention that studying flow instability by temporal and spatial theory and artificially patching the two together is fraught with danger. This has been demonstrated by the authors and colleagues that there is really no need to make this artificial demarcation into spatial and temporal routes. The best course of action is to adopt the spatio-temporal route to avoid this ambiguity. While studying instability of mixed convection boundary layer, authors came across a perplexing situation, when a singularly cooled plate demonstrated simultaneous presence of temporal and spatial instability [10]. The problem was resolved because of the availability of DNS results obtained by high-accuracy method, which showed the flow to follow primarily temporal growth, and not the spatial route (the growth should be strictly stated to be via spatio-temporal route). This emphasizes the need to perform spatio-temporal analysis, which has been practiced and yielded new insights, which form the contents of this treatise.

However, despite Taylor demonstrating the strength of linear stability of flow inside concentric cylinder [12], TS wave could not be detected experimentally. This created doubts about the existence of TS wave and, in turn, on the relevance of viscous linear theory. This problem was resolved by Dryden's group, by demonstrating for the first time the existence of TS wave by vibrating a ribbon inside the boundary layer, with the results announced after the Second World War [3, 5]. This experimental approach may be termed as frequency response of the fluid dynamical system. It would be appropriate to note that exciting the flow at any frequency will not lead to flow instability, as the vibrating diaphragm experiment of Taylor [13] at 2 Hz demonstrated. The frequencies to be excited were present in the works of Tollmien and Schlichting [8, 15], but Taylor's experimental vibration was at too low a frequency [13]. This aspect of how to experimentally study eigenvalue problem is not clearly understood. In a recent work, researchers have explained the relationship between frequency response and instability for an associated problem of bifurcation [6].

Viewing fluid flows as dynamical systems, it is thus unwise to ignore the receptivity aspect. Receptivity was emphasized by Schubauer and Skramstad in reporting their classic experimental results [7]. Thereafter, reported TS waves as early marker of flow instability was readily embraced by the research community to conclude prematurely that TS waves are the cause for flow transition. It is worthwhile remembering that vortical excitation validated instability theory, while the acoustic excitation did not! Also, instability theory does not require any specific excitation, except in prescribing the qualitative nature of boundary conditions. The main feature of this monograph is to relate receptivity with flow instability and show how different routes of excitation lead to different types of disturbance evolution. This has been achieved here primarily by DNS to explain theoretical aspects of the flow.

Viscous instability results also suffered credibility due to the use of a parallel flow assumption for the equilibrium flow, along with linearization. This criticism was partly silenced by experimental verification provided [7], but whether TS waves can cause transition or not was not known for a long time, till the definitive routes of transition were identified for 2D zero pressure gradient flow [9]. It has been shown by the present authors that when the Navier–Stokes equation is computed in 2D framework to mimic the experiment [7] for moderate- to high-frequency wall excitation, created TS wave packet remains passive, while the spatio-temporal wave-front (STWF) causes transition. The STWF is the first wave front that is created upon starting off the excitation, as in an experiment, which has the property to regenerate other STWF in its wake [9]. It is interesting to note that STWF was originally proposed in research on electromagnetic wave propagation [2]. In fluid dynamics, the major success in tracking STWF from the spatio-temporal dynamics of Orr–Sommerfeld equation was reported [11], following the developed technique of Bromwich contour integral method. While these have been obtained for the frequency response of disturbance evolution, corresponding role of STWF during impulse response has been reported only recently [1], which also shows that the same physical mechanism of STWF explains the features of geophysical phenomena like tsunami and rogue waves.

This book has evolved into an account of the research interests of the authors over the years. Efforts are made to keep the treatment at an elementary level requiring rudimentary knowledge of calculus, Laplace–Fourier transform, and complex analysis, which should be equally amenable to undergraduate students, as well as serious researchers in the field of hydrodynamics and mixed convection. This monograph shows to readers that without good computing, this subject will be poorer in linking spatio-temporal growth, instability at low frequencies, and the actual physical mechanisms of transition. In providing computational results from receptivity to a fully developed turbulent stage of 2D and 3D disturbance flows, we also definitely provide the basis of experimental approach for transition to turbulence. In doing so, we state that turbulence is deterministic in its origin, as it is implicitly assumed in transition experiments [3, 5, 7]. Trained with high-accuracy computing methods, users of this monograph will be able to further contribute in this rich field of nonlinear dynamics.

Thus, the emphasis of the monograph is on DNS of transitional and turbulent flows, basis of which are laid out in Chap. 2. The applications of DNS technique are used to explain receptivity and instability in hydrodynamics and mixed convection flows in Chap. 3. The power of DNS is demonstrated by using results to explain vortex-induced instability in a nonlinear framework to explain receptivity and instability in Chap. 4. This is done by developed disturbance enstrophy transport equation [16, 17]. This provides a nonlinear framework based on Navier–Stokes equation without making any assumption on the equilibrium flow. Another nonlinear framework to study instability in this chapter is by using enstrophy-based proper orthogonal decomposition. The last two chapters are to demonstrate the power and accuracy presented by the DNS techniques described in earlier chapters to show that STWF is the precursors of transition to turbulence, for both 2D and 3D disturbance flow simulations. The resulting turbulent flows are as have been created experimentally [3, 5, 7]. The transition to turbulence is shown to be experimentally obtained for the integral properties and presented in textbooks. In closing this discussion, we note that the subject has matured very rapidly in recent times with the advent of very high-accuracy methods of computing, which will lead to an explosive growth of activities in the subject field. The contents of the monograph can be adopted as text for a high-level course on DNS and transition to turbulence.

The contents of this monograph are based on the doctoral thesis work of one of the authors and the other graduate students who have been associated with the first author over the last three decades. The contents have been specifically enriched by faculty colleagues in NUS, Singapore, specifically by Profs. Y. T. Chew, T. T. Lim, B. C. Khoo, K. S. Yeo, and S. Chang and Prof. Pierre Sagaut (UPMC), Prof. Julio Soria (Monash University), Prof. W. Schneider (TUW, Vienna), Prof. K. R. Sreenivasan (ICTP and NYU), Prof. M. Klocker (University of Stuttgart), Dr. J. M. Kendall (Jet Propulsion Lab, USA), Dr. B. R. Noack (Pprime, University of Poitier, France), Prof. M. Deville and Prof. F. Gallaire (EPFL, Switzerland), Prof. S. Girimaji (TAMU, USA), Prof. P. J. Strykowski (University of Minnesota), and Prof. A. Tumin (University of Arizona). Much of our work on receptivity has been influenced by the works of Prof. H. Fasel (University of Arizona). The first author acknowledges the influence of Prof. M. Gaster and Prof. D. G. Crighton (University of Cambridge) for encouraging him to develop a theory of receptivity. Many of the contributors of this book are now faculty colleagues in different parts. It is our pleasant duty to acknowledge the contributions by Sandeep Nijhawan, Manish Ballav, A. P. Sinha, Vivek Rana, Manoj T. Nair, K. Venkatasubbaiah, A. K. Rao, Manojit Chattopadhyay, Z. Y. Wang, A. Dipankar, S. Sarkar, S. De, Yogesh Bhumkar, Vijay Vedula, Manoj Rajpoot, S. Unnikrishnan, R. Bose, N. A. Sreejith, Ashish Bhole, and Soumyo Sengupta. In a very recent development, the authors have been looking at "exact" nonlinear theories of instability for incompressible flows, based on disturbance mechanical energy and disturbance enstrophy, with the former covered in the text here. However, we refrain from providing the details on the theory based on disturbance enstrophy and is currently being probed for different types of flow fields and would be described elsewhere. The authors acknowledge Aditi Sengupta and V. K. Suman for helping with the development of this instability theory based on disturbance enstrophy [17].

The authors would like to acknowledge the competent help provided in typing this material by Mrs. Baby Gaur and Mrs. Shashi Shukla. Specifically, Mrs. Gaur takes exceptional care in typing the text and preparing figures at all times. We also acknowledge various helps provided to us by Mr. Mukesh Kumar.

Kanpur, India   Tapan K. Sengupta
Jammu & Kashmir, India   Swagata Bhaumik

References

1. Bhaumik, S., & Sengupta, T. K. (2017). Impulse response and spatio-temporal wave-packets: The common feature of rogue waves, tsunami and transition to turbulence. *Physics of Fluids, 29*, 124103.
2. Brillouin, L. (1960). *Wave propagation and group velocity.* New York: Academic Press.
3. Dryden, H. L. (1955). Fifty years of boundary-layer theory and experiment. *Science, 121* (3142), 375–380.
4. Heisenberg, W. (1924). Üeber stabilität und turbulenz von flüssigkeitsströmen. *Annalen der Physik, 74*, 577–627.
5. Klebanoff, P. S., Tidstrom, K. D., & Sargent, L. M. (1962). The three-dimensional nature of boundary-layer instability. *Journal of Fluid Mechanics, 12*, 1–34.
6. Lestandi, L., Bhaumik, S., Avatar, G. R. K. C., Mejdi, A., & Sengupta, T. K. (2018). Multiple Hopf bifurcations and flow dynamics inside a 2D singular lid driven cavity. *Computers & Fluids.*
7. Schubauer, G. B., & Skramstad, H. K. (1947). Laminar boundary layer oscillations and the stability of laminar flow. *Journal of the Aeronautical Sciences, 14*(2), 69–78.
8. Schlichting, H. (1933). Zur entstehung der turbulenz bei der plattenströmung. *Nachrichten von der Gesellschaft der Wissenschaften zu Göttingen, Mathematisch-Physikalische Klasse,* 181–208.
9. Sengupta, T. K., & Bhaumik, S. (2011). Onset of turbulence from the receptivity stage of fluid flows. *Physical Review Letters, 107*(15), 154501.
10. Sengupta, T. K., Bhaumik, S., & Bose, R. (2013). Direct numerical simulation of transitional mixed convection flows: Viscous and inviscid instability mechanisms. *Physics of Fluids, 25,* 094102.
11. Sengupta, T. K., Rao, A. K., & Venkatasubbaiah, K. (2006). Spatio–temporal growing wave fronts in spatially stable boundary layers. *Physical Review Letters, 96*(22), 224504.
12. Taylor, G. I. (1923). Stability of a viscous liquid contained between two rotating cylinders. *Philosophical Transactions of the Royal Society (London), A223,* 289–343.
13. Taylor, G. I. (1939). Some recent developments in the study of turbulence. In J. P. Den Hartog & H. Peters (Eds.), *Proceedings of the Vth International Conference on Applied Mechanics.*
14. Tietjens, O. (1925). Beiträge zur enstehung der turbulenz. *ZAMM, 5,* 200–217.
15. Tollmien, W. (1931). Über die enstehung der turbulenz. I, English translation. *NACA TM 609.*
16. Sengupta, T. K., Sharma, N. & Sengupta, A. (2018). Non-linear instability analysis of the two-dimensional Navier-Stokes equation: The Taylor-Green vortex problem. *Physics of Fluids* (Accepted).
17. Sengupta, A., Suman, V. K., Sengupta, T. K. & Bhaumik, S. (2018). An enstrophy-based linear and non-linear receptivity theory. *Physics of Fluids* (Accepted).

Contents

Symbols Description

Chapter 1

f	Function in physical plane
F	Bilateral Laplace transform of f
k^*	Wavenumber
p	Pressure
Re	Reynolds number
Re_{cr}	Critical Reynolds number
Re_{tr}	Transitional Reynolds number
T	Temperature
U_1, U_2	Velocity in Kelvin-Helmholtz instability problem
U_j	Streamwise velocity
v	Specific volume
x, y, t	External streamline fixed coordinates
z_s	Interface displacement
α	Wavenumber in x-direction
β	Wavenumber in y-direction
δ^*	Displacement thickness
ε	Small parameter
$\eta, \hat{\eta}$	Interface displacements
γ	Wavenumber vector inclination with x-axis
ν	Kinematic viscosity
$\bar{\omega}$	Circular frequency
ϕ_j	Velocity potentials
ρ_1, ρ_2	Density in Kelvin-Helmholtz instability problem

Chapter 2

a_1, a_2	Wave amplitude
A_j	Quotient for RK method
c	physical phase speed
c_N	Numerical phase speed
D_ω	Divergence of vorticity
D_v	Divergence of velocity field
e	Internal energy per unit mass
$E(k, \varepsilon)$	Energy spectrum
F_b	Turbulence burst frequency
G	Spectral amplification factor
H	Shape factor
h	Grid-spacing
h_1, h_2	Scale factors of transformation
k	Complex wavenumber
k_c	Cutoff wavenumber
l	Landau coefficient
N_c	CFL number
p_m	Mechanical pressure
Re	Reynolds number
TF	Transfer function
u_j	Velocity components
u, v, w	Cartesian velocity components
U_c	Convection velocity
V_g	Group velocity
V_{gN}	Numerical group velocity
x_j	Coordinates
β_0	Fixed frequency
β_1, β_2	Frequency bandwidth of excitation
δ^*	Displacement thickness
δ_{ij}	Kronecker delta function
η_K	Kolmogorov length scale
Γ-form	Error analysis by space-time discretization
κ	Thermal conductivity
λ	Bulk viscosity
$\bar{\lambda}$	Wavelength
μ	Molecular viscosity
∇	Gradient operator
$\bar{\omega}$	Circular frequency
Ω_f	Frequency ranges in temporal scale
Φ_0	Dissipation function
Π-form	Error analysis by spatial discretization
Ψ	Stream function

ρ	Density
Σ	Mean strain
σ	Stress
τ_{ij}	Stress tensor
$\vec{V}$	Velocity vector
ξ, η	Transformed coordinates

Chapter 3

A, A_m	Amplitude of wall excitation
Br	Bromwich contour
c^*	Complex conjugate of phase speed
c_i	Temporal growth rates
c_r	Phase speed, real part
Gr	Grashof number
F	Excitation frequency (viscous scale)
$H(t)$	Heaviside function
K	Buoyancy parameter
K_i, K_a	Isothermal and adiabatic, buoyancy parameter
L	Length
Pr	Prandtl number
q	Instantaneous quantity
Ri	Richardson number
v_d	Wall-normal excitation velocity
U_c	Convective velocity
u_d	Streamwise disturbance velocity
U_e	Boundary layer edge velocity
$U_e(X)$	Non-uniform edge velocity
ω_d	Disturbance vorticity
w_{ex}	Exciter width
X	Streamwise distance
α_1	Amplitude control parameter of wall exciter
ω_{0j}	Circular frequency
α_r, α_i	Wavenumber, real and imaginary part
β	Complex frequency
β_0	Fixed frequency
$(\beta)_{cr}$	Critical circular frequency
Re_{δ^*}	Reynolds number based on displacement thickness
$(Re_\delta^*)_{cr}$	Critical Reynolds number
$\delta(\tilde{x})$	Dirac-delta function
δ^*	Displacement thickness
δT_L	Temperature scale
ΔT	Temperature scale
Γ	Contour

Γ_F	Fjortoft integral, Theorem II
∇^2	Laplacian operator
$\omega_0, \bar{\omega}_0$	Circular frequency
ϕ	Amplitude of v-velocity
Φ_v	Viscous dissipation
ψ, ω	Stream function, vorticity
ρU_∞^2	Pressure scale
θ	Momentum thickness
Θ	Temperature
Θ'	Wall-normal derivative of temperature
Θ_w	Wall temperature
U_∞, T_∞	Free stream temperature, velocity denoted
ξ, ζ	Transformed orthogonal coordinate

Chapter 4

a_m, a_j	Amplitude functions
d	Core diameter of convecting vortex
E	Total mechanical energy
E_d	Disturbance mechanical energy
H	Constant height of convecting vortex
R	Regular POD modes
Re_δ^*	Reynolds number
R_{ij}	Cross-correlation
u, v, w	Velocity components
u', v'	Disturbance velocity components
U_c	Convective speed
u_d	Disturbance velocity components
$\vec{V}, \omega$	Velocity–vorticity vectors
$(\beta)_L$	Circular frequency
β_x, β_y	Grid stretching parameter
ΔT	Non-dimensional frequency
Γ	Convecting vortex strength
ε_j	Normalization factor in POD
ε	Dissipation function
v	Kinematic viscosity
$\hat{\Omega}$	Average enstrophy
$\Omega_1, \Omega_2, \Omega_3$	Higher moments of enstrophy
Ω_d	Disturbance enstrophy
ϕ_i	Deterministic vectors for POD
ϕ_j	POD eigenfunctions
ψ, ω	Stream function, vorticity
ξ, ζ	Transformed plane coordinates

Chapter 5

F	Non-dimensional excitation frequency
k	Wavenumber
k_m	Dominant wavenumber
m	Falkner–Skan pressure gradient parameter
u	Streamwise component of velocity
U_c	Convective speed of disturbance at outflow
u_d	Streamwise disturbance velocity amplitude
u_{dm}	Maximum amplitude of u_d
α_1	Amplitude control function of excitation
f'	Similarity functions
ψ, ω	Stream function and vorticity
ψ_{wp}	Perturbation stream function
U_∞	Free stream speed

Chapter 6

C_f	Skin friction coefficient
D_v	Divergence of velocity
E_d	Disturbance mechanical energy
F_f	Non-dimensional frequency
h_1, h_2, h_3	Scale factors of transformation
k_x, k_1	Streamwise wavenumber
x_1, x_2	Wall-excitation stretch
u_d, v_d, w_d	Disturbance velocity components
β_x, β_y	Grid stretching parameters
δ^* and θ	Displacement and momentum thickness
λ_2	Second eigenvalue of the velocity strain matrix
$\overrightarrow{V}$	Velocity vector
$\overrightarrow{\Omega}$	Vorticity vector
Ω_ξ, Ω_η and Ω_ζ	Vorticity components
(ξ, η, ζ)	Transformed plane coordinates

List of Figures

List of Tables

Chapter 1
DNS of Wall-Bounded Turbulent Flow: An Introduction

1.1 Introduction

This book covers the topic of direct numerical simulation (DNS) of wall-bounded turbulent flow by first principle approach. It is mandatory to track the flow from laminar state to fully developed turbulent state, and which has been solved very recently for specific cases. While one can attempt DNS of Navier-Stokes equation (NSE) for fully developed turbulent flow, there would be ambiguity about specifying initial and boundary conditions. Moreover, such an exercise even if it is successful, then also it may not answer most of the fundamental questions related to the evolution of the flow field. First, by erroneous numerical methods and artificial excitations, one may achieve fully developed turbulence, but that approach should be avoided. Such simulations would not enlighten one in following the physical processes during transition. A detailed review and critique of some of the methods used for DNS and large eddy simulation (LES) of transitional and turbulent flows is given in [24]. This is of fundamental relevance for the contents of this book. Here, we present a more fundamental approach that studies the topic by considering the flow evolution from laminar to fully developed state by a thoroughly analyzed numerical method. The aim is to computationally reproduce the physical phenomena as recorded in classical transition experiments by Schubauer and Skramstad [19] for 2D transition and Klebanoff et al. [10] for the 3D routes. These experiments were designed after lots of soul searching in the fluid dynamic community, when experimentalists failed to detect some wave-solutions predicted for instability studies by Heisenberg, Tollmien and Schlichting. The reader is directed to the introductory discussion in [23] about the historic development of the subject of instability studies and a brief account will be provided later.

The qualitative features of stability of zero pressure gradient flow over a flat plate has been studied in many complimentary ways, after the governing viscous fluid flow equations were written down in the form of NSE in the first half of the nineteenth century. The correctness of this equation is now well accepted in the continuum regime of fluid flow. Apart from the fact that correct governing equations

© Springer Nature Singapore Pte Ltd. 2019
T. K. Sengupta and S. Bhaumik, *DNS of Wall-Bounded Turbulent Flows*,
https://doi.org/10.1007/978-981-13-0038-7_1

are needed, one has also to ensure that correct boundary and initial conditions are used in the solution process. One notes that the no-slip boundary condition is a modeling approximation and has never been proven rigorously. Batchelor [2] has noted for Newtonian fluid flow that *the absence of slip at a rigid wall is now amply confirmed by direct observation and by the correctness of its many consequences under normal conditions.* Despite this, we do not question the correctness of the no-slip condition for continuum flows.

However, it became immediately apparent to Stokes that the closed-form solution of NSE for circular pipe flow did not match with the experimental observation. This led all the leading fluid dynamicists of the day to understand that mere existence and uniqueness of the governing equation is not adequate. It is equally important that such solutions should be stable, if the solution is to represent a physical flow condition. This is due to the fact that such physical fluid dynamical systems are always visited upon by omnipresent background disturbances. It is often likely that such disturbances will destabilize the nominal solution, as obtained by solving the governing equation. This nominal solution is called the equilibrium solution, as it is obtained from the solution of the governing equation describing the dynamic equilibrium of participating forces and moments. Obtaining the equilibrium solution and testing it for its instability is central to most fluid dynamical studies for practical systems [23]. The instability is related to growth of background disturbances, which eventually is responsible for creation of turbulence.

To present the state of art on the subject topic, we focus upon the most fundamental canonical problem, namely transition of fluid flow over a flat plate subjected to zero pressure gradient. This zero pressure gradient flow can be simplified by boundary layer assumption into an ordinary differential equation by similarity transformation, as was performed by Blasius and the resultant velocity is given by the Blasius profile [22]. In the present book, equilibrium flow will be obtained from the solution of NSE, and not from the approximate boundary layer solution.

1.2   Why Deterministic Study Is More Relevant than Stochastic Approaches?

The reason for numerically simulating deterministic experiments here, is on purpose. In real life situation, the flow becomes turbulent due to stochastic inputs, which may be difficult to quantify. Although some investigators have attempted using random noise or trip to trigger transition, results of such simulations do not unambiguously show route(s) to turbulence [18, 20, 33]. It is worthwhile noting that no proof exists that turbulence is unique and hence making a flow turbulent by artificial numerical means may simulate a broadband spectrum flow field, but whether it is the right one is not certain. Also in many practical applications, the nature and extent of transitional flow matters and knowledge about such flows are essential. For example, in turbomachineries one notices a very large streamwise extent of transitional flows in the turbine stages. To calculate fluid dynamic and thermal loads for such devices, one needs exact knowledge of the flow field. This point of view is espoused in the presentation here.

To understand transitional and turbulent flows, it is therefore paramount to understand the flow physics accurately, by the applications of computational tools. Also viewing the flow to be a dynamical system, where one can correlate causes with effects; this helps us develop understanding of transitional and turbulent flows. In this context, one appreciates the role of *kernel* experiments in exploratory studies of fluid flow [31]. It is in this spirit, experiments at NBS helped the fluid dynamic community understand the role of instability theory via the experiments in [10, 19] for the canonical zero pressure gradient boundary layer.

For example, all earlier studies viewed it as a problem of linear hydrodynamic instability. As the background disturbances are not always distinguishable due to the infinitesimally small amplitudes, it is quite natural to study the instabilities by linearizing the disturbance equations, and solving the resultant eigenvalue problem. The disturbance equations are often obtained from the governing equation, by splitting the flow variable into equilibrium quantity and a disturbance component. Formal application of perturbation theory to derive this equation is well described in [6, 23]. Despite this, in the presented approach, we will avoid linearization of the governing equation and study fully the nonlinear instability and receptivity of the flow.

1.2.1   Historic Developments

At the beginning, leading scientists of the day in fluid dynamics, like Rayleigh, Kelvin and Helmholtz proposed that one can study the onset of turbulence problem by studying disturbance growth by considering the disturbance field to be inviscid, even if the equilibrium state whose instability has to be studied, is obtained by including viscous term(s) in governing equations. This was justified from the point of view that the action of viscosity is essentially dissipative and will attenuate the disturbance. In contrast, the inviscid equation will provide more critical estimate of flow instability via disturbance growth. Instead of providing a quantitative measure of instability of flows (whose growth in the streamwise direction was neglected), Rayleigh enunciated a theorem which provided a necessary condition (and not sufficient!) for instability to be that the velocity profile must have an inflection point inside the viscous region. This is known as the Rayleigh's inflection point theorem for the growth of disturbance in time or temporal instability.

However at the onset of such an attempt for the zero pressure gradient flow over flat plate, it was realized that this flow does not suffer temporal instability, as the corresponding velocity profile does not have an inflection point. This failure of temporal inviscid instability led researchers to look for growing disturbance in space, a natural extension of the temporal theory. Also for this canonical flow, it was noted that the growing disturbances convect in space. However, it is easy to show that Rayleigh's stability equation does not account for spatial inviscid growth of disturbances.

This necessitated inclusion of viscous terms in disturbance equation, which was proposed by Orr and Sommerfeld [16, 32] independently, and is now known as the Orr-Sommerfeld equation. It is now becoming more and more apparent, that the action of viscous diffusion is not merely attenuating the solution, it can by itself contribute to instability, as noted in combustion. Moreover, viscous actions can also add to phase shift of components and under suitable conditions, such phase shift can make the disturbance grow. Recently a theory of instability is being proposed which is based on enstrophy. In many work on turbulence, researchers have identified enstrophy with dissipation, based on work for homogeneous periodic flow.

It was Heisenberg [7] under the guidance of Sommerfeld, attempted to solve the Orr-Sommerfeld equation for complex wavenumber (for variation in the streamwise direction) as the eigenvalue, for prescribed Reynolds number (based on displacement thickness of the equilibrium boundary layer) and real frequency of excitation. Subsequently, Tollmien [34] and Schlichting [21], under the guidance of Prandtl, also solved this eigenvalue problem analytically. All of these researchers predicted the instability via growing wave-like solutions in the streamwise direction. This is now known as Tollmien-Schlichting (TS) *wave* and has been the pacemaker of instability research for decades. We emphasize that the TS *wave* has been predicted for an equilibrium flow, which does not grow in streamwise direction. This may hold true for fully developed channel flow, but is strictly not correct for boundary layers. Right from the beginning, it was assumed that the analysis can be performed locally for boundary layers, with the current boundary layer thickness considered to be held constant in the streamwise direction, for the purpose of solving the Orr-Sommerfeld equation. In an actual growing boundary layer, obtained eigenvalue will continually change for the complex wavenumber. The real part of this complex wavenumber indicates the wavelength of the TS *wave*, while the imaginary part indicates growth or attenuation rate of the TS *wave*. Thus, in conjunction with the local parallel flow assumption, the TS *wave* solution would indicate a wave-packet, instead of a wave, whose streamwise extent is determined by the imaginary part of the wavenumber. This approach did not find immediate acceptance, as it was difficult to detect TS *wave* experimentally. This situation prevailed till the experimental reporting of TS *wave* by Schubauer and Skramstad [19] for 2D zero pressure gradient boundary layer representing the equilibrium flow. The success of this experiment, in the light of continuing failure by others, was related to shifting of attention from natural transition to forced excitation of the equilibrium flow by vibrating a ribbon inside the boundary layer. To accentuate the relationship of cause and effect, a tunnel was designed for extremely negligible background disturbances and then providing deterministic excitation at a fixed frequency, as was the case of instability studies. Whereas the natural disturbances are not monochromatic disturbances, these are not expected to create TS *wave* of eigenvalue analysis.

Along with the successful reporting of 2D TS *wave* for 2D equilibrium flow, natural extension of deterministic route of 2D transition was investigated, where the imposed deterministic excitation was made three-dimensional (3D) and resultant transition was investigated by Klebanoff et al. [10]. This was considered important, as in most turbulent flows (excepting some flows, including atmospheric dynamics), three-dimensionality is important due to vortex stretching. This work was followed by many other forms of 3D instability mechanisms, which were also considered important for transition to turbulence for the canonical zero pressure gradient boundary layer [8].

1.3   Present State of Art in the Field

The reader is reminded that the experiment in [19] only demonstrated the existence of TS *wave* and did not show that TS *wave* eventually causes transition. This was an implicit assumption right from the beginning of instability studies, which presupposes that any instability waves eventually lead to causing turbulence. Over the last two decades, it has been shown that apart from TS *wave*, there exists the spatio-temporal wave-front (STWF). This has been made possible first by solving the Orr-Sömmerfeld equation in spatio-temporal framework [26, 28, 29]. This was performed to show how a so-called *linearly stable* boundary layer supports massive wave-packets with sharp wave-fronts, something like tsunami or rogue waves. The existence of STWF has also been established from the solution of NSE. Furthermore, it has been shown subsequently that such a STWF via non-linear growth causes 2D turbulence [25, 30]. The STWF results from finite start-up time of the exciter in experiments/computations. However, once the STWF is set-up, it has been shown that this structure has self regeneration mechanism [25]. One of the assuring features of this finding is that the energy spectrum displays k^{-3} variation, as has been predicted by Kraichnan [11] and Batchelor [1] for 2D turbulence.

After the experimental verification of TS *wave* [19], and reporting of experimental results for 3D transition routes (see [8] for a review), computing these scenario was reported by only a few groups of researchers. We will discuss the merits and drawbacks of these efforts in greater details. In some of these, governing equation itself has been changed; in some the numerical methods are flawed; while in some of these the initial and boundary conditions were not correct. Problem arises when generic codes have been adopted to simplify the simulation efforts and in the process unphysical routes of transition to turbulence have been documented for very small computational domains. Instead of noticing 3D TS *wave* and subsequent secondary or nonlinear stages of transition in the experiments of [10], the readers have been told that the simulations represent bypass transition (see for example, [18, 36–38] for various reviews). While in most of the cited works, numerical artifacts created flow fields, which look like turbulent flows. Authors in [3, 4] have solved the problem for two primary routes of 3D transition described in the literature for wall excitation in [10] without any numerical artifacts. Of specific interest will be a comparison between the work in [18] with [4]. In the present book, all the essential details of of the work in [3, 4] are provided with explanation.

1.4   Different Transition Routes

So far we have discussed transition and turbulence, which is caused by wall excitation. This type of transition is stated to be caused by wall modes [23] in the linearized analysis, and in the same reference, transition caused by convecting vortices
in the free stream is said to be due to free stream modes. Experimentally, Lieb et
al. [13] have shown how these two types of modes are coupled and the theoretical
aspect of this coupling is given in [23] for the Orr-Sommerfeld equation. In the context
of nonlinear instability, solution of NSE displays a k^{-3} spectrum, when the Fourier
transform of streamwise disturbance velocity is plotted as a function of streamwise
wavenumber, for the inhomogeneous 2D flow past a flat plate subjected to periodically passing train of vortices at a fixed height. Thus one cannot distinguish between
wall and free stream excitation for 2D turbulence. The problem of vortex-induced
instability caused by a single convecting 2D vortex at a fixed speed and height over
the plate can cause instability at a streamwise location, where growing TS *wave* is
not seen by wall excitation [14, 27]. Thus, the vortex-induced instability is a sub-
critical phenomenon (discussed in Chap. 4). It was Morkovin [15] who coined the
term bypass transition, for all those cases where transition is not associated with the
creation of TS *wave*. Originally, it was noted that Couette flow, Poiseuille flow in
a pipe were noted to be stable by linear spatial theory and it was thought that such
flows, along with bluff body flows, suffer transition by bypass route. Even the critical Reynolds number obtained by linear theory for channel flow ($Re_{cr} = 5772$) is
far above the value for which the flow is noted to become turbulent (at $Re_{tr} \approx 1000$).
Thus, it becomes apparent that apart from the zero pressure gradient boundary layer,
most flows do not even show the existence of TS *wave*. With varied pressure gradient,
as in flow over aerofoil, one does not notice monochromatic TS *wave* also. Finally,
even for zero pressure gradient boundary layer, one does not notice TS *wave*, when
the amplitude of wall excitation is significantly higher. However, the original con-
notation of flow instability is with respect to vanishingly small excitation. Thus, we
should explain bypass transition, when this can be attributed to physical processes and
not due to numerical means by which stable laminar flow is tripped to turbulent state
[18, 37]. The TS *wave* is thus noted only with extremely carefully designed experiment and the resultant turbulence provides the true test of computational ability of
capturing turbulence by first being able to reproduce these deterministic disturbances
[10, 19]. In this respect the work reported in [3, 4] has shown the correct methodology
for DNS of turbulent flow. This approach is specifically described in this book.

1.5   Role of Equilibrium Flow in DNS

In discussing about the role of analytical solution, Landau and Lifshitz [12] noted that
*the flow that occurs in nature must not only follow the equations of fluid dynamics, but
also be stable*. If solutions are not *observable*, then the corresponding equilibrium

flows are not *stable*. This also tells us about the implication of flow *instability* in the context of observation of Stokes with respect to pipe flow experimental results, following a series of simplifying assumptions. Moreover, the flow was considered to be laminar, while the experimental results corresponded to fully developed turbulent pipe flow. The mismatch between the two results were attributed to flow instability in [12]. The laminar flow steady solution can be viewed as the equilibrium solution, and the following instability caused the growth of infinitesimally small perturbations present in the surroundings. This sensitive dependence on the disturbance environment makes the subject of instability very interesting, as well as challenging in understanding the state of the flow. The smallness of background disturbances allows one to study the problem of growth of these by small perturbation theory. This greatly helps, if the governing nonlinear equations can be solved for the equilibrium solution with ease and then its stability can be studied by linearizing the governing equation for the perturbation field. In theoretical instability approach, this leads to an eigenvalue problem, as we have already discussed above for flow past a plate experiencing zero pressure gradient. One of the drawback of eigenvalue analysis is that this does not require any information on input, that triggers instability. At the same time, Schubauer and Skramstad [19] have clearly noted in their experiment, that the vibrating ribbon placed inside the boundary layer can create TS *wave*, while an acoustic exciter placed outside the boundary layer did not cause any instability. This brings in the concept of receptivity, by which one can say that wall-bounded shear layer are receptive to vortical excitation (more strongly when it is placed inside it) and not so receptive to acoustic excitation, when it is applied from outside. Naturally, the amplitude of excitation should be a factor in determining the causation of instability. Interestingly, this was noted by Osborne Reynolds and he recorded it unambiguously [17].

The dye experiment performed [17] is perhaps the first recorded experimental observations of flow instability. Reynolds took pipes of different diameters fitted with a trumpet shaped mouth-piece or bell-mouth, which accelerated the flow on entry to the pipe. Such acceleration creates a favorable pressure gradient, attenuating background disturbances accompanying the oncoming water in the pipe. He performed experiments in night, to avoid noise from daytime vehicular traffic. Reynolds observed that the rapid diffusion of dye with surrounding fluid depends on the non-dimensional parameter Va/v, with V as the center-line velocity in the pipe of diameter a, and v is the kinematic viscosity. This non-dimensional parameter is the Reynolds number (Re). Reynolds found that the flow can be kept orderly or laminar up to $Re = 12, 830$ in his set up. He noted this value to be extremely sensitive to the disturbances in the flow before it enters the tube and he noted prophetically that *this at once suggested the idea that the condition might be one of instability for disturbance of a certain magnitude and stable for smaller disturbances.* The relationship of instability with disturbance amplitude is often attributed to non-linear instability. The implicit assumption in this point of view is that linear instability does not require any knowledge for the amplitude and spectrum of the input disturbances. This must be viewed as a serious shortcomings of linear theory, which will be used only for qualitative understanding of flow instability.

When it comes to computational efforts in understanding instability and transition to turbulence, one of the strengths lies in being able to obtain equilibrium flow without making too many simplifying assumptions. On the other hand, no computations can be made without the attendant numerical error. Thus, knowledge and ability to reduce errors must be one of the central themes of DNS and LES. Since a steady equilibrium flow becomes unsteady, before becoming fully turbulent, obtaining a steady flow accurately is equivalent to obtaining the equilibrium flow correctly. At the same time, we all understand that there are many flows, for range of parameter values, which cannot be computed without the flow displaying inherent unsteadiness. A simple example is a flow past a circular cylinder for Reynolds number in excess of 65 or so, for which the computed flow display alternately shed vortices. Vortex shedding itself is a manifestation of instability of steady laminar flow for a critical Reynolds number. It is well to remember that neither computationally or experimentally, one obtains a single fixed critical Reynolds number. In computations and experiments, background disturbances in experiment and numerical errors act as seed of instabilities for any flow, provided sufficient numerical accuracy is maintained. Numerical errors are due to discretization of continuum equations and for computing space-time dependent problems, additional error arises due to coupling of space and time discretizations considered together. Such a relationship for space and time scales appear as dispersion relation for linear constant coefficient partial differential equations. The coupled space-time discretization error is thus, termed as dispersion error. It has been variously investigated for different model equation, for the purpose of calibrating space-time discretization considered together. For other flows, there may be other sources of errors, like aliasing error, Gibbs' phenomenon etc. which can also provide the seed to destabilize a given flow. Thus, following the approach used in experiments on transition to turbulence, computationally also, one must design numerical methods which systematically removes error sources, and thereby we improve our ability to obtain equilibrium flows. Such equilibrium flows should be studied for their receptivity to different generic types of errors, which mimic acoustic, vortical and entropic disturbances.

1.6   What Is Instability?

To study a physical problem analytically, one first obtains governing equations which model the phenomenon adequately. If the auxiliary equations pertaining to initial and boundary conditions are well posed, then obtaining the solution is straightforward. Mathematically, one is concerned with the *existence* and *uniqueness* of the solution. Yet not every solution of equations of motion, even if it is exact, is observable in nature. This is at the core of many physical phenomena where *observability* of solution is of fundamental importance. If the solution is not *observable*, then the corresponding basic flow is not *stable*. Here, the implication of *stability* is in the context of the solution with respect to infinitesimally small perturbations.

In studying the stability of a problem with respect to ambient disturbances, it is hardly ever possible that one can incorporate all the contributing factors in a given physical scenario for posing a physical problem. Occasionally these neglected *causes* can be incorporated by *process noise* and results are made to correlate with the physical situation. This is possible when the *causes* are statistically independent and then it follows upon using the Central Limit Theorem.

1.7   Temporal and Spatial Instability

Instability of an autonomous system is strictly for time-dependent systems that display growth of disturbances with time. This may also mean that either we are studying the stability of a flow at a fixed spatial location or the full system displays identical variation in time for each and every spatial location. Fluid flow instabilities are treated as if the disturbance growth is either in space or in time. This approach is merely for expediency and not based on any assessment of reality. It is ideally suited to treat the growth in a spatio-temporal framework, but no theoretical framework exists, except by the use of the Bromwich contour integral method [23]. This method is used in a linear framework and is capable of tracking growth in space and time simultaneously. This method will be described and results shown in Chaps. 3 and 5. A disturbance originating from a fixed location in space can grow as it convects downstream. Such a disturbance is termed unstable if it grows as it moves downstream. This is the convective instability. This type of instability is considered for wall-bounded shear layers, exemplified by the classic vibrating ribbon experiment of [19]. This experiment was performed in a very quiet facility to create TS waves by vibrating a ribbon inside a flat plate boundary layer, a detailed description of which is provided in Chap. 3. This experiment was the first to show the existence of a viscous unstable wave which was predicted earlier theoretically in [7, 21, 34]. Hence, the existence of TS waves was doubted before the experimental results in [19] were published. Additionally, this experiment was also the first to display receptivity of wall-bounded shear layers to vortical disturbances created inside a shear layer, while showing the inadequacy of acoustic excitation in creating TS waves. One of the major aims of the book is to show that the classical picture of convective instability is not correct, and the transition is shown to be created by STWF.

In many flows, it can happen that the disturbance can grow first in time at a fixed location, before it is convected downstream. Such growth of disturbances both in space and time are seen in many free shear layers and bluff-body flows. If we subject the equilibrium solution of such an unstable fluid dynamical system to a localized impulse, then the response field spreads both upstream and downstream of the location with respect to the local flow where it originated, while growing in amplitude. Such instabilities have been termed *absolute instabilities*. Here, an additional distinction needs to be made between convective and absolute instabilities. On application of an impulse, both situations display disturbances in upstream and downstream directions. However, in a convectively unstable system, the growth of the disturbance is predominantly in one direction, while for an absolutely unstable system the growth will be omni-directional.

1.8 Some Instability Mechanisms

Here two simple cases of instabilities are considered to emphasize the concepts described above. We begin by distinguishing the difference between static and dynamic instability by considering the stability of the atmosphere as an example.

When a parcel of air in the atmosphere is moved rapidly from an equilibrium condition and its tendency to come back to its undisturbed position is noted, then we term the atmosphere statically stable. The movement of the packet is considered as impulsive, to preclude any heat transfer from the parcel to the ambience. This tendency of static stability, when if exists, is due to the buoyancy force caused by the density differential due to temperature variation with height and such a body force acts upon the displaced air parcel. In static stability studies, we do not look for detailed time-dependent motion of the parcel following the displacement (as the associated accelerations are considered negligible). We refer the reader to detailed dynamical instability of atmosphere studied in [23].

1.8.1 Kelvin–Helmholtz Instability

This arises when two layers of fluids (may not be of the same species or density) are in relative motion. Thus, this is an interfacial instability and the resultant flow features due to imposed disturbance will be much more complicated due to relative motion. The physical relevance of this problem was seized upon by Helmholtz [35], who observed that the interface as a surface of separation tears the flow *asunder*. Sometime later Kelvin [9] posed this problem as one of instability and solved it. We follow this latter approach here. The basic equilibrium flow is assumed to be inviscid and incompressible - as two parallel streams having distinct density and velocity - flowing one over the another, is depicted in Fig. 1.1.

Before any perturbation is applied, the interface is located at $z = 0$ and subsequent displacement of this interface is expressed parametrically as

$$z_s = \hat{\eta}(x, y, t) = \varepsilon\eta(x, y, t) \tag{1.7}$$

where ε is a small parameter, defined to perform a linearized perturbation analysis. One can view the interface itself as a shear layer of vanishing thickness. For the considered inviscid irrotational flow, velocity potentials in the two domains are given by

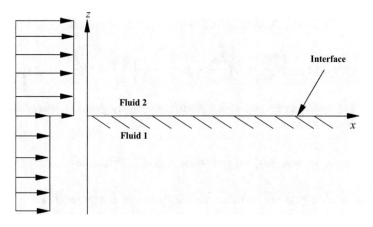

Fig. 1.1 Kelvin–Helmholtz instability at the interface of two flowing fluids

$$\tilde{\phi}_j(x, y, z, t) = U_j x + \varepsilon \phi_j(x, y, z, t) \tag{1.8}$$

The governing equations in either of the flow domains are given by

$$\nabla^2 \tilde{\phi}_j = 0 \tag{1.9}$$

The potentials must satisfy the following far-stream boundary conditions,

$$\phi_j s \quad \text{are} \quad \text{bounded} \quad \text{as} \quad z \to \pm \infty \tag{1.10}$$

Another set of boundary conditions is applied at the interface, which is the no-fluid through the interface condition, i.e.,

$$\frac{\partial \hat{\eta}}{\partial t} - \frac{\partial \tilde{\phi}_j}{\partial z} = -\frac{\partial \hat{\eta}}{\partial x} \frac{\partial \tilde{\phi}_j}{\partial x} - \frac{\partial \hat{\eta}}{\partial y} \frac{\partial \tilde{\phi}_j}{\partial y} \tag{1.11}$$

In addition, in the absence of surface tension, pressure must be continuous across the interface. Upon linearization, the interface boundary condition of Eq. (1.11) simplifies to

$$\frac{\partial \eta}{\partial t} + U_j \frac{\partial \eta}{\partial x} - \frac{\partial \phi_j}{\partial z} = 0 \quad \text{for} \quad j = 1, 2 \tag{1.12}$$

where $\tilde{\phi}_j$ and ϕ_j are as related in Eq. (1.8). Defining the pressure on either flow domain by the unsteady Bernoulli's equation, one can write

$$p_j = C_j - \rho_j \left\{ \frac{\partial \tilde{\phi}_j}{\partial t} + \frac{1}{2} (\nabla \tilde{\phi}_j)^2 + g \hat{\eta} \right\} \tag{1.13}$$

Simplifying and retaining up to $0(\varepsilon)$ terms, we get the following conditions

$$0(1) \text{ condition}: \ C_1 - \frac{1}{2}\rho_1 U_1^2 = C_2 - \frac{1}{2}\rho_2 U_2^2 \tag{1.14a}$$

$$0(\varepsilon) \text{ condition}: \ \rho_1\left\{\frac{\partial\phi_1}{\partial t} + U_1\frac{\partial\phi_1}{\partial x} + g\eta\right\} = \rho_2\left\{\frac{\partial\phi_2}{\partial t} + U_2\frac{\partial\phi_2}{\partial x} + g\eta\right\} \tag{1.14b}$$

One can consider a very general interface displacement given in terms of a bilateral Laplace transform as

$$\eta(x, y, t) = \int\int F(\alpha, \beta, t)\, e^{i(\alpha x + \beta y)} d\alpha\, d\beta \tag{1.15}$$

Correspondingly, the perturbation velocity potential is expressed as

$$\phi_j(x, y, z, t) = \int\int Z_j(\alpha, \beta, z, t)\, e^{i(\alpha x + \beta y)}\, d\alpha\, d\beta \tag{1.16}$$

Writing $k^2 = \alpha^2 + \beta^2$ and using Eq. (1.16) in Eq. (1.9), one gets the solution that satisfies the far-stream boundary conditions of Eq. (1.10) as

$$Z_j = f_j(\alpha, \beta, t)\, e^{\pm kz} \quad \text{for} \quad j = 1 \text{ and } 2 \tag{1.17}$$

Using Eq. (1.15) in the interface boundary condition of Eq. (1.12) one gets

$$\dot{F} + i\alpha U_1 F - k f_1 = \dot{F} + i\alpha U_2 F + k f_2 = 0 \tag{1.18}$$

where the dots denote differentiation with respect to time. If we denote the density ratio $\rho = \rho_2/\rho_1$, then the linearized pressure continuity condition of Eq. (1.14b) gives

$$\frac{\partial\phi_1}{\partial t} - \rho\frac{\partial\phi_2}{\partial t} + U_1\frac{\partial\phi_1}{\partial x} - \rho U_2\frac{\partial\phi_2}{\partial x} + (1-\rho)g\eta = 0 \tag{1.19}$$

Using Eqs. (1.15) and (1.16) in the above one gets

$$\dot{f_1} - \rho\dot{f_2} + i\alpha U_1 f_1 - i\alpha\rho U_2 f_2 + (1-\rho)g F = 0 \tag{1.20}$$

Eliminating f_1 and f_2 from Eq. (1.20) using Eq. (1.18), one gets, after simplification,

$$(1+\rho)\ddot{F} + 2i\alpha(U_1 + \rho U_2)\dot{F} - \{\alpha^2(U_1^2 + \rho U_2^2) - (1-\rho)gk\}F = 0 \tag{1.21}$$

This ordinary differential equation for the time variation of the interface displacement F can be understood better in terms of its Fourier transform defined by

$$F(., t) = \int \hat{F}(., \bar{\omega})\, e^{i\bar{\omega}t}\, d\bar{\omega} \tag{1.22}$$

One obtains the following dispersion relation by substitution of Eq. (1.22) in (1.21) as

$$- \bar{\omega}^2(1 + \rho) - 2\alpha\bar{\omega}(U_1 + \rho U_2) + (1 - \rho)gk - \alpha^2(U_1^2 + \rho U_2^2) = 0 \quad (1.23)$$

This provides the characteristic exponents in Eq. (1.22) as

$$\bar{\omega}_{1,2} = -\frac{\alpha(U_1 + \rho U_2)}{(1 + \rho)} \mp \frac{\sqrt{gk(1 - \rho^2) - \alpha^2\rho(U_1 - U_2)^2}}{(1 + \rho)} \quad (1.24)$$

Based on this dispersion relation the following sub-cases are considered:

CASE 1: If the interface is disturbed in the spanwise direction only, i.e., $\alpha = 0$, then

$$\bar{\omega}_{1,2} = \mp\sqrt{g\beta\frac{(1 - \rho)}{(1 + \rho)}} \quad (1.25)$$

Thus, the streaming velocities U_1 and U_2 do not affect the response of the system. If in addition, $\rho > 1$, i.e., a heavier liquid is over a lighter liquid, then the buoyancy force causes temporal instability (if β is considered real) as is the case for Rayleigh-Taylor instability (see Chandrasekhar [5]).

CASE 2: For a general interface perturbation if $gk(1 - \rho^2) - \alpha^2\rho(U_1 - U_2)^2 < 0$, then the interface displacement will grow in time. This condition can be alternately stated as a condition for instability as $(U_1 - U_2)^2 > \frac{gk}{\alpha^2}\left(\frac{1-\rho^2}{\rho}\right)$.

Thus, for a given shear at the interface given by $(U_1 - U_2)$ and for a given oblique disturbance propagation direction at the interface indicated by the wavenumber vector k, instability would occur for all wavenumbers k^*, given by

$$k^* > \left(\frac{k^*}{\alpha}\right)^2 \frac{g}{(U_1 - U_2)^2}\left(\frac{\rho_1}{\rho_2} - \frac{\rho_2}{\rho_1}\right)$$

Note that the wavenumber vector makes an angle γ with the x-axis, such that $\cos\gamma = \frac{\alpha}{k^*}$ and the above condition can be conveniently written as

$$k^* > \frac{g}{(U_1 - U_2)^2 \cos^2\gamma}\left(\frac{\rho_1}{\rho_2} - \frac{\rho_2}{\rho_1}\right) \quad (1.26)$$

The lowest value of wavenumber ($k^* = k_{min}$) would occur for two-dimensional disturbances, i.e., when $\cos\gamma = 1$ and this minimum is given by

$$k_{min}^* = \frac{g}{(U_1 - U_2)^2}\left(\frac{\rho_1}{\rho_2} - \frac{\rho_2}{\rho_1}\right) \quad (1.27)$$

CASE 3: Consider the case of shear only of the same fluid in both domains, i.e., $\rho = 1$. The characteristic exponents then simplify to

$$\bar{\omega}_{1,2} = -\alpha \frac{U_1 + U_2}{2} \mp \frac{i\alpha}{2}(U_1 - U_2) \tag{1.28}$$

The presence of an imaginary part with a negative sign implies temporal instability for all wave lengths. Also, note that since the group velocity and phase speed in the y direction are identically zero, therefore the Kelvin–Helmholtz instability for pure shear always will lead to two-dimensional instability.

References

1. Batchelor, G. K. (1969). Computation of the energy spectrum in homogeneous two-dimensional decaying turbulence. *Physics of Fluids, 12*(suppl. II), 233–239.
2. Batchelor, G. K. (1988). *An introduction to fluid dynamics*. UK: Cambridge University Press.
3. Bhaumik, S. (2013). Direct numerical simulation of inhomogeneous transitional and turbulent flows. Ph. D. Thesis, I. I. T. Kanpur.
4. Bhaumik, S., & Sengupta, T. K. (2014). Precursor of transition to turbulence: Spatiotemporal wave front. *Physical Review E, 89*(4), 043018.
5. Chandrasekhar, S. (1960). *Radiative transfer* (p. 393). New York: Dover Publications Inc.
6. Drazin, P. G., & Reid, W. H. (1981). *Hydrodynamic stability*. UK: Cambridge University Press.
7. Heisenberg, W. (1924). Über stabilität und turbulenz von flüssigkeitsströmen. *Annalen der Physik (Leipzig), 379*, 577–627 (Translated as 'On stability and turbulence of fluid flows'. NACA Tech. Memo. Wash. No 1291 1951)
8. Kachanov, Y. S. (1994). Physical mechanisms of laminar-boundary-layer transition. *Annual Review of Fluid Mechanics, 26*, 411–482.
9. Kelvin, L. (1871). Hydrokinetic solutions and observations. *Philosophical Magazine, 4*(42), 362–377.
10. Klebanoff, P. S., Tidstrom, K. D., & Sargent, L. M. (1962). The three-dimensional nature of boundary-layer instability. *Journal of Fluid Mechanics, 12*, 1–34.
11. Kraichnan, R. H. (1967). Inertial ranges in two-dimensional turbulence. *Physics of Fluid, 10*(7), 1417–1423
12. Landau, L. D., & Lifshitz, E. M. (1959). *Fluid mechanics* (Vol. 6). London: Addison - Wesley. Pergamon Press.
13. Leib, S. J., Wundrow, D. W., & Goldstein, M. E. (1999). Effect of free stream turbulence and other vortical disturbances on a laminar boundary layer. *Journal of Fluid Mechanics, 380*, 169–203.
14. Lim, T. T., Sengupta, T. K., & Chattopadhyay, M. (2004). A visual study of vortex-induced subcritical instability on a flat plate laminar boundary layer. *Experiments in Fluids, 37*, 47–55.
15. Morkovin, M. V. (1991). Panoramic view of changes in vorticity distribution in transition instabilities and turbulence. D.C. Reda, H.L. Reed & R. Kobayashi (Eds.), *Transition to turbulence* (Vol. 114, pp. 1–12) ASME FED Publication.
16. Orr, W. M. F. (1907). The stability or instability of the steady motions of a perfect liquid and of a viscous liquid. Part I: A perfect liquid. Part II: A viscous liquid. *Proceedings of the Royal Irish Academy, A27*, 9–138.
17. Reynolds, O. (1883). An experimental investigation of the circumstances which determine whether the motion of water shall be direct or sinuous and of the law of resistance in parallel channels. *Philosophical Transactions of the Royal Society, 174*, 935–982.

18. Sayadi, T., Hamman, C. W., & Moin, P. (2013). Direct numerical simulation of complete H-type and K-type transitions with implications for the dynamics of turbulent boundary layers. *Journal of Fluid Mechanics, 724*, 480–509.
19. Schubauer, G. B., & Skramstad, H. K. (1947). Laminar boundary layer oscillations and the stability of laminar flow. *Journal of Aerospace Sciences, 14*(2), 69–78.
20. Schlatter, P., & Örlü, R. (2012). Turbulent boundary layers at moderate Reynolds numbers: Inflow length and tripping effects. *Journal of Fluid Mechanics, 710*, 5–34.
21. Schlichting, H. (1933). Zur entstehung der turbulenz bei der plattenströmung. *Nach. Gesell. d. Wiss. z. Gött., MPK, 42*, 181–208.
22. Schlichting, H. (1979). *Boundary layer theory* (7th ed.). New York: McGraw Hill.
23. Sengupta, T. K. (2012). *Instabilities of flows and transition to turbulence*. Florida, USA: CRC Press, Taylor & Francis Group.
24. Sengupta, T. K. (2015). A critical assessment of simulations for transitional and turbulent flows. In T. K. Sengupta, S. K. Lele, K. R. Sreenivasan & P. A. Davidson (Eds.), *In the IUTAM Symposium Proceedings on Advances in Computation, Modeling and Control of Transitional and Turbulent Flows* (pp. 491–532). World Scientific Publishing Company.
25. Sengupta, T. K., & Bhaumik, S. (2011). Onset of turbulence from the receptivity stage of fluid flows. *Physical Review Letters, 154501*, 1–5.
26. Sengupta, T. K., Ballav, M., & Nijhawan, S. (1994). Generation of Tollmien-Schlichting waves by harmonic excitation. *Physics of Fluids, 6*(3), 1213–1222.
27. Sengupta, T. K., De, S., & Sarkar, S. (2003). Vortex-induced instability of an incompressible wall-bounded shear layer. *Journal of Fluid Mechanics, 493*, 277–286.
28. Sengupta, T. K., Rao, A. K., & Venkatasubbaiah, K. (2006). Spatio-temporal growing wave fronts in spatially stable boundary layers. *Physical Review Letters, 96*(22), 224504.
29. Sengupta, T. K., Rao, A. K., & Venkatasubbaiah, K. (2006). Spatio-temporal growth of disturbances in a boundary layer and energy based receptivity analysis. *Physics of Fluids, 18*, 094101.
30. Sengupta, T. K., Bhaumik, S., & Bhumkar, Y. (2012). Direct numerical simulation of two-dimensional wall-bounded turbulent flows from receptivity stage. *Physical Review E, 85*(2), 026308.
31. Smith, C. R. (1993). Use of 'Kernel' experiments for modeling of near-wall turbulence. In R.M.C. So, C.G. Speziale & B.E. Launders (Eds.), *Near wall turbulent flows* (pp. 33–42). Amsterdam, Holland: Elsevier.
32. Sommerfeld, A. (1908). Ein Beitrag zur hydrodynamiscen Erklarung der turbulenten Flussigkeitsbewegung. In *Proceedings of the 4th International Congress of Mathematicians, Rome* (pp. 116–124).
33. Spalart, P. (1988). Direct numerical study of leading edge contamination. *AGARD CP, **438***, 5.1–5.13.
34. Tollmien, W. (1931). Über die enstehung der turbulenz. I, English translation. *NACA TM 609*.
35. von Helmholtz, H. (1868). On discontinuous movements of fluids. *Philosophical Magazine, 36*(4), 337–346.
36. Wu, X. (2017). Inflow turbulence generation methods. *Annual Review of Fluid Mechanics, 49*, 23–49.
37. Wu, X., & Moin, P. (2009). Direct numerical simulation of turbulence in a nominally-zero-pressure-gradient flat-plate boundary layer. *Journal of Fluid Mechanics, 630*, 5–41.
38. Wu, X., Moin, P., & Hickey, J. P. (2014). Boundary layer bypass transition. *Physics of Fluids, 26*, 091104.

Chapter 2
DNS of Navier–Stokes Equation

2.1 Fluid Dynamical Equations

The governing equations of fluid mechanics are all based on identical fundamental classical principles of dynamics, namely conservation of mass, momentum and energy. Based on these principles, Claude-Louis Navier and George Gabriel Stokes, derived the governing equation for viscous fluids by applying Newton's second law to fluid motion, along with the assumption that stresses arising in the fluids are due to diffusing viscous effects and the pressure gradient. These governing equations are known as Navier–Stokes Equation (NSE). It is to be noted that there is no fixed version of NSE, an appropriate version and the formulation has to be chosen based on the need and the characteristics of the fluid-dynamical problem to be considered. For example, for low-speed applications without heat-transfer, the appropriate version is the incompressible NSE, without considering the energy equation. For such equations, it can be shown that the kinetic energy of the fluid in a control volume is automatically conserved from the momentum equation. For flows dominated by heat-transfer, an additional energy equation is to considered which essentially determines the temperature at each point. Temperature gradient induces gradient in fluid-density, which in turn affects the flow if gravitational effects are also present. In incompressible NSE, such buoyancy effects due to temperature induced density-gradient are generally modeled via Boussinesq approximation [87]. In contrast, when the flow speed is on the higher side and the compressibility effects are quite dominant, the appropriate governing equations are the compressible verisons of NSE, where all the conservation constraints (mass, momentum and energy) are to be considered. Additional equations are also required for compressible NSE to close the system of equations which is given by the thermodynamic equation of state relating fluid temperature, density and pressure. To derive the appropriate governing equations, following general steps are to be followed:

© Springer Nature Singapore Pte Ltd. 2019
T. K. Sengupta and S. Bhaumik, *DNS of Wall-Bounded Turbulent Flows*,
https://doi.org/10.1007/978-981-13-0038-7_2

(i) First, appropriate physical principles have to be decided, which are required to be satisfied, e.g., if one focuses upon the incompressible flow without heat transfer, then only conservation principles of interest are mass and momentum.

(ii) Based on (i), an appropriate model for the flow problem has to be chosen.

(iii) Finally, based on (i) and (ii), appropriate governing mathematical equations should be derived. Once, appropriate governing equations are derived, suitable computational tools are to be used to numerically solve these equations, as generally these equations do not admit any closed–form analytical solution (save some very few simplified cases). Development and use of highly accurate numerical methodologies are very important, particularly for transitional and turbulent flows. These computations require special approaches, which are distinctly different from the conventional CFD methodologies one usually comes across for engineering applications.

For proper characterization of transitional flows, one has to accurately track disturbance evolution in a base or equilibrium flow, whose properties would determine the characteristics of amplification and propagation of perturbation. The equilibrium flow can be obtained by considering the full nonlinear NSE or some simplified form of it. Historically, even inviscid irrotational assumptions are also made to obtain an equilibrium flow. However, recent efforts have shown the importance of using full nonlinear NSE to obtain the equilibrium flows, especially in the context of analyzing its receptivity to imposed excitations. Typically, there are two ways by which traditionally transitional flows are analyzed. First, through eigenvalue approach by considering corresponding linearized NSE. In this approach, which have been the focus of most of the earlier studies in the field, one is not required to characterize the disturbance environment. In contrast, to study various aspects of the response of a dynamical system to a specific class of disturbances, one adopts what has been identified as receptivity analysis. Receptivity analysis can also be performed in the linearized framework and has been advanced in [26, 30, 31] via the Bromwich contour integral method. Irrespective of the amplitude of perturbations, one can also use the full NSE for receptivity study. This is undertaken here in Chaps. 3–6. Thus, it is imperative that readers appreciate the nuances of solution techniques of various space-time dependent equations.

Unlike in solid mechanics problems, fluid dynamical systems are characterized by a relatively larger number of degrees of freedom, as the later is characterized by very low physical dissipation. However, it cannot be completely ignored from the whole flow domain, as the delicate balance of dissipation and energy transfer to other spectral components determines the characteristics of transitional and turbulent flows. In the spectral plane the maximum dissipation occurs at higher wavenumbers corresponding to smallest eddies. Thus, for simulating transitional flows one would have to be very careful in resolving very high wavenumbers. The importance of this is often lost on CFD practitioners, who insist on performing large eddy simulations (LES) of transitional flows, by considering only the energy spectrum.

The conservation principles for mass, momentum and energy for fluid flows are illustrated next.

2.1.1 Equation of Continuity

This follows from the conservation of mass, which for a *control volume* essentially states that net rate of creation/destruction of mass inside it is balanced by the amount of mass flow rate through the *control surfaces*. For unsteady compressible flows, this leads to

$$\frac{D\rho}{Dt} + \rho \, \nabla \cdot \vec{V} = \frac{\partial \rho}{\partial t} + \nabla \cdot (\rho \, \vec{V}) = 0 \tag{2.1}$$

where D/Dt stands for the material derivative, i.e.,

$$\frac{D}{Dt} = \left(\frac{\partial}{\partial t} + \vec{V} \cdot \nabla \right)$$

For incompressible flows, density is treated as constant i.e., $\rho =$ constant and subsequently Eq. (2.1) simplifies to

$$\nabla \cdot (\vec{V}) = 0 \tag{2.2}$$

Any vector field that satisfies the equivalent condition given by Eq. (2.2) is called the *divergence-free* or *solenoidal* field.

2.1.2 Momentum Conservation Equation

Similarly, this stems from the principle of conservation of translational momentum following Newton's second law of motion applied to a control volume. Following a *control volume* in motion is essentially like considering *a control-mass system* for which the mass does not change with time. Hence, the rate of change of momentum inside such a *control-mass system* is basically the mass multiplied by the acceleration experienced by the fluid. Therefore, net rate of change of momentum inside the *control volume* is balanced by summation of net force exerted on the control volume and the total rate of of momentum crossing (entering or leaving) the corresponding *control surface*. The constituent forces acting on the elementary *control mass* are body forces acting on the control volume (which do not depend on the geometry of the body, e.g., gravitational/buoyancy/electromagnetic forces) and the surface forces acting directly on the *control surfaces* caused by *normal* and *shear* stresses. For an elementary volume element of size $(dx \, dy \, dz)$, if the local density is ρ, then the body force acting in the x-direction can be expressed as

$$F_x = \rho f_x (dx \, dy \, dz) \tag{2.3}$$

where f_x is the associated acceleration in the x-direction. If τ_{ij} denotes the stress tensor acting in the j-direction on a plane whose normal is in the i-direction, then

one can account for all the contributory stresses giving rise to surface force in the
x-direction as

$$\left(-\frac{\partial p}{\partial x} + \frac{\partial \tau_{xx}}{\partial x} + \frac{\partial \tau_{yx}}{\partial y} + \frac{\partial \tau_{zx}}{\partial z}\right) dx\, dy\, dz \tag{2.4}$$

Net force term acting on the fluid element along x-direction is, therefore, the vector
summation of Eqs. (2.3) and (2.4). As the mass of the fluid element is $(\rho\, dx\, dy\, dz)$
and the acceleration of the moving element is given by its substantive or the material
derivative of the velocity vector, the momentum conservation equation for the fluid
element along x-direction can, therefore, be expressed as

$$\rho\frac{Du}{Dt} = \left(-\frac{\partial p}{\partial x} + \frac{\partial \tau_{xx}}{\partial x} + \frac{\partial \tau_{yx}}{\partial y} + \frac{\partial \tau_{zx}}{\partial z}\right) + \rho f_x \tag{2.5}$$

Similarly y- and z-components of the momentum equation can also be obtained
as

$$\rho\frac{Dv}{Dt} = \left(-\frac{\partial p}{\partial y} + \frac{\partial \tau_{xy}}{\partial x} + \frac{\partial \tau_{yy}}{\partial y} + \frac{\partial \tau_{zy}}{\partial z}\right) + \rho f_y \tag{2.6}$$

$$\rho\frac{Dw}{Dt} = \left(-\frac{\partial p}{\partial z} + \frac{\partial \tau_{xz}}{\partial x} + \frac{\partial \tau_{yz}}{\partial y} + \frac{\partial \tau_{zz}}{\partial z}\right) + \rho f_z \tag{2.7}$$

Equations (2.5)–(2.7) are known as the Cauchy equations, which in vector form
can be written as

$$\rho\frac{D\vec{V}}{Dt} = -\nabla p + \nabla \cdot \tau_{ij}\, \delta_{ij} + \rho\vec{F} \tag{2.8}$$

where δ_{ij} is the Kronecker delta function. By using the vector form of the continuity
equation, the x-component of the Cauchy equations in the *conservation form*, can
be expressed as

$$\frac{\partial(\rho u)}{\partial t} + \nabla \cdot (\rho\, u \vec{V}) = \left(-\frac{\partial p}{\partial x} + \frac{\partial \tau_{xx}}{\partial x} + \frac{\partial \tau_{yx}}{\partial y} + \frac{\partial \tau_{zx}}{\partial z}\right) + \rho f_x \tag{2.9}$$

Similarly, one can also express the other two components of the momentum equation in *conservative form*. For these system of equations to be useful, one needs
information about the general stress system in the flow field. For Newtonian fluids,
Stokes [109] has shown that stress is related to the strain rates as

$$\tau_{ij} = \lambda\delta_{ij}\frac{\partial v_k}{\partial x_k} + \mu\left(\frac{\partial v_i}{\partial x_j} + \frac{\partial v_j}{\partial x_i}\right) - p\delta_{ij} \tag{2.10}$$

where μ is the molecular viscosity, λ is the second coefficient of viscosity and p is
the thermodynamic pressure or hydrostatic stress. If p_m is the mechanical pressure,

then by definition $p_m = \frac{\tau_{ii}}{3}$. Stokes [109] again hypothesized a relationship between μ and λ by relating thermodynamic and mechanical pressure from

$$3\lambda + 2\mu = 0 \tag{2.11}$$

However, in recent times, various researchers have shown that this hypothesis have serious shortcomings [104]. Stokes' hypothesis states that bulk viscosity $\bar{\kappa} = \lambda + \frac{2}{3}\mu$ is set to zero for any flow following the assumption that for compressible flows, thermodynamic pressure and mechanical pressures are identical [15, 26, 27, 35]. However, several researchers have criticized this assumption [15, 26, 27, 35, 69, 103]. Emanuel [26] and Buresti [15] have weakly justified Stokes' hypothesis for mono-atomic rigid molecule gases. However, Liebermann [57], Karim and Rosenhead [47] and Rosenhead [79] have opined that λ is independent of μ and also, it can be orders of magnitude higher in amplitude with a reverse sign. Emanuel [26] and Gad-el-Hak [27] have also questioned the Stokes' hypothesis and its applicability for flows such as re-entry into planetary atmosphere. Cramer [19] have shown that many common fluids including diatomic gases can have bulk viscosities ($\bar{\kappa}$) thousand times larger than λ, prompting Rajagopal [69] to suggest that new approach in obtaining NSE for all fluids should be adopted without using the simple relationship given by Eq. (2.11), between λ and μ.

Using the constitutive relation given in Eq. (2.10), one can write the final form of the NSE for compressible flow as

$$\frac{\partial(\rho u)}{\partial t} + \frac{\partial(\rho u^2)}{\partial x} + \frac{\partial(\rho uv)}{\partial y} + \frac{\partial(\rho uw)}{\partial z} = -\frac{\partial p}{\partial x} +$$
$$\frac{\partial}{\partial x}\left(\lambda \nabla \cdot \vec{V} + 2\mu \frac{\partial u}{\partial x}\right) + \frac{\partial}{\partial y}\left(\mu\left[\frac{\partial v}{\partial x} + \frac{\partial u}{\partial y}\right]\right) + \frac{\partial}{\partial z}\left(\mu\left[\frac{\partial u}{\partial z} + \frac{\partial w}{\partial x}\right]\right) + \rho f_x \tag{2.12}$$

$$\frac{\partial(\rho v)}{\partial t} + \frac{\partial(\rho uv)}{\partial x} + \frac{\partial(\rho v^2)}{\partial y} + \frac{\partial(\rho vw)}{\partial z} = -\frac{\partial p}{\partial y} +$$
$$\frac{\partial}{\partial y}\left(\lambda \nabla \cdot \vec{V} + 2\mu \frac{\partial v}{\partial y}\right) + \frac{\partial}{\partial x}\left(\mu\left[\frac{\partial v}{\partial x} + \frac{\partial u}{\partial y}\right]\right) + \frac{\partial}{\partial z}\left(\mu\left[\frac{\partial v}{\partial z} + \frac{\partial w}{\partial y}\right]\right) + \rho f_y \tag{2.13}$$

$$\frac{\partial(\rho w)}{\partial t} + \frac{\partial(\rho uw)}{\partial x} + \frac{\partial(\rho wv)}{\partial y} + \frac{\partial(\rho w^2)}{\partial z} = -\frac{\partial p}{\partial z} +$$
$$\frac{\partial}{\partial z}\left(\lambda \nabla \cdot \vec{V} + 2\mu \frac{\partial w}{\partial z}\right) + \frac{\partial}{\partial x}\left(\mu\left[\frac{\partial w}{\partial x} + \frac{\partial u}{\partial u}\right]\right) + \frac{\partial}{\partial y}\left(\mu\left[\frac{\partial v}{\partial z} + \frac{\partial w}{\partial y}\right]\right) + \rho f_z \tag{2.14}$$

For incompressible flows, these equations can be further simplified. As for incompressible flows $\nabla \cdot \vec{V} = 0$, the terms associated with the bulk viscosity coefficient $\bar{\kappa}$ drop out in Eqs. (2.12)–(2.14). Further, for incompressible flows with very mild or almost absent temperature variation, μ can be treated as a constant. Using these assumptions, the vector form of the NSE for incompressible flow is obtained as

$$\frac{\partial \vec{V}}{\partial t} + (\vec{V} \cdot \nabla) \vec{V} = -\frac{\nabla p}{\rho} + \nu \nabla^2 \vec{V} + \vec{F} \tag{2.15}$$

This is also called the equations in *primitive variables* or the *primitive variable formulation* of incompressible NSE.

2.1.3   Energy Conservation Equation

Here, this is the first law of thermodynamics stated for a control volume system which is: The rate of change of energy inside the *control volume* must be the summation of heat transfer across the *control surface* and the work done by body and surface forces. Detailed derivations can be noted in many text books on fluid mechanics and CFD; see, e.g., [81]. The constituent terms of the energy equation are illustrated next.

Terms Due to Work Done by Body and Surface Forces:

The rate of work done by body forces for the *control volume* of mass $\rho\ (dx\ dy\ dz)$ is given in a mixed form as

$$\left[-\nabla \cdot (p \vec{V}) + \frac{\partial}{\partial x_j} (u_i \tau_{ji}) \right] dx\ dy\ dz + \rho\ \vec{F} \cdot \vec{V}\ dx\ dy\ dz \tag{2.16}$$

Terms Due to Heat Transfer:

The net flux of heat is due to volumetric heating such as absorption or emission of radiation and heat transfer across the control surface due to thermal conduction. If one defines the rate of volumetric heat addition per unit mass as $\dot{q}$, then the volumetric heating of the element is

$$= \rho\ \dot{q}\ dx\ dy\ dz \tag{2.17}$$

The directional conductive heat transfer through the control surfaces is related to corresponding temperature gradient, using Fourier's law by

$$\dot{q}_j = -\kappa \frac{\partial T}{\partial x_j}$$

where κ is the thermal conductivity. Hence the total heat interaction term is obtained by using Newton's law to provide

$$\left[\rho \, \dot{q} + \frac{\partial}{\partial x_j} \left(\kappa \frac{\partial T}{\partial x_j} \right) \right] dx \, dy \, dz \qquad (2.18)$$

Terms Due to the Rate of Change of Energy:

For a moving fluid element, $E = e + \frac{V^2}{2}$, where E, e and $\frac{V^2}{2}$ represent total, internal and kinetic energy per unit mass. The time rate of change of E is given by the substantive derivative as

$$\rho \frac{D}{Dt} \left(e + \frac{V^2}{2} \right) dx \, dy \, dz \qquad (2.19)$$

The Final Form:

The final form of the energy equation is obtained by collating all the terms illustrated above to obtain the *non-conservation* or *convective* form of the energy equation as

$$\rho \frac{D}{Dt} \left(e + \frac{V^2}{2} \right) = \left[\rho \, \dot{q} + \frac{\partial}{\partial x_j} \left(\kappa \frac{\partial T}{\partial x_j} \right) \right] - \nabla \cdot (p \vec{V}) + \frac{\partial}{\partial x_j} (u_i \tau_{ji}) + \rho \, \vec{F} \cdot \vec{V}$$
$$(2.20)$$

2.1.4 Alternate Forms of the Energy Equation

Occasionally, the energy equation is written in terms of internal energy only. This is obtained as

$$\rho \frac{De}{Dt} = \rho \, \dot{q} + \frac{\partial}{\partial x_j} \left(\kappa \frac{\partial T}{\partial x_j} \right) - p \nabla \cdot \vec{V} + \lambda (\nabla \cdot \vec{V})^2 + 2\mu \left[\left(\frac{\partial u}{\partial x} \right)^2 + \left(\frac{\partial v}{\partial y} \right)^2 + \left(\frac{\partial w}{\partial z} \right)^2 \right]$$

$$+ \mu \left[\left(\frac{\partial u}{\partial y} + \frac{\partial v}{\partial x} \right)^2 + \left(\frac{\partial u}{\partial z} + \frac{\partial w}{\partial x} \right)^2 + \left(\frac{\partial v}{\partial z} + \frac{\partial w}{\partial y} \right)^2 \right] \qquad (2.21)$$

The energy equation in conservation form is obtained from the above by noting

$$\rho \frac{De}{Dt} = \frac{\partial}{\partial t} (\rho e) + \nabla \cdot (\rho e \vec{V})$$

Terms involving λ and μ constitute the dissipation term Φ_0 in Eq. (2.21) .

2.1.5 Vorticity Transport Equation for Incompressible Flows

In fluid flows, vorticity, $\vec{\omega}$ is defined as the curl of the velocity field, i.e., $\vec{\omega} = \nabla \times \vec{V}$. One of the problems of primitive variable formulation is the absence of unique and definitive boundary conditions for pressure and this is often avoided by using an alternative formulation that does not have explicit pressure dependent terms. For NSE corresponding to incompressible flows, one can eliminate it by taking a curl of Eq. (2.15) and using the kinematic relation between vorticity and velocity. This gives the *non-conservative form* of vorticity transport equation (VTE) as

$$\frac{\partial \vec{\omega}}{\partial t} + (\vec{V} \cdot \nabla)\vec{\omega} = (\vec{\omega} \cdot \nabla)\vec{V} + \nu \nabla^2 \vec{\omega} \tag{2.22}$$

Note that the first term on the right hand side of Eq. (2.22) is the vortex stretching term, which is identically zero for 2D flows. Using the vector identity $\nabla \times (\vec{A} \times \vec{B}) = \vec{A}(\nabla \cdot \vec{B}) - \vec{B}(\nabla \cdot \vec{A}) + (\vec{B} \cdot \nabla)\vec{A} - (\vec{A} \cdot \nabla)\vec{B}$ and divergence-free condition on velocity and vorticity ($D_v = \nabla \cdot \vec{V} = 0$ and $D_\omega = \nabla \cdot \vec{\omega} = 0$), one gets the *Laplacian form* of VTE as,

$$\frac{\partial \vec{\omega}}{\partial t} + \nabla \times (\vec{\omega} \times \vec{V}) = \nu \nabla^2 \vec{\omega} \tag{2.23}$$

The viscous term in the Laplacian form of VTE can be further modified by noting that

$$\nabla^2 \vec{\omega} = \nabla(\nabla \cdot \vec{\omega}) - \nabla \times (\nabla \times \vec{\omega})$$

Thus, $\nabla^2 \vec{\omega} = -\nabla \times (\nabla \times \vec{\omega})$, since $\nabla \cdot \vec{\omega} = 0$.

Modifying the right hand side of Eq. (2.23) using the above relation, one gets the *rotational form* of VTE as,

$$\frac{\partial \vec{\omega}}{\partial t} + \nabla \times (\vec{\omega} \times \vec{V}) + \nu \nabla \times (\nabla \times \vec{\omega}) = 0 \tag{2.24a}$$

which can also be written in a concise form, by denoting $\vec{H} = (\vec{\omega} \times \vec{V} + \nu \nabla \times \vec{\omega})$ and thus,

$$\frac{\partial \vec{\omega}}{\partial t} + \nabla \times \vec{H} = 0 \tag{2.24b}$$

One notes that ideally the vorticity field should always be divergence free, i.e., $D_\omega = \nabla \cdot \vec{\Omega} = 0$, which comes from the kinematic relation between velocity and vorticity. For 2D flows, this condition is always satisfied ($D_\omega = \nabla \cdot \vec{\omega} = 0$). However, for numerical computation of 3D flows in derived variable formulation, this is not automatically guaranteed. For wall-bounded flows, vorticity is continually generated at the no-slip wall and if the generated wall vorticity is not made divergence free, this will be a continual source of error, if left unchecked. Therefore, one needs to chose judiciously the form of VTE for 3D simulations. This can be performed by deriving the governing equation for D_ω for the three forms of VTE as given in Eqs. (2.22)–(2.24a).

For the non-conservative form of the VTE (Eq. (2.22)), the evolution equation of D_ω is given as

$$\frac{\partial D_\omega}{\partial t} + \vec{V} \cdot (\nabla \cdot D_\omega) - \vec{\omega} \cdot (\nabla \cdot D_v) = \frac{1}{Re} \nabla^2 D_\omega \qquad (2.25a)$$

One readily identifies that this evolution equation of D_ω represents a unsteady convection-diffusion equation with a source term given as $S_\omega = \vec{\omega} \cdot (\nabla \cdot D_v)$. In order to have the source term $S_\omega = 0$, one must have $D_v = 0$ at all the points inside the computational domain. The necessary and essential conditions to ensure $D_\omega = 0$ in the computational domain for all times are, (i) $D_\omega = 0$ at $t = 0$ in the complete domain, (ii) $D_\omega = 0$ on the boundary of the domain for all $t > 0$ and (iii) the divergence of velocity D_v is zero for all $t > 0$ in the full domain. The evolution equation of D_ω for the Laplacian form of VTE (Eq. (2.23)), represents a unsteady diffusion equation given as

$$\frac{\partial D_\omega}{\partial t} = \frac{1}{Re} \nabla^2 D_\omega \qquad (2.25b)$$

Hence, the satisfaction of only first two conditions are necessary for D_ω to be zero for all times inside the computational domain. For the rotational form of VTE given by Eq. (2.24a), the evolution equation of D_ω is given simply by,

$$\frac{\partial D_\omega}{\partial t} = 0 \qquad (2.25c)$$

which requires only the satisfaction of the first condition to ensure that $D_\omega = 0$ for $t \geq 0$, inside the computational domain. Hence, one notes that the rotational form of VTE is much superior to the other two forms and requires very much less stringent condition to preserve the solenoidality condition on vorticity vector for 3D flows.

In *velocity-vorticity formulation* of the incompressible NSE, one augments the transport equations by deriving an auxiliary relation from the kinematic definition of vorticity. For example, taking a curl of vorticity expressed in the inertial frame, one obtains

$$\nabla^2 \vec{V} = -\nabla \times \vec{\omega} \qquad (2.26)$$

Any form of VTE given in Eqs. (2.22)–(2.24a) along with the velocity Poisson equation, Eq. (2.26), constitute the governing *velocity-vorticity* equations in the derived variable formulation. In *velocity–vorticity* formulation of incompressible NSE, the satisfaction of the divergence free condition on velocity (i.e., $D_v = 0$) is subject to the accuracy up to which Eq. (2.26) is solved [8]. Therefore, some researchers have not used Eq. (2.26) for computing the velocity field. For example, authors in [31–33, 65] have directly used the continuity equation and the kinematic definition of vorticity, along with the VTE. However, such approaches lead to an over–determined system of linear algebraic set of equations and consequently, special treatment is required to solve these numerically.

Another approach is the use of vector potential $\vec{\psi}$, instead of the velocity vector. Vector potential $\vec{\psi}$ is defined such that the velocity field $\vec{V}$ is the curl of $\vec{\psi}$ i.e.,

$$\vec{V} = \nabla \times \vec{\psi} \qquad (2.27)$$

One notes that for any given velocity field $\vec{V}$, if an uniquely defined $\vec{\psi}$ is obtained then $D_v = 0$. Taking curl of Eq. (2.27) and also imposing the restriction that $\vec{\psi}$ is divergence free, i.e., $\nabla \cdot \vec{\psi} = 0$, one gets the equation for $\vec{\psi}$ as

$$\nabla^2 \vec{\psi} = -\vec{\omega} \qquad (2.28)$$

In 2D, only relevant component of both vector potential and vorticity is the one which is normal to the plane of the flow and consequently, these equations are simplified to yield *stream function-vorticity* formulation ((ψ, ω)-formulation) for 2D flows. For 2D flows, the (ψ, ω)-formulation possess certain significant advantages over the primitive variable formulation. These are due to (i) lesser number of unknowns as compared to three variables in ($p, \vec{V}$)-formulation; (ii) exact satisfaction of mass conservation everywhere in the flow field; (iii) obtaining the primary quantity, namely the vorticity, directly from the governing equation and not by numerical differentiation of computed quantities and (iv) removing the ambiguity regarding the prescription of a suitable boundary condition for pressure. However, it is not straight-forward to specify boundary conditions in terms of $\vec{\psi}$ for 3D problems, in general. A general mathematical formulation of the boundary conditions on $\vec{\psi}$ for 3D flows are given in Hirasaki and Hellums [40, 41]. Solution of 3D NSE using vector potential ($\vec{\psi}$)- vorticity ($\vec{\omega}$) formulation for the flows are solved in Wong and Reizes [129] for 3D duct flows for $Re = 10$ and 100; for cubic lid-driven cavity (LDC) at $Re = 500$ and 3200 in Weinan and Liu [126]; for computing nonlinear stability of rotating Hagen–Poiseuille flow (RHPF) in Ortega-Casanova and Fernandez-Feria [64]; in Holdeman [43] for cubic LDC problem at $Re = 400$ and fully developed flow inside 3D open duct problems using finite element method. The difficulty in prescribing the boundary condition in $\vec{\psi}$ for multiply-connected domains are addressed in Wong and Reizes [130], where an alternative is proposed to determine the boundary condition, by introducing an over-relaxation factor. This approach was shown to accelerate the convergence of several stable free and forced convection cases inside annular cavities.

Equation (2.28) is derived based on the assumption that $\nabla \cdot \vec{\psi} = 0$. However, it is not guaranteed for the solution of $\vec{\psi}$ obtained from Eq. (2.28). For $\vec{\psi}$ obtained from Eq. (2.28) to be divergence-free, the associated vorticity field also needs to be solenoidal. If the computed $\vec{\psi}$ is not solenoidal, then following Helmholtz decomposition, one can express $\vec{\psi}$ as

$$\vec{\psi} = \nabla \Phi + \vec{\tilde{\psi}}$$

where $\vec{\Psi}$ is a solenoidal vector field such that at the boundary of the domain the normal components of $\vec{\Psi}$ and $\vec{\psi}$ are identical. Using vector identity, $\nabla \times \nabla\Phi = 0$, one readily notes from this relation that $\vec{V} = \nabla \times \vec{\Psi} = \nabla \times \vec{\psi}$. Therefore, even if the computed vector potential field is not divergence-free, it does not affect the computation of the associated velocity field.

2.1.6 Derived Variable Formulation for 2D Incompressible Navier–Stokes Equation

For 2D incompressible flows, stream-function and vorticity has only one component, which is perpendicular to the plane of the flow. The non-dimensional governing 2D Navier–Stokes equation in (ψ, ω)-formulation is given as

$$\frac{\partial \omega}{\partial t} + \frac{\partial \psi}{\partial y} \frac{\partial \omega}{\partial x} - \frac{\partial \psi}{\partial x} \frac{\partial \omega}{\partial y} = \frac{1}{Re} \nabla^2 \omega \tag{2.29}$$

$$\nabla^2 \psi = -\omega \tag{2.30}$$

Equations (2.29) and (2.30) are the VTE and stream-function equation (SFE), respectively. Often one need to solve these equations in transformed computational domain, so that effects of grid-point clustering at suitable locations can be taken care of. In the transformed computational (ξ, η)-plane, Eqs. (2.29) and (2.30) convert to the following form [81, 83]

$$h_1 h_2 \frac{\partial \omega}{\partial t} + \frac{\partial \psi}{\partial \eta} \frac{\partial \omega}{\partial \xi} - \frac{\partial \psi}{\partial \xi} \frac{\partial \omega}{\partial \eta} = \frac{1}{Re}\left[\frac{\partial}{\partial \xi}\left(\frac{h_2}{h_1}\frac{\partial \omega}{\partial \xi}\right) + \frac{\partial}{\partial \eta}\left(\frac{h_1}{h_2}\frac{\partial \omega}{\partial \eta}\right)\right] \tag{2.31}$$

$$\frac{\partial}{\partial \xi}\left(\frac{h_2}{h_1}\frac{\partial \psi}{\partial \xi}\right) + \frac{\partial}{\partial \eta}\left(\frac{h_1}{h_2}\frac{\partial \psi}{\partial \eta}\right) = -h_1 h_2 \omega \tag{2.32}$$

In Eqs. (2.31) and (2.32), h_1 and h_2 refer to the scale-factors along transformed ξ- and η-directions for an orthogonal grid. These are given as,

$$h_1 = \sqrt{\left(\frac{\partial x}{\partial \xi}\right)^2 + \left(\frac{\partial y}{\partial \xi}\right)^2}$$

$$h_2 = \sqrt{\left(\frac{\partial x}{\partial \eta}\right)^2 + \left(\frac{\partial y}{\partial \eta}\right)^2} \tag{2.33}$$

The contra-variant components of the velocity vector are given by

$$u = \frac{1}{h_2} \frac{\partial \psi}{\partial \eta}$$

$$v = -\frac{1}{h_1} \frac{\partial \psi}{\partial \xi} \tag{2.34}$$

In Eqs. (2.29) and (2.31), we used the nondimensional Reynolds number which is defined as

$$Re = \frac{U_{ref} L_{ref}}{\nu}$$

where, U_{ref}, L_{ref} and ν are reference integral velocity and length scales and kinematic viscosity, respectively. For the velocity-vorticity formulation (($\vec{V}$, ω)-formulation), VTE can be written in either conservative or nonconservative formulation. The non-conservative formulation of VTE in transformed computational (ξ, η)-plane can be obtained from Eqs. (2.31) as

$$h_1 h_2 \frac{\partial \omega}{\partial t} + h_2 u \frac{\partial \omega}{\partial \xi} + h_1 v \frac{\partial \omega}{\partial \eta} = \frac{1}{Re} \left[\frac{\partial}{\partial \xi} \left(\frac{h_2}{h_1} \frac{\partial \omega}{\partial \xi} \right) + \frac{\partial}{\partial \eta} \left(\frac{h_1}{h_2} \frac{\partial \omega}{\partial \eta} \right) \right] \tag{2.35}$$

while the corresponding conservative form of VTE is given as

$$h_1 h_2 \frac{\partial \omega}{\partial t} + \frac{\partial (h_2 u \omega)}{\partial \xi} + \frac{\partial (h_1 v \omega)}{\partial \eta} = \frac{1}{Re} \left[\frac{\partial}{\partial \xi} \left(\frac{h_2}{h_1} \frac{\partial \omega}{\partial \xi} \right) + \frac{\partial}{\partial \eta} \left(\frac{h_1}{h_2} \frac{\partial \omega}{\partial \eta} \right) \right] \tag{2.36}$$

The continuity equation (i.e., $\nabla \cdot \vec{V} = 0$) in the (ξ, η)-plane is given as

$$\frac{1}{h_1 h_2} \left[\frac{\partial (h_2 u)}{\partial \xi} + \frac{\partial (h_1 v)}{\partial \eta} \right] = 0 \tag{2.37}$$

In the ($\vec{V}$, ω)-formulation, attendant velocity components are obtained from the velocity Poisson equation (Eq. (2.26)) and for u- and v-components, these equations in the transformed plane are given as

$$\left(\frac{1}{h_1} \frac{\partial}{\partial \xi} \left[\frac{1}{h_1 h_2} \frac{\partial (h_2 u)}{\partial \xi} \right] + \frac{1}{h_2} \frac{\partial}{\partial \eta} \left[\frac{1}{h_1 h_2} \frac{\partial (h_1 u)}{\partial \eta} \right] \right) +$$
$$\left\{ \frac{1}{h_1} \frac{\partial}{\partial \xi} \left[\frac{1}{h_1 h_2} \frac{\partial (h_1 v)}{\partial \eta} \right] - \frac{1}{h_2} \frac{\partial}{\partial \eta} \left[\frac{1}{h_1 h_2} \frac{\partial (h_2 v)}{\partial \xi} \right] \right\} = -\frac{1}{h_2} \frac{\partial \omega}{\partial \eta}$$
$$\left(\frac{1}{h_1} \frac{\partial}{\partial \xi} \left[\frac{1}{h_1 h_2} \frac{\partial (h_2 v)}{\partial \xi} \right] + \frac{1}{h_2} \frac{\partial}{\partial \eta} \left[\frac{1}{h_1 h_2} \frac{\partial (h_1 v)}{\partial \eta} \right] \right) + \tag{2.38}$$
$$\left\{ \frac{1}{h_2} \frac{\partial}{\partial \eta} \left[\frac{1}{h_1 h_2} \frac{\partial (h_2 u)}{\partial \xi} \right] - \frac{1}{h_1} \frac{\partial}{\partial \xi} \left[\frac{1}{h_1 h_2} \frac{\partial (h_1 u)}{\partial \eta} \right] \right\} = \frac{1}{h_1} \frac{\partial \omega}{\partial \xi}$$

One notes from Eq. (2.38) that unlike SFE in Eq. (2.32), the Poisson Equations for velocity components (Eq. (2.38)) contain mixed-derivatives. Consequently the discretized matrix for Eq. (2.38) would contain more terms along any row which generally slows down the convergence of these equations, when an iterative solver is used to obtain the solution of the resultant linear algebraic equations. However, when ξ- and η-directions are along physical x- and y-directions, respectively, so that $h_1 = h_1(\xi)$ and $h_2 = h_2(\eta)$, Eq. (2.38) can be simplified to

$$\frac{\partial}{\partial \xi}\left(\frac{h_2}{h_1}\frac{\partial u}{\partial \xi}\right) + \frac{\partial}{\partial \eta}\left(\frac{h_1}{h_2}\frac{\partial u}{\partial \eta}\right) = -h_1\frac{\partial \omega}{\partial \eta}$$
$$\frac{\partial}{\partial \xi}\left(\frac{h_1}{h_2}\frac{\partial v}{\partial \xi}\right) + \frac{\partial}{\partial \eta}\left(\frac{h_2}{h_1}\frac{\partial v}{\partial \eta}\right) = h_2\frac{\partial \omega}{\partial \xi}$$

The above equation is in self-adjoint form, similar to Eq. (2.32) for stream-function ψ.

2.2 Spatial and Temporal Scales for Transitional and Turbulent Flows

Transitional and turbulent flows are inherently unsteady, exhibiting wide-range of space and time scales. For turbulent flows, the existence of broadband spectra is a consequence of nonlinearity in the governing equations where transfer of energy takes place from one wavenumber (or frequency) component to neighboring wavenumber (or frequency) components. For 3D flows, the vortex-stretching terms (the first term on the right-hand side of Eq. (2.22)) are responsible for transferring energy from large to small scales. As these features are intrinsic to NSE, therefore these are also noted for later stages of transitional flows. For 2D flows also, there can exist an enstrophy cascade (see Davidson [23] for details), which causes a reverse migration of energy from small to large scale, a process known as the inverse cascade process.

Scales of turbulent flows are related to eddy sizes in the evolving flow field. The largest scale is associated with the integral dimension of the fluid dynamical system, denoted by l, at which the flow is fed with energy. For *homogeneous and isotropic turbulent flows*, Kolmogorov has shown that (see Tennekes and Lumley [114]) the smallest excited length scale (known as the *Kolmogorov scale*) is given by

$$\eta_K = (v^3/\varepsilon)^{\frac{1}{4}} \tag{2.39}$$

If we define u as the representative velocity-scale in the large scale (associated with kinetic energy per unit mass), then we can define a corresponding Reynolds number given by

$$Re = \frac{ul}{v}$$

Thus the largest and the smallest length scales of turbulent flows are related by

$$\frac{l}{\eta_K} = (Re)^{\frac{3}{4}} \tag{2.40}$$

For turbulent flows, most energetic structures occur at the lower wavenumbers scales. Consequently, the peak in the energy spectrum is obtained at lower wavenumbers. However, the dissipation peak is located at a higher wavenumber *as compared to the peak of energy spectrum*. This is due to the fact that the dissipation is given by $\nu||\nabla u||_2^2$. This can be shown from the energy budget of the disturbance field. In general, the energy spectrum depends on the wavenumber (k), dissipation (ε) and kinematic viscosity (ν).

The above description elaborates how wide range of scales are excited particularly in turbulent flows, at high Reynolds numbers. For computations at high Reynolds numbers via DNS, one needs to resolve these wide-range of scales. If the cut-off wavenumber is represented by k_c (related to η_K), then Eq. (2.40) can also be written as,

$$k_c l \approx Re^{\frac{3}{4}} \tag{2.41}$$

This equation is often used to state grid requirements for DNS. For 3D flows, this shows that the resolution requirement scales as $(Re^{3/4})^3$ or roughly about Re^2. In deriving Kolmogorov's scaling theory, it is said that there exist length scales shorter than those are directly excited (l), but larger than the Kolmogorov scale (η_K), for which the energy spectrum is independent of the viscous dissipation mechanism. At these intermediate scales – the *inertial subrange* – the structure of the energy spectrum ($E(k)$) is determined solely by nonlinear energy transfer (via the stretching term given by the first term on the right-hand side of Eq. (2.22)) by a cascade process and the overall energy flux is shown to depend as

$$E(k, \varepsilon) = C_k k^{-\frac{5}{3}} \varepsilon^{\frac{2}{3}} \tag{2.42}$$

The existence of an *inertial subrange* suggests some form of universality of flow structure, at this length scale. This is exploited in LES where the flow is computed by resolving all the way up to the *inertial subrange* and anything smaller than this is modeled via sub-grid scale (SGS) stress models.

An inhomogeneous turbulent flow-field is generally identified as having large-scale low-frequency coherent structures and seemingly random high-frequency turbulent fluctuations. These organized coherent structures in turbulent flows carry about 20% of total turbulent kinetic energy (TKE) and hence their role in determining turbulent flow dynamics cannot be underestimated. In fact, several researchers have identified the significant role these coherent structures play in determining various physical phenomena like noise radiation [113], low frequency oscillation of the separation zone in turbulent shock-boundary layer interactions [115], mixing and entrainment in turbulent free-shear layers and wakes [76]; swirling jets [66], low-frequency buffeting for transonic flow past an airfoil [21] and many other similar applications. Therefore, accurate prediction of the various dynamical features of these large-scale coherent structures are essential to obtain vital information of the flow-physics. However, the dynamics of these structures are also strongly dependent on the random incoherent fluctuations. Reynolds and Hussain [76] have showed that one has to fairly accurately model the drain of energy from the coherent structures to incoherent fluctuation to predict the dynamics of the former.

For wall-bounded flows, these organized structures show up as peaks and valleys in the near-wall region and in terms of wall units, they have lengths between 100 and 2000 units in the streamwise direction and have a spacing of about 50 units in the spanwise direction. These high energetic events occur at a height of about 20 to 50 units (i.e., in and around the buffer layer) − all these are statistical estimates. Additionally, these near-wall events are interspersed by bursting of these structures. After bursts, new intermediate scale motions ensue in the buffer layer; those are also streamwise and/or hair-pin vortices. Thus, the unsteadiness of the turbulent boundary layer is characterized by bursting frequency, even for an attached shear layer. For example, in the zero pressure gradient boundary layer, this critical frequency is roughly between 20 and 100% of the turbulence burst frequency (F_b), where $F_b = U_\infty/5\delta$, with δ as the shear layer thickness. For adverse pressure gradient flows, this critical frequency is between 6 and 28% of F_b.

In Fig. 2.1, various temporal scales excited in typical engineering flows are displayed for different speed regimes. For high Reynolds number flows, only the mean frequencies of large eddies are shown. It is quite clear that ranges of non-dimensional frequencies span over three orders of magnitude and numerical methods must resolve these scales for high Reynolds number flows.

2.3   Numerical Methods for Developing DNS/LES

Previous discussion highlights the importance of resolving broadband spatial and temporal scales for accurate computations of transitional and turbulent flows. Therefore, for any numerical computation of transitional and turbulent flows, the computing methodologies should be *spectrally accurate*. Moreover, the numerical methods should also preserve the spectral relationship between spatial and temporal scales i.e, the numerical methods should preserve the physical *dispersion relationship*. In computing the fluid dynamics governing equations accurately, therefore, both space-time dependence of the problem has to be considered together. This aspect of computing is very often overlooked, where spatial and temporal discretization is often decoupled, which can lead to serious dispersion errors [95, 101].

2.3.1   Waves − Building Blocks of a Disturbance Field

We begin by first highlighting the aspects of disturbance evolution which takes a flow from the laminar to the turbulent state. Irrespective of the mechanism being linear or non-linear, it is always possible to explain disturbance as being composed of Fourier–Laplace transforms. For space-time dependent systems, such disturbances

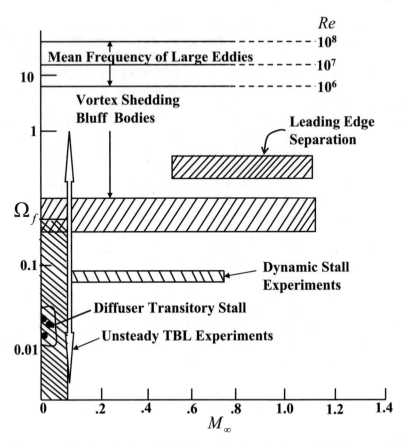

Fig. 2.1 Ranges of temporal scales excited in flows of engineering interest for different speeds and the corresponding frequency ranges (Ω_f). Also, indicated is the band for mean frequencies of large eddies in high Reynolds number flows

can often be viewed as plane waves. Waves can be dispersive or non-dispersive, governed by hyperbolic or non-hyperbolic partial differential equations. The prototype of hyperbolic waves is often taken as the one dimensional convection equation

$$\frac{\partial u}{\partial t} + c \frac{\partial u}{\partial x} = 0 \tag{2.43}$$

where u can represent any disturbance quantity. Consider the propagation of disturbances subject to initial conditions

$$u(x, 0) = f(x) \quad \text{for} \quad -\infty < x < \infty \tag{2.44}$$

The unbounded spatial domain makes this a Cauchy problem. In Eq. (2.44), $f(x)$ is considered continuous functions. The exact solution of Eq. (2.43) is given as

$$u(x, t) = f(x - ct)$$

The nature of this solution indicates that the initial disturbance propagates to the right at a convection speed of c, while maintaining identical amplitude and shape at all time instants. The building blocks of any arbitrary aggregation of plane waves can be understood by defining certain wave parameters for a single-periodic function

$$u(x, t) = a \sin\left[\frac{2\pi}{\lambda}(x - ct)\right] \tag{2.45}$$

One can identify this as a specific solution for Eq. (2.43). In the above, the quantity in the square brackets represents the phase of the wave and a represents the amplitude of the wave. The quantity λ is the wavelength, since u does not change when x is changed by λ, with t held fixed. One defines wavenumber k $(= \frac{2\pi}{\lambda})$, which provides the number of full waves in a length 2π. Thus, the representation of Eq. (2.45) can be alternately written as

$$u(x, t) = a \sin[k(x - ct)] \tag{2.46}$$

Keeping one's gaze fixed at a single point, the least time after which $u(x, t)$ retains the same value determines the time period T, and this is also the time required for the wave to travel one wavelength: $T = \frac{\lambda}{c}$. The number of oscillations at a point per unit time is the frequency given by $f_0 = \frac{1}{T}$. One can define a circular frequency $\bar{\omega}$ by noting

$$\bar{\omega} = kc \tag{2.47}$$

Thus, c $(= \bar{\omega}/k)$ has a dimension of speed and is appropriately called the phase speed, the rate at which the phase of the wave propagates. Such movement is not always physical and most often illusory. Equation (2.47) is known as the physical dispersion relation for obvious reasons.

One consequence of the dispersion property is the group velocity, V_g. The physical implications of the group velocity V_g is that it is the speed with which energy travels in a system displaying a wide-band spectrum. This has been recognized by researchers across many disciplines of science and engineering and studied in [3, 13, 46, 58, 127]. Rayleigh [73] laid the foundation for group velocity as opposed to phase speed – although it was discussed earlier in [38]. The carrier waves use the phase speed for phase variation, while the group velocity is associated with the propagation of amplitude. According to Brillouin [13], group velocity is the velocity of energy propagation and this was identified as signal speed by Rayleigh [74]. For linear dispersive systems following some conservation law, the group velocity is defined as

$$V_g = \frac{d\bar{\omega}}{dk} \tag{2.48}$$

Therefore, for 1D convection equation, the dispersion relation shows

$$\frac{d\bar{\omega}}{dk} = V_g = c \tag{2.49}$$

This signifies that the phase speed and group velocity are indistinguishable for non-dispersive systems like the one given in Eq. (2.43). However, this is not true for any general dispersive system. For such a system, the phase speed and the group velocity are not identical. For example, for deep water waves, the dispersion relation is given as [127]

$$\bar{\omega} = \sqrt{gk} \tag{2.50}$$

where g is the acceleration due to gravity. For such waves, the group velocity and the phase speed are given as

$$V_g = \frac{d\bar{\omega}}{dk} = \frac{1}{2}\sqrt{\frac{g}{k}}$$

$$c = \frac{\bar{\omega}}{k} = \sqrt{\frac{g}{k}} \tag{2.51}$$

Therefore, the group velocity is half the phase speed for deep water waves. Similarly, consider propagation of transverse elastic waves in a non-ideal string, for which the dispersion relation is given as

$$\bar{\omega}^2 = \frac{T}{\mu}k^2 + \alpha k^4 \tag{2.52}$$

where T is the tension in the string, μ is the string's mass per unit length and the parameter α depends on the stiffness of the string. One notes that for this system also the group velocity and the phase speed are not identical. A more detailed discussion from physical and mathematical stand-point on group velocity can be found in [82, 83, 127].

2.3.2   Resolution of Spatial Discretization

NSE basically governs space-time evolution of the primary variables. As noted in all of its forms described till now, it contains various forms of spatial derivatives. Therefore, the resolution of the discretization scheme to estimate these derivatives would invariably affect the accuracy of the numerical solution of NSE. Here, we talk about the resolution of the spatial discretization schemes for the estimation of first and second derivatives. We first begin with the numerical discretization of the first derivative.

Let, $u'_j = \left(\frac{\partial u}{\partial x}\right)_j$ represents the first derivative of the variable $u = u(x)$ at the jth-node. Let, all the nodes be equispaced with h representing the distance between

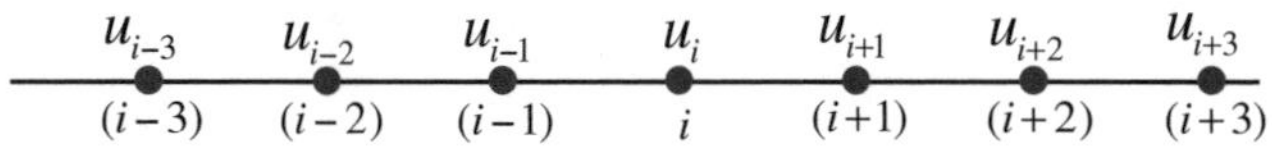

$$\underset{(i-3)}{\overset{u_{i-3}}{\bullet}} \quad \underset{(i-2)}{\overset{u_{i-2}}{\bullet}} \quad \underset{(i-1)}{\overset{u_{i-1}}{\bullet}} \quad \underset{i}{\overset{u_i}{\bullet}} \quad \underset{(i+1)}{\overset{u_{i+1}}{\bullet}} \quad \underset{(i+2)}{\overset{u_{i+2}}{\bullet}} \quad \underset{(i+3)}{\overset{u_{i+3}}{\bullet}}$$

Fig. 2.2 Schematic of grid-spacing. All the nodes are equispaced and the distance between adjacent grid-points is h

adjacent nodes, as shown in Fig. 2.2. The 2nd-order central difference (CD_2) scheme to evaluate u'_j is given as

$$u'_j = \frac{u_{j+1} - u_{j-1}}{2h} \tag{2.53}$$

Similarly, the 4th-order accurate CD_4-scheme for spatial discretization of the first derivative is given as

$$u'_j = \frac{-u_{j+2} + 8u_{j+1} - 8u_{j-1} + u_{j-2}}{12h} \tag{2.54}$$

To evaluate the resolution of a particular scheme, we express the function in terms of its Fourier–Laplace transform by

$$u_j = \int U(k)e^{ikx_j} dk \tag{2.55}$$

From Eq. (2.55), one notes that ideally

$$(u'_j)_{exact} = \int ikU(k)e^{ikx_j} dk \tag{2.56}$$

For CD_2-scheme in Eq. (2.53), u'_j can be written in spectral form as

$$(u'_j)_{CD_2} = \int \frac{e^{ikh} - e^{-ikh}}{2h} U(k)e^{ikx_j} dk \tag{2.57}$$

which gives

$$(u'_j)_{CD_2} = \int i\frac{\sin(kh)}{h} U(k)e^{ikx_j} dk = \int ik_{eq}U(k)e^{ikx_j} dk \tag{2.58}$$

where k_{eq} is the equivalent wavenumber. The term k_{eq}/k basically defines the resolution of any spatial discretization scheme. For CD_2 and CD_4 schemes, this term is given from Eqs. (2.53) and (2.54) as

$$\left(\frac{k_{eq}}{k}\right)_{CD_2} = \frac{\sin(kh)}{kh} \tag{2.59}$$

$$\left(\frac{k_{eq}}{k}\right)_{CD_4} = \left[\frac{(4 - \cos(kh))}{3}\right]\frac{\sin(kh)}{kh} \tag{2.60}$$

Note that for both Eqs. (2.59) and (2.60), $(k_{eq}/k) \longrightarrow 1$ as $kh \longrightarrow 0$. This is the *Consistency condition*. This must be true for any spatial discretization scheme for as $kh \longrightarrow 0$, one approaches the continuum from the discrete limit, and this implies $(k_{eq}/k) \longrightarrow 1$. Similar expression of resolution can also be derived for other higher-order discretization schemes, like CD_6 and CD_8 schemes. Note from Eqs. (2.59) and (2.60) for CD_2 and CD_4 schemes, (k_{eq}/k) is a real quantity. This is true for any central discretization scheme in uniform grids.

For upwinded spatial discretization schemes, (k_{eq}/k) is a complex quantity. For example, for first-order upwind scheme UD_1,

$$u'_j = \frac{u_{j+1} - u_j}{h} \tag{2.61}$$

and (k_{eq}/k) for UD_1 scheme is given as

$$\left(\frac{k_{eq}}{k}\right)_{UD_1} = \frac{\sin(kh)}{kh} + i\frac{(\cos(kh) - 1)}{kh} = \left(\frac{k_{eq}}{k}\right)_{real} + i\left(\frac{k_{eq}}{k}\right)_{img} \tag{2.62}$$

where, $(k_{eq}/k)_{real}$ and $(k_{eq}/k)_{img}$ are the real and imaginary part of (k_{eq}/k). Similarly for second order upwind UD_2 scheme,

$$u'_j = \frac{3u_j - 4u_{j-1} + u_{j-2}}{2h} \tag{2.63}$$

the complex resolution can be obtained as

$$\left(\frac{k_{eq}}{k}\right)_{UD_2} = \frac{\sin(kh)(2 - \cos(kh))}{kh} - i\frac{(\cos(kh) - 1)^2}{kh} \tag{2.64}$$

In Fig. 2.3, real and imaginary parts of (k_{eq}/k) are plotted as a function of modified wavenumber, kh, for indicated spatial discretization schemes. Here, $(k_{eq}/k)_{real}$ denotes the resolution of the particular spatial discretization scheme. On the other hand, value of $(k_{eq}/k)_{img}$ signifies the added numerical diffusion of the scheme. This can be understood considering the semi-discrete form (where time-integration is considered as exact and spatial discretization is as given for the chosen numerical schemes) of the 1D convection equation (Eq. (2.43)). As for any numerical scheme

$$\left(\frac{\partial u}{\partial x}\right)_j = \int ik_{eq}U(k, t)e^{ikx_j}dk \tag{2.65}$$

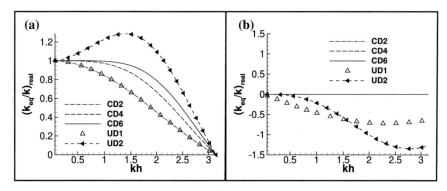

Fig. 2.3 **a** $(k_{eq}/k)_{real}$ and **b** $(k_{eq}/k)_{img}$ plotted for CD_2, CD_4, CD_6, UD_1 and UD_2 spatial discretization schemes as a function of kh

Expressing the variable $u(x, t)$ in terms of its Fourier–Laplace amplitude $U(k, t)$, Eq. (2.43) can be written as

$$\int \left(\frac{\partial U}{\partial t} + ik_{eq}U(k, t) \right) e^{ikx_j} dk = 0 \tag{2.66}$$

Therefore, for any wavenumber component k, the equation for $U(k, t)$ is given as

$$\left(\frac{\partial U}{\partial t} + ik_{eq}U(k, t) \right) = 0 \tag{2.67}$$

The exact solution of Eq. (2.67) can be obtained as

$$U(k, t) = U(k, 0) \, e^{-i(k_{eq}/k)_{real}kt} \, e^{(k_{eq}/k)_{img}kt} \tag{2.68}$$

Equation (2.68) clearly shows that $U(k, t)$ grows or decays exponentially with time, when $(k_{eq}/k)_{img}$ is positive or negative, respectively. It can be further shown that (by considering the spatial discretization schemes for second derivatives, which is illustrated next), effect of upwinding converts the 1D-convection equation (Eq. (2.43)) to

$$\frac{\partial u}{\partial t} + c\frac{\partial u}{\partial x} = \beta \frac{\partial^2 u}{\partial x^2} \tag{2.69}$$

where $\beta > 0$ if $(k_{eq}/k)_{img} < 0$. Thus, the role of upwinding is to induce a diffusive term when $(k_{eq}/k)_{img}$ is negative. A positive value of $(k_{eq}/k)_{img}$ is referred to as *anti-diffusion*, which is potentially destabilizing for the overall scheme via such spatial discretization.

Next, the resolution of the spatial discretization schemes for the evaluation of second derivatives are provided. The second order central difference scheme for the evaluation of second derivatives is given as

$$u''_j = \frac{u_{j+1} - 2u_j + u_{j-1}}{h^2} \tag{2.70}$$

where double-prime denotes the spatial second derivative. We refer this scheme as CD_2^2 to distinguish it from the second order central difference scheme (Eq. (2.53)) for the first derivative. Following Eq. (2.65), one can express the exact and the numerically obtained second derivative as

$$\left(\frac{\partial^2 u_j}{\partial x^2}\right)_{exact} = \int -k^2 U(k) e^{ikx_j} dk \tag{2.71}$$

$$u''_j = \left(\frac{\partial^2 u_j}{\partial x^2}\right)_{num} = \int -k_{eq}^2 U(k) e^{ikx_j} dk \tag{2.72}$$

where k_{eq} is the modified wavenumber for the evaluation of the second derivative. Therefore, similar to the cases for the numerical evaluation of first derivatives, here also (k_{eq}/k) would denote the resolution of the scheme to approximate second derivatives. The spectral resolution of the scheme of (2.70), in terms of the modified wavenumber k_{eq}, is given as

$$\left(\frac{k_{eq}}{k}\right)^2 = \frac{\sin^2(kh/2)}{(kh/2)^2} \tag{2.73}$$

It is noted that the scheme given by Eq. (2.70) is formally second order accurate in terms of the truncation error, as obtained from the Taylor series expansion. One can also approximate the second derivative by the following approximation while maintaining second order accuracy in terms of the truncation error as

$$u''_j = \frac{u_{j+2} - 2u_j + u_{j-2}}{4h^2} \tag{2.74}$$

We refer this scheme as CD_{2a}^2 to distinguish it from the CD_2^2 scheme given in Eq. (2.70). Though both the schemes are formally second order accurate, the distinction between these become clear when one plots the corresponding spectral resolution as a function of wavenumber. For scheme (2.74), the spectral resolution in terms of $(k_{eq}/k)^2$ is given as

$$\left(\frac{k_{eq}}{k}\right)^2 = \frac{\sin^2(kh)}{(kh)^2} \tag{2.75}$$

In Fig. 2.4, $(k_{eq}/k)_{real}$ is plotted for both the schemes as a function of kh. It becomes readily apparent from this figure that CD_2^2 scheme (Eq. (2.70)) has higher resolution as compared to the CD_{2a}^2 scheme (Eq. (2.74)), specially at higher wavenumbers.

Fig. 2.4 $(k_{eq}/k)_{real}$ plotted for the CD_2^2- and CD_{2a}^2-schemes as a function of kh. These schemes are given by Eqs. (2.70) and (2.74), respectively

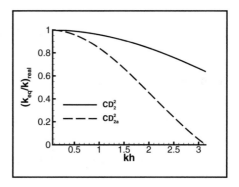

The resolution (k_{eq}/k) for CD_{2a}^2 scheme becomes zero at the Nyquist limit i.e., at $kh = \pi$. In contrast, $(k_{eq}/k) = (2/\pi)$ at $kh = \pi$ for the CD_2^2-scheme given by Eq. (2.70).

2.3.3   High-Accuracy Compact Schemes for Evaluation of First Derivatives

In the previous subsection, we have described various explicit schemes to discretize first and second derivatives. Here, we introduce *compact schemes*, in which the derivatives are treated as unknowns as function of the variables. The derivatives at the grid points are expressed as an auxiliary implicit equation involving the unknown. There are two essential features: high spectral accuracy and relatively compact stencil, to relate derivatives with the unknown. These methods provide near-spectral accuracy and robustness, at the same time being less costly than the spectral scheme. Historically, these methods owe their origin to the entire class of centered explicit spatial discretization using Padé schemes for ODE, described in [50]. Application of such schemes to PDEs can be found in many places and for CFD, these can be found in [53, 81].

The derivatives following the compact schemes are obtained by solving a system of linear algebraic equations. The compact representation of these schemes indicates that the resultant matrix is strictly tri-diagonal or penta-diagonal, so that the linear algebraic equations can be easily solved. The general stencil following compact schemes to evaluate the nth derivative of the variable u_j at the jth-node for a uniform grid of spacing h can be given as

$$\sum_{k=-N_1}^{N_2} \alpha_k u_{j+k}^{(n)} = \frac{1}{h^n} \sum_{l=-M_1}^{M_2} \beta_l u_{j+l} \tag{2.76}$$

where $u_j^{(n)}$ represents the nth derivative of u_j. It is apparent that the resultant matrix is band-limited with bandwidth defined by N_1 and N_2 points towards left and right of the jth-node. Let us consider the general stencil of a compact scheme for evaluation of first derivative (denoted by the superscript $'$) with $N_1 = N_2 = 2$ and $M_1 = M_2 = 3$. Such a scheme can be expressed as

$$\alpha_{+2}u'_{j+2} + \alpha_{+1}u'_{j+1} + \alpha_0 u'_j + \alpha_{-1}u'_{j-1} + \alpha_{-2}u'_{j-2} = \frac{1}{h}\Big(\beta_{+3}u_{j+3} + \beta_{+2}u_{j+2}+$$

$$\beta_{+1}u_{j+1} + \beta_0 u_j + \beta_{-1}u_{j-1} + \beta_{-2}u_{j-2} + \beta_{-3}u_{j-3}\Big) \tag{2.77}$$

One notes that if $\alpha_{+2} = \alpha_{-2} = 0$ then the resultant matrix for Eq. (2.77) would be tridiagonal and one can easily employ Thomas' tridiagonal matrix algorithm (TDMA) to solve the above system of equations to obtain u'_j. Using Taylor series expansion of both right- and left-hand side expressions, one can derive the consistency and other conditions. By matching the coefficient of u_j, the first consistency condition (corresponding to first order accuracy) can be obtained as

$$(\beta_{+3} + \beta_{+2} + \beta_{+1} + \beta_0 + \beta_{-1} + \beta_{-2} + \beta_{-3}) = 0 \tag{2.78}$$

Similarly, by matching the coefficient of u'_j from left- and right-hand sides, the second condition (corresponding to second order accuracy) can be obtained as

$$(\alpha_{+2} + \alpha_{+1} + \alpha_0 + \alpha_{-1} + \alpha_{-2})$$

$$= 3(\beta_{+3} - \beta_{-3}) + 2(\beta_{+2} - \beta_{-2}) + (\beta_{+1} - \beta_{-1}) \tag{2.79}$$

The general condition for $(p + 1)$th-order of accuracy (where $p \geq 2$), in terms of truncation error, can be given as

$$2^{p-1}(\alpha_{+2} + (-1)^{p-1}\alpha_{-2}) + (\alpha_{+1} + (-1)^{p-1}\alpha_{-1}) = \frac{3^p}{p}(\beta_{+3} + (-1)^p\beta_{-3})$$

$$+\frac{2^p}{p}(\beta_{+2} + (-1)^p\beta_{-2}) + \frac{1}{p}(\beta_{+1} + (-1)^p\beta_{-1}) \tag{2.80}$$

Equation (2.77) contains 12 unknown coefficients (α_{+2}, ..., α_{-2}, β_{+3}, ..., β_{-3}) to be determined and therefore, 11th-order of accuracy can be obtained from the compact scheme for spatial discretization, described in Eq. (2.77). However, in practice hardly such higher order accurate compact schemes are used. Instead, to obtain a better accuracy scheme, some of the higher order conditions are sacrificed and additional conditions are imposed based on the resolution (k_{eq}/k) in the spectral plane, as followed to develop high-accuracy compact schemes like $OUCS3$ in [92, 96]. One also notes from Eq. (2.77) that if $\alpha_{+k} = \alpha_{-k}$, $(\beta_{+l} + \beta_{-l}) = 0$ and $\beta_0 = 0$, where $k = 1, 2$ and $l = 1, 2, 3$, then one essentially obtains a central spatial discretization scheme. Following tridiagonal compact scheme is proposed for the evaluation of first derivative:

$$\alpha u'_{j+1} + u'_j + \alpha u'_{j-1} = c \frac{u_{j+3} - u_{j-3}}{6h} +$$
$$b \frac{u_{j+2} - u_{j-2}}{4h} + a \frac{u_{j+1} - u_{j-1}}{2h} \tag{2.81}$$

If $c = 0$, then for 4th-order accuracy one must have

$$a = \frac{2}{3}(\alpha + 2), \quad b = \frac{1}{3}(4\alpha - 1)$$

while $\alpha = 1/3$ makes the scheme 6th-order accurate. If $c \neq 0$ then sixth order accuracy can be obtained if

$$a = \frac{1}{6}(\alpha + 9), \quad b = \frac{1}{15}(32\alpha - 9), \quad c = \frac{1}{10}(-3\alpha + 1)$$

If $\alpha = 3/8$, then a formally eighth-order accurate scheme can be obtained. If the scheme given by Eq. (2.81) is applied to a periodic problems, its resolution in the spectral plane is given as

$$\left(\frac{k_{eq}}{k}\right) = \frac{1}{kh} \frac{a \sin(kh) + (b/2) \sin(2kh) + (c/3) \sin(3kh)}{1 + 2\alpha \cos(kh)} \tag{2.82}$$

The advantage gained in terms of accuracy, while using pentadiagonal compact schemes, can be effectively leveraged if spectrally optimized tridiagonal compact schemes are developed.

In [92, 96], high-accuracy optimized upwind compact scheme $OUCS3$ is described, which has the stencil at interior nodes given as

$$p_{j-1} u'_{j-1} + u'_j + p_{j+1} u'_{j+1} = \frac{1}{h} \sum_{n=-2}^{2} q_n u_{j+n} \tag{2.83}$$

where, $p_{j\pm1} = D \pm \frac{\eta_*}{60}$; $q_{\pm2} = \pm\frac{F}{4} + \frac{\eta_*}{300}$; $q_{\pm1} = \pm\frac{E}{2} + \frac{\eta_*}{30}$; $q_0 = -\frac{11\eta_*}{150}$; $D = 0.3793894912$; $E = 1.57557379$ and $F = 0.183205192$. Here, η is the upwind parameter and a value of zero implies the corresponding central scheme. This scheme is second order accurate. Additional coefficients of the scheme are found by maximizing the resolution in the spectral plane. The elaborate methodology of spectral optimization by which these coefficients are obtained for non-periodic stencils are provided in [92, 96]. For periodic problems, this essentially implies that one maximizes the following objective function

$$I_\gamma = \int_0^\gamma \left| 1 - \left(\frac{k_{eq}}{k}\right) \right| d(kh) \tag{2.84}$$

where $0 \leq \gamma \leq \pi$. In [96], $\gamma = \pi$ was chosen. Despite only second order accuracy of the scheme (Eq. (2.83)), its resolution in the spectral plane is significantly higher

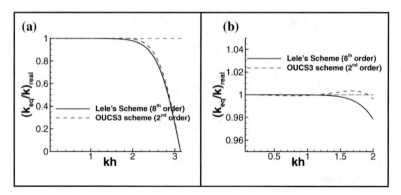

Fig. 2.5 (k_{eq}/k) plotted as a function of kh for the interior stencil of OUCS3 (Eq. (2.83)) and the eighth order compact schemes proposed by Lele (Eq. (2.81)). Here, we consider only the central stencil for periodic problems. For OUCS3, we take the central stencil, i.e., $\eta = 0$ in Eq. (2.83)

than explicit schemes like CD_2, CD_4 and CD_6. In Fig. 2.5, (k_{eq}/k) for compact schemes proposed by Lele [53] with eighth order accuracy and OUCS3 scheme (Eq. (2.83)) with second order accuracy are compared for interior stencil applied to periodic problems. For OUCS3 scheme, we have used the central stencil, i.e., $\eta = 0$ in Eq. (2.83). Figure 2.5 shows that all these schemes maintain $(k_{eq}/k) \simeq 1$ up to significantly larger range of kh. For OUCS3 scheme, $(k_{eq}/k) \simeq 1$, up to $kh \approx 2.3$, after which it starts to fall off and becomes zero at the Nyquist limit. Even though the scheme proposed in Lele [53] is eighth order accurate, its spectral resolution is slightly inferior than the central OUCS3 stencil, which is only second order accurate in terms of truncation error.

2.3.4 Boundary Closure Schemes for Compact Schemes to Evaluate First Derivative

So far we have only discussed about the compact schemes applied to periodic problems. For non-periodic problems, one needs stencils at near boundary points ($j = 1$, 2, $(N - 1)$ and N) to close the implicit system of equations, Eq. (2.81) or (2.83).

Adams [1] used the following stencils for non-periodic problem

$$2u'_1 + 4u'_2 = \frac{1}{h}(-5u_1 + 4u_2 + u_3) \quad \text{at } j = 1 \tag{2.85}$$

$$u'_1 + 4u'_2 + u'_3 = \frac{3}{h}(u_3 - u_1) \quad \text{at } j = 2 \tag{2.86}$$

$$u'_{j-1} + 3u'_j + u'_{j+1} = \frac{1}{12h}(-u_{j-2} - 28u_{j-1}$$
$$+ 28u_{j+1} + u_{j+2}) \quad \text{for } 3 \leq j \leq (N - 2) \tag{2.87}$$

Similar stencils can be obtained at $j = N - 1$ and $j = N$, following the stencils given by Eqs. (2.85) and (2.86), respectively. In the above, near-boundary point stencil at $j = 2$ has fourth order accuracy, while the interior points have sixth order formal accuracy.

From, Eqs. (2.85)–(2.87), it is clear that for non-periodic problems, use of compact stencil to evaluate first derivative result in solving the following linear algebraic equation [81, 83, 96]

$$[A]\{u'\} = \frac{1}{h}[B]\{u\} \tag{2.88}$$

where both $[A]$ and $[B]$ are $(N \times N)$ square matrices. Denoting $[C] = [A]^{-1}[B]$, one obtains

$$u' = \frac{1}{h}[C]u \tag{2.89}$$

It is pertinent to note that though, $[A]$ and $[B]$ are sparse matrices (here, $[A]$ is tridiagonal and $[B]$ contains non-zero elements only at three diagonal columns around the main diagonal entry), the matrix $[C]$ is in general a non-sparse matrix. Equation (2.89) implies that the derivative at the jth node is evaluated as

$$u'_j = \frac{1}{h} \sum_{l=1}^{N} C_{jl}\, u_l$$

where $u_l = u(x_l) = \int U(k)\, e^{ikx_l}\, dk$ is the function value at the lth node. Using spectral representation, one can alternately write the numerical derivative as

$$u'_j = \int \frac{1}{h} \sum C_{jl}\, U(k)\, e^{ik(x_l - x_j)}\, e^{ikx_j}\, dk \tag{2.90}$$

From Eq. (2.90), one obtains

$$\left(\frac{[k_{eq}]_j}{k}\right) = -\frac{i}{kh} \sum_{l=1}^{N} C_{jl}\, e^{ik(x_l - x_j)} \tag{2.91}$$

Although in physical plane computations C_{jl}'s are real, $([k_{eq}]_j/k)$ is in general complex, with real and imaginary parts representing numerical phase and added numerical diffusion, respectively, as illustrated before. It is determined by the numerical method fixing the entries of $[C]$. Following Eq. (2.91), one can obtain the real and imaginary parts of the complex resolution, (k_{eq}/k), at every grid point ranging from $j = 1$ to N. This is the global spectral analysis (GSA) of implicit compact schemes for non-periodic problems introduced in [81, 83, 96].

It can be shown by plotting (k_{eq}/k) that implicit boundary closure schemes cause numerical anti-diffusion at near-boundary points.

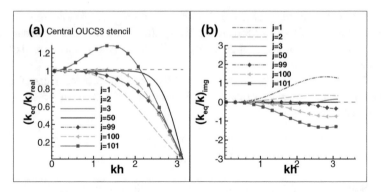

Fig. 2.6 **a** Real and **b** imaginary part of (k_{eq}/k) plotted for OUCS3 scheme with explicit boundary closure schemes given by Eqs. (2.92) and (2.93). For the interior OUCS3 stencil here we consider $\eta = 0$ in Eq. (2.83)

The problem of anti-diffusion at near-boundary points for compact schemes with implicit boundary closure can be addressed by prescribing explicit boundary closure schemes at those points, i.e., at $j = 1, 2, (N - 1)$ and N. In [96], special explicit boundary closure schemes have been derived while deriving $OUCS1, OUCS2, OUCS3$ schemes, whose interior stencil is given by Eq. (2.83). Following boundary closure schemes have been proposed in [96],

$$u'_1 = \frac{1}{2h}\left[-3u_1 + 4u_2 - u_3\right] \quad \text{at } j = 1 \tag{2.92}$$

$$u'_2 = \frac{1}{h}\left[\left(\frac{2\beta_*}{3} - \frac{1}{3}\right)u_1 - \left(\frac{8\beta_*}{3} + \frac{1}{2}\right)u_2 + \left(4\beta_* + 1\right)u_3\right.$$
$$\left. - \left(\frac{8\beta_*}{3} + \frac{1}{6}\right)u_4 + \frac{2\beta_*}{3}u_5\right] \quad \text{at } j = 2 \tag{2.93}$$

with β_* as a floating parameter. Similar closure schemes for $j = (N - 1)$ and N can also be written. The non-zero parameter β_* represents added fourth order diffusion, which reduces anti-diffusion at near-boundary points. In [96], it is suggested that for better global numerical properties, one requires $\beta_* = -0.025$ for $j = 2$ and $\beta_* = 0.09$ for $j = N - 1$. Figure 2.6 shows real and imaginary parts of (k_{eq}/k) for OUCS3 scheme with explicit boundary closure schemes given by Eqs. (2.92) and (2.93). One readily notes that use of explicit boundary closure scheme have reduced the problem of anti-diffusion at near-boundary points significantly.

2.3.5 Compact Schemes for Second Derivative Evaluation

Similar to the evaluation of first derivatives by compact schemes, one can also use such schemes to evaluate second derivatives. Lele [53] proposed the following compact stencils for the evaluation of second derivative in an uniform grid

$$\beta_2 u''_{j+2} + \alpha_2 u''_{j+1} + u''_j + \alpha_2 u''_{j-1} + \beta_2 u''_{j-2} =$$
$$c\frac{u_{j+3} - 2u_j + u_{j-3}}{9h^2} + b\frac{u_{j+2} - 2u_j + u_{j-2}}{4h^2} + a\frac{u_{j+1} - 2u_j + u_{j-1}}{h^2} \quad (2.94)$$

Comparing the coefficient of u''_j from right- and left-hand side of Eq. (2.94), one gets the consistency condition corresponding to second order accuracy as

$$(a + b + c) = (1 + 2\alpha_2 + \beta_2) \quad (2.95)$$

From the Taylor series expansion of Eq. (2.94), one gets a generalized equation for the $(2p + 2)$th-order term (with $p = 1, 2, 3, 4$) as

$$(3^{2p}c + 2^{2p}b + a) = \frac{(2p + 2)!}{(2p)!}(2^{2p}\beta_2 + \alpha_2) \quad (2.96)$$

If $\beta_2 = c = 0$, one obtains a tridiagonal compact scheme for the evaluation of second derivative as

$$\alpha_2 u''_{j+1} + u''_j + \alpha_2 u''_{j-1} = b\frac{u_{j+2} - 2u_j + u_{j-2}}{4h^2} +$$
$$a\frac{u_{j+1} - 2u_j + u_{j-1}}{h^2} \quad (2.97)$$

For sixth order accuracy of the stencil, Eq. (2.97), one requires

$$(a + b) = (1 + 2\alpha_2) \quad \text{for second order accuracy}$$
$$(a + 4b) = 12\alpha_2 \quad \text{for fourth order accuracy}$$
$$(a + 16b) = 30\alpha_2 \quad \text{for sixth order accuracy} \quad (2.98)$$

Solving Eqs. (2.98), one gets $\alpha_2 = 2/11$, $a = 12/11$ and $b = 3/11$ for sixth order accuracy. Therefore, stencil Eq. (2.97) and above values of a, b and α_2, defines a sixth order scheme for periodic problems. Similarly a single parameter, fourth order accurate, tridiagonal compact scheme can be derived for

$$a = \frac{4}{3}(1 - \alpha_2) \quad (2.99)$$

$$b = \frac{1}{3}(10\alpha_2 - 1) \quad (2.100)$$

Following Eq. (2.72), the resolution of the scheme Eq. (2.97) for periodic problems can be expressed in terms of the modified wavenumber k_{eq} as

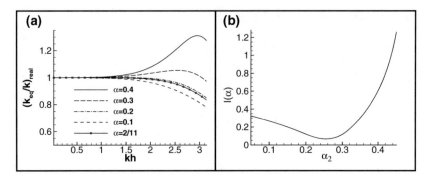

Fig. 2.7 a (k_{eq}/k) for the periodic tridiagonal compact stencil, Eq. (2.97) plotted for different values of α_2. $\alpha_2 = 2/11$ yields a sixth order accurate scheme. **b** Integral objective function I given by Eq. (2.102) plotted as a function of α_2 for $\gamma = \pi$

$$\left(\frac{k_{eq}}{k}\right)^2 = \frac{1}{(1 + 2\alpha_2 \cos(kh))} \frac{(b \sin^2(kh) + 4a \sin^2(kh/2))}{(kh)^2} \tag{2.101}$$

Therefore, an optimized tridiagonal compact scheme, for second derivative evaluation and applicable to periodic problems, in uniform grid can be derived, if one obtains α_2 by minimizing the following objective function

$$I = \int_0^{\gamma} \left| 1 - \left(\frac{k_{eq}}{k}\right)^2 \right| d(kh) \tag{2.102}$$

where $\gamma \leq \pi$. In Fig. 2.7a, $(k_{eq}/k)^2$ for the periodic tridiagonal compact stencil Eq. (2.97) is plotted for different values of α_2. As noted earlier, $\alpha_2 = 2/11$ yields a sixth order accurate scheme. Figure 2.7b shows the integral objective function I given by Eq. (2.102), as a function of α_2 for $\gamma = \pi$. Figure 2.7b shows that I is minimum when $\alpha_2 = 0.26$. Therefore, the fourth order accurate stencil Eq. (2.97) with $\alpha_2 = 0.26$ provides better spectral accuracy, than corresponding sixth order accurate stencil with $\alpha_2 = 2/11$ for periodic problems.

For non-periodic problems, one requires boundary closure stencils at near-boundary points. Lele [53] prescribed following boundary closure stencils for the compact scheme

$$u_1'' + 11u_2'' = \frac{1}{h^2}(13u_1 - 27u_2 + 15u_3 - u_4) \quad \text{at } j = 1 \tag{2.103}$$

$$\frac{1}{10}u_1'' + u_2'' + \frac{1}{10}u_3'' = \frac{12}{10h^2}(u_3 - 2u_2 + u_1) \quad \text{at } j = 2 \tag{2.104}$$

Similar stencils can be derived at $j = (N - 1)$ and $j = N$. One notes that Lele [53] proposed implicit boundary closure. As noted in Sect. 2.3.4 for the first derivative, implicit boundary closure schemes lead to anti-diffusion at near boundary points and spreads over a wider range of near-boundary points. For the scheme given in Eq. (2.94) with $\beta_2 = c = 0$, one can significantly remove numerical problems, if explicit second-order accurate schemes are used at the boundary nodes. The composite compact stencil can then be expressed as

$$\alpha_2 u''_{j+1} + u''_j + \alpha_2 u''_{j-1} = \frac{b}{4h^2} (u_{j+2} - 2u_j + u_{j-2})$$

$$+ \frac{a}{h^2} (u_{j+1} - 2u_j + u_{j-1}) \quad \text{for } 3 \leq j \leq (N - 3) \qquad (2.105)$$

$$u''_j = \frac{1}{h^2} (u_{j+1} - 2u_j + u_{j-1}) \quad \text{for } j = 2 \text{ and } j = (N - 1) \qquad (2.106)$$

$$u''_1 = \frac{1}{h^2} (2u_1 - 5u_2 + 4u_3 - u_4) \quad \text{for } j = 1 \qquad (2.107)$$

$$u''_N = \frac{1}{h^2} (2u_N - 5u_{N-1} + 4u_{N-2} - u_{N-3}) \quad \text{for } j = N \qquad (2.108)$$

Equation (2.88) expresses the linear algebraic equation corresponding to compact stencils for first derivative evaluation applied to non-periodic problems in equivalent matrix-vector form. Similar matrix-vector form for non-periodic compact stencils for second derivatives can be written as

$$[A]\{u''\} = \frac{1}{h^2} [B]\{u\} \qquad (2.109)$$

where both $[A]$ and $[B]$ are $(N \times N)$ square matrices. Denoting $[D] = [A]^{-1}[B]$, one obtains

$$\{u''\} = \frac{1}{h^2} [D]\{u\} \qquad (2.110)$$

Though, $[A]$ and $[B]$ are sparse matrices ($[A]$ is tridiagonal and $[B]$ contains non-zero elements only along three diagonals including the main diagonal entry), the matrix $[D]$ is not a sparse-matrix. From Eq. (2.110) the derivative at the jth node is evaluated as

$$u''_j = \frac{1}{h^2} \sum_{l=1}^{N} D_{jl} u_l$$

Using spectral representation, one can alternately write the second derivative as [81, 83],

$$u''_j = \int \frac{1}{h^2} \sum D_{jl} U(k) e^{ik(x_l - x_j)} e^{ikx_j} dk \qquad (2.111)$$

Following Eq. (2.111), one can perform a full domain spectral analysis for non-periodic compact stencils and express $(k_{eq}/k)^2$ at jth-node as

$$\left(\frac{[k_{eq}]_j}{k}\right)^2 = -\frac{1}{(kh)^2} \sum_{l=1}^{N} D_{jl}\, e^{ik(x_l - x_j)} \tag{2.112}$$

Therefore, $([k_{eq}]_j/k)^2$ is in general complex, with real and imaginary parts representing numerical resolution error and artificial numerical dispersion, respectively.

Next, the combined compact differencing (CCD) scheme is described [18]. This scheme has implicit sixth order stencils, where both first and second derivatives are evaluated simultaneously. The CCD scheme [18] for uniform grid-points are given as

$$u'_j + \alpha_{1*}(u'_{j+1} + u'_{j-1}) + \beta_{1*}h(u''_{j+1} - u''_{j-1}) = \frac{a_{1*}}{2h}(u_{j+1} - u_{j-1}) \tag{2.113}$$

$$u''_j + \alpha_{2*}(u''_{j+1} + u''_{j-1}) + \frac{\beta_{2*}}{2h}(u'_{j+1} - u'_{j-1}) = \frac{a_{2*}}{h^2}(u_{j+1} - 2u_j + u_{j-1}) \tag{2.114}$$

for sixth order accuracy $\alpha_{1*} = 7/16, \beta_{1*} = -1/16, a_{1*} = 15/8, \alpha_{2*} = -1/8, \beta_{2*} = 9/4$ and $a_{2*} = 3$. To obtain spectral resolution of the CCD scheme (both first and second derivatives), one has to express u_j by its Fourier–Laplace transform, as explained before. Let, $U(k)$, $U_1(k)$ and $U_2(k)$ be the Fourier–Laplace amplitude of u_j, u'_j and u''_j, respectively. Then considering a periodic problem, the equations for $U_1(k)$ and $U_2(k)$ in terms of $U(k)$ can be obtained from Eqs. (2.113) and (2.114) as

$$(1 + 2\alpha_{1*}\cos(kh))U_1(k) + 2i\beta_{1*}h\sin(kh)U_2(k) = 2i\frac{a_{1*}}{2h}\sin(kh)U(k) \tag{2.115}$$

$$2i\frac{\beta_{2*}}{2h}\sin(kh)U_1(k) + (1 + 2\alpha_{2*}\cos(kh))U_2(k) = \frac{a_{2*}}{h^2}(1 + 2\cos(kh))U(k) \tag{2.116}$$

solving Eqs. (2.115) and (2.116) for $U_1(k)$ and $U_2(k)$, one obtains

$$\frac{U_1(k)}{U(k)} = \frac{2i\sin(kh)}{2h}\frac{(a_{1*} - 2a_{2*}\beta_{1*}) + 2(a_{1*}\alpha_{2*} - 2a_{2*}\beta_{1*})\cos(kh)}{(1 + 2\alpha_{1*}\cos(kh))(1 + 2\alpha_{2*}\cos(kh)) + 2\beta_{1*}\beta_{2*}\sin^2(kh)} \tag{2.117}$$

$$\frac{U_2(k)}{U(k)} = \frac{1}{h^2}\frac{a_{1*}\beta_{2*}\sin^2(kh) + a_{2*}(1 + 2\cos(kh))(1 + 2\alpha_{1*}\cos(kh))}{2\beta_{1*}\beta_{2*}\sin^2(kh) + (1 + 2\alpha_{1*}\cos(kh))(1 + 2\alpha_{2*}\cos(kh))} \tag{2.118}$$

One can define k_{1eq} and k_{2eq}, the modified wavenumbers for the evaluation of first and second derivatives, respectively, following

$$(U_1(k)/U(k)) = ik_{1eq} \quad \text{and} \quad (U_2(k)/U(k)) = -k_{2eq}^2$$

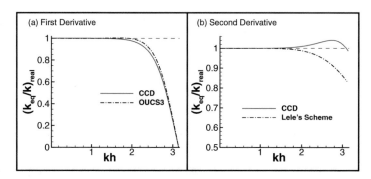

Fig. 2.8 **a** (k_{1eq}/k) and **b** $(k_{2eq}/k)^2$ plotted for the first and second derivatives corresponding to CCD scheme (Eqs. (2.113) and (2.114)) for periodic problems. Stencils of OUCS3-scheme Eq. (2.83) (first derivative) and the scheme due to Lele Eq. (2.97) (for second derivative) are also included in frames **a** and **b**, respectively

Therefore, the resolution of the periodic CCD scheme for the evaluation of first and second derivatives are given as

$$\left(\frac{k_{1eq}}{k}\right) = \frac{\sin(kh)}{kh} \frac{(a_{1*} - 2a_{2*}\beta_{1*}) + 2(a_{1*}\alpha_{2*} - 2a_{2*}\beta_{1*})\cos(kh)}{(1 + 2\alpha_{1*}\cos(kh))(1 + 2\alpha_{2*}\cos(kh)) + 2\beta_{1*}\beta_{2*}\sin^2(kh)}$$
(2.119)

$$\left(\frac{k_{2eq}^2}{k}\right)^2 = \frac{1}{(kh)^2} \frac{a_{1*}\beta_{2*}\sin^2(kh) + a_{2*}(1 + 2\cos(kh))(1 + 2\alpha_{1*}\cos(kh))}{2\beta_{1*}\beta_{2*}\sin^2(kh) + (1 + 2\alpha_{1*}\cos(kh))(1 + 2\alpha_{2*}\cos(kh))}$$
(2.120)

In Fig. 2.8, (k_{1eq}/k) (frame (a)) and $(k_{2eq}/k)^2$ (frame (b)) are plotted for the first and second derivatives, as obtained for CCD scheme given by Eqs. (2.113) and (2.114) for periodic problems. In the figures, corresponding results for OUCS3-scheme (Eq. (2.83) for first derivative) and for the scheme due to Lele (Eq. (2.97) for second derivative) are also included. The superior resolution characteristics of CCD-scheme is evident from this figure. The $(k_{eq}/k)^2$ for the second derivative shows slight overshoot close to the Nyquist limit. For NSE simulations with uniform grid-points, this is very effective because (a) it results in very high accuracy in terms of discretization of diffusion terms and (b) high value of $(k_{eq}/k)^2$ at high wavenumber range signifies adding numerical diffusion selectively at these high values of kh. The last aspect is computationally very helpful, as it leads to the damping of the amplitudes of high wavenumber components, thereby controlling several numerical sources of error, like spurious numerical errors due to aliasing [107].

For non-periodic problems, the CCD-scheme also requires boundary closure schemes. Chu and Fan [18] proposed the following stencils at $j = 1$ and $j = N$, while using CCD stencil (Eqs. (2.113) and (2.114)) from $j = 2$ to $(N - 1)$.

$$u_1' + 2u_2' - hu_2'' = \frac{1}{h}\left(-\frac{7}{2}u_1 + 4u_2 - \frac{1}{2}u_3\right) \tag{2.121}$$

$$u_1'' + 5u_2'' - \frac{6}{h}u_2' = \frac{1}{h^2}(9u_1 - 12u_2 + 3u_3) \tag{2.122}$$

$$u_N' + 2u_{N-1}' - hu_{N-1}'' = \frac{1}{h}\left(\frac{7}{2}u_N - 4u_{N-1} + \frac{1}{2}u_{N-2}\right) \tag{2.123}$$

$$u_N'' + 5u_{N-1}'' - \frac{6}{h}u_{N-1}' = \frac{1}{h^2}(9u_N - 12u_{N-1} + 3u_{N-2}) \tag{2.124}$$

In terms of the matrix-vector form, the linear algebraic equations, Eqs. (2.113)–(2.114) and (2.121)–(2.124), are obtained as

$$[A]\{u'\} + h[B]\{u''\} = \frac{1}{h}[C]\{u\} \tag{2.125}$$

$$\frac{1}{h}[D]\{u'\} + [E]\{u''\} = \frac{1}{h^2}[F]\{u\} \tag{2.126}$$

where $[A]$ and $[E]$ are symmetric tridiagonal matrices, whose diagonal elements are unity, whereas $[B]$ and $[D]$ are skew-symmetric tridiagonal matrices, whose diagonal elements are all zeros. Solving Eqs. (2.125) and (2.126), one obtains

$$\{u'\} = \frac{1}{h}[A - BE^{-1}D]^{-1}[C - BE^{-1}F]\{u\} \tag{2.127}$$

$$\{u''\} = \frac{1}{h^2}[E - DA^{-1}B]^{-1}[F - DA^{-1}C]\{u\} \tag{2.128}$$

Following Eqs. (2.127) and (2.128) a full domain spectral analysis can be carried out, as shown previously through Eqs. (2.91) and (2.112), respectively, for first and second derivatives. As noted earlier, for CCD also, implicit boundary closure schemes lead to numerical problems at near-boundary points. To overcome this difficulty explicit boundary closure schemes are proposed in [98] at $j = 2$ and $(N - 1)$, while the CCD-stencil is applied for $3 \le j \le (N - 2)$. This scheme with explicit boundary closure is termed as NCCD (acronym for *New CCD*) scheme. The explicit boundary stencils proposed in [98] at $j = 2$ are given as

$$u_2' = \frac{1}{h}\left[\left(\frac{2\beta_2}{3} - \frac{1}{3}\right)u_1 - \left(\frac{8\beta_2}{3} + \frac{1}{2}\right)u_2 + \left(4\beta_2 + 1\right)u_3\right.$$
$$\left. - \left(\frac{8\beta_2}{3} + \frac{1}{6}\right)u_4 + \frac{2\beta_2}{3}u_5\right] \tag{2.129}$$

$$u_2'' = \frac{1}{h^2}(u_1 - 2u_2 + u_3) \tag{2.130}$$

where $\beta_2 = -0.025$. Note that the closure scheme, Eq. (2.129), corresponds to the closure scheme of OUCS3 (see Eq. (2.93)). Similar stencils can be obtained at

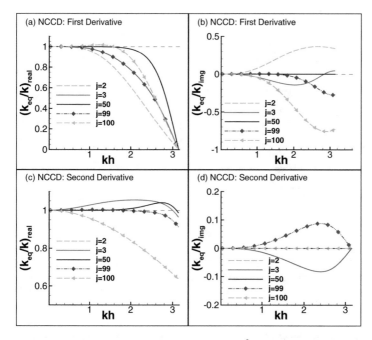

Fig. 2.9 Real and imaginary parts of $(k_{1eq}/k)_j$ and $(k_{2eq}/k)_j^2$ are plotted at indicated grid points for NCCD scheme [98] with explicit boundary closure schemes given by Eqs. (2.129)–(2.130)

$j = (N - 1)$, for which $\beta_{N-1} = 0.09$ is prescribed. One notes that because of the use of such explicit stencils at $j = 2$ and $(N - 1)$, no stencil is required at $j = 1$ or N. In Fig. 2.9, $(k_{1eq}/k)_j$ and $(k_{2eq}/k)_j$ are plotted at indicated grid points for NCCD scheme [98] (with explicit boundary closure schemes given by Eqs. (2.129) and (2.130)), respectively. One clearly notes from this figure that the numerical problems (in terms of added numerical anti-diffusion and dispersive effects) are mitigated to a considerable degree in the NCCD scheme. Moreover, the NCCD scheme also substantially improves the resolution of the second derivative at $j = 2$ and $(N - 1)$, compared to the original CCD scheme.

2.3.6 Compact Schemes for Interpolation and First Derivative Evaluation in Staggered Grids

Sometimes while solving NSE numerically, one needs staggered arrangement of variables, to achieve higher accuracy of the solution. For the velocity-vorticity formulation of NSE on orthogonal curvilinear grids, staggered grid arrangement of variables results in less error, as compared to solutions using non-staggered or collocated arrangement of variables [44]. In [70], it was shown while analyzing linearized

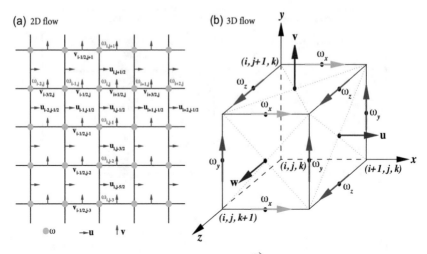

Fig. 2.10 **a** Staggered representation of variables in 2D ($\vec{V}$, ω)-formulation. Here, ω refers to the component of vorticity perpendicular to the 2D plane. **b** Staggered arrangement in 3D of u-, v- and w-velocity components and ω_x-, ω_y- and ω_z-vorticity components shown on an elementary cell

shallow water equations, that staggered arrangement of relevant variables provides best numerical accuracy.

If staggered grid is used for the solution of NSE, as in [8, 9], then one needs to interpolate and evaluate derivatives at the mid-point locations. For example, if the variables are defined at integral points, i.e., at jth nodes, then using staggered arrangement of variables, interpolation or evaluation of derivatives are required at $(j + 1/2)$th nodes. As an example of staggered arrangement of variables in Fig. 2.10b, staggered arrangement of velocity and vorticity components are shown for 3D problems, while using velocity-vorticity formulation of NSE [7–9]. The velocity components are defined at the center of the sides (for 2D, Fig. 2.10a) or at the center of the faces of the elementary cube in the transformed computational plane (for 3D, Fig. 2.10b). For 2D problems, the only relevant component of vorticity is ω_z, i.e., the vorticity component perpendicular to the 2D plane and in Fig. 2.10a, it is defined at the corner points of each elementary square cell. For 3D problems, however, all the three components of vorticity are to be considered and these components are defined at the center of each sides of the cell as shown in Fig. 2.10b.

The interpolation in staggered grids are carried out by the optimized compact mid-point interpolation scheme [62]. This interpolation scheme is given as

$$\alpha_I \hat{u}_{j-1} + \hat{u}_j + \alpha_I \hat{u}_{j+1} = \frac{a_I}{2}(u_{j-\frac{1}{2}} + u_{j+\frac{1}{2}}) + \frac{b_I}{2}(u_{j-\frac{3}{2}} + u_{j+\frac{3}{2}}) \qquad (2.131)$$

where $\hat{u}_j$'s are the interpolated values at the jth-location obtained from the known $u_{j\pm n/2}$ values at the $(j \pm n/2)$th-locations. For 4th order accuracy, one should have $a_I = \frac{1}{8}(9 + 10\alpha_I)$ and $b_I = \frac{1}{8}(6\alpha_I - 1)$, with α_I as a free parameter. Furthermore, a choice of $\alpha_I = \frac{3}{10}$ yields 6th order accuracy for the interpolation scheme. However, here $\alpha_I = 0.42$ is obtained by optimizing the integrated phase error of the scheme in the spectral plane to achieve better DRP properties as given in [7]. Using Fourier–Laplace transform one can express, $\hat{u}_j = \int \hat{U}(k) e^{ikx_j} dk$ and $u_j = \int U(k) e^{ikx_j} dk$. Then resolution properties and the performance of an interpolation scheme can be characterized by a transfer function given by $TF = \hat{U}(k)/U(k)$, which is the ratio of the Fourier amplitude of the interpolated function to that of the original function. From Eq. (2.131), this transfer function is obtained as,

$$TF(kh) = \left[\frac{a_I \cos(kh/2) + b_I \cos(3kh/2)}{1 + 2\alpha_1 \cos(kh)} \right] \qquad (2.132)$$

with uniform grid spacing h. Then to minimize the error in the spectral plane, one can define an objective function given as,

$$I_{interp} = \int_0^\gamma \left| 1 - TF(kh) \right| d(kh) \qquad (2.133)$$

which is nothing but the error integrated over the range of kh up to γ, which is less than or equal to the Nyquist limit. The TF given above is shown plotted as a function of kh in Fig. 2.11a, for the indicated values of α_I. The objective function I_{interp} is shown plotted in Fig. 2.11b, as a function of α_I for indicated values of γ. It is noted clearly that this objective function is minimum at $\alpha_I = 0.42$ for $\gamma = \pi$. This optimized fourth order scheme has been used here, for all reported computations. For non-periodic problems, as the flow inside a square lid-driven cavity, the interpolation scheme given by Eq. (2.131) has to be supplemented by the following boundary stencil given as

$$\hat{u}_1 = \frac{1}{2}(u_{\frac{1}{2}} + u_{\frac{3}{2}})$$
$$\hat{u}_N = \frac{1}{2}(u_{N-\frac{1}{2}} + u_{N+\frac{1}{2}}) \qquad (2.134)$$

Such explicit boundary closure does not induce any numerical instability at near-boundary points.

The evaluation of first derivative in staggered grids can be performed by a scheme, which is a variation of the scheme given in [62] for derivative evaluation in staggered grid. The general interior stencil for first derivative (indicated by primed quantity) is given by

$$\alpha_{II} u'_{j-1} + u'_j + \alpha_{II} u'_{j+1} = \frac{b_{II}}{3h}(u_{j+3/2} - u_{j-3/2}) + \frac{a_{II}}{h}(u_{j+1/2} - u_{j-1/2}) \qquad (2.135)$$

where u'_j s defined at the jth-location are the first derivatives of the known function $u_{j\pm n/2}$, defined at the $(j \pm n/2)$th-locations. Here h is the uniform grid-spacing. In [62], a fourth-order, single parameter (α_{II}) version was considered with $b_{II} =$

Fig. 2.11 a Transfer function of the staggered interpolation scheme given by Eq. (2.131) plotted as a function of kh for indicated values of α_I. **b** Integrated phase error plotted as a function of α_I for the stencil for indicated values of γ

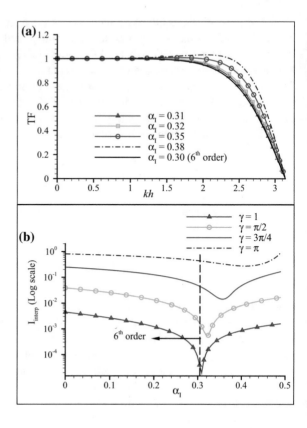

$$k_{eq} = \frac{2}{h}\left[\frac{a_{II}\sin(kh/2) + (b_{II}/3)\ \sin(3kh/2)}{1 + 2\alpha_{II}\cos(kh)}\right] \quad (2.136)$$

Therefore, one can define an objective function to be optimized as

$$I_{comp} = \int_0^{\gamma}\left|1 - \frac{k_{eq}}{k}\right|\ d(kh) \quad (2.137)$$

$(22\,\alpha_{II} - 1)/8$ and $a_{II} = (9 - 6\,\alpha_{II})/8$ and a sixth order version with $\alpha_{II} = 9/62$ was used. However to obtain better spectral properties, an optimized version of the integrated phase error in the spectral plane for Eq. (2.135) has been developed in [7]. Using Fourier–Laplace transform, one can define an equivalent wavenumber k_{eq} [98, 107] for the compact scheme in Eq. (2.135) as

This objective function I_{comp} is plotted as a function of α_{II} in Fig. 2.12b for the indicated values of γ, while in Fig. 2.12a, k_{eq}/k as a function of kh is plotted for different indicated values of α_{II}. It is noted from Fig. 2.12 that the optimum value of $\alpha_{II} = 0.216$ is obtained for $\gamma = \pi$. For non-periodic problems, a second order

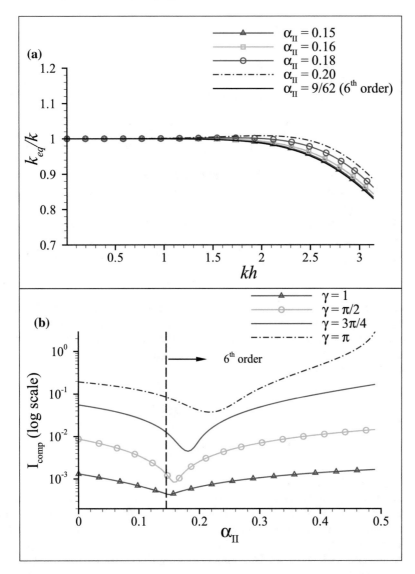

Fig. 2.12 **a** Spectral resolution of the scheme given by Eq. (2.135), shown for indicated values of α_{II}. **b** Integrated phase error plotted as a function of α_{II} for the stencil for the indicated values of γ

stencil can be used at $j = 1$ and N to avoid numerical problems. Such explicit closure schemes can be given as

$$u'_1 = \frac{1}{h}(u_{\frac{3}{2}} - u_{\frac{1}{2}})$$

$$u'_N = \frac{1}{h}(u_{N+\frac{1}{2}} - u_{N-\frac{1}{2}}) \tag{2.138}$$

2.3.7 Discretization of Self-Adjoint Terms

When NSE is solved in the transformed plane (as discussed in Sect. 2.1.6), one notes
the appearance of self-adjoint terms corresponding to viscous diffusion. For example,
consider the following two terms that appear in Eq. (2.31),

$$\frac{\partial}{\partial \xi}\left(\frac{h_2}{h_1}\frac{\partial \omega}{\partial \xi}\right) \quad \text{and} \quad \frac{\partial}{\partial \eta}\left(\frac{h_1}{h_2}\frac{\partial \omega}{\partial \eta}\right)$$

These self-adjoint terms can be generally represented as $(gf')'_j$ at the jth node in
terms of the functions g_j and f_j. Here, the function g_j is one of the grid transformation
terms as $(h_1/h_2)_j$ and $(h_1/h_2)_j$. For most of the 2D receptivity cases reported here,
these terms are discretized by second order central difference scheme CD_2 as,

$$(gf')'_j = \frac{1}{h}\left[\left(\frac{g_{j+1}+g_j}{2}\right)\left(\frac{f_{j+1}-f_j}{h}\right) - \left(\frac{g_j+g_{j-1}}{2}\right)\left(\frac{f_j-f_{j-1}}{h}\right)\right]$$

(2.139)

A new compact scheme for the discretization of these self-adjoint terms has been
developed in [88], which improves the effectiveness of diffusion discretization over
conventional CD_2-scheme given in Eq. (2.139). This scheme is named in [88] as the
self-adjoint compact difference scheme (SACD). Denoting $D_j = (gf')'_j$, the scheme
proposed in [88] is given as

$$\alpha_I D_{j-1} + D_j + \alpha_I D_{j+1} = \frac{\beta_I}{h^2}\left[g_{j-1/2}f_{j-1}\right.$$

$$\left. + \left(g_{j-1/2}+g_{j+1/2}\right)f_j + g_{j+1/2}f_{j+1}\right]$$

(2.140)

where $\beta_I = 1 + 2\alpha_I$, $g_{j-1/2} = (g_{j-1}+g_j)/2$ and $g_{j+1/2} = (g_j+g_{j+1})/2$. The
scheme given in Eq. (2.140) requires boundary closure scheme. For the variable
f_j defined in the nodes ranging from $j = 1$ to N, one only requires to evaluate the
diffusion terms in nodes ranging from $j = 2$ to $N - 1$. The boundary closure of the
SACD scheme is given by the conventional CD_2 scheme given by Eq. (2.139).

The spectral analysis of the SACD scheme is done by taking the Fourier–Laplace
transform of the functions, f and g. Here, the function f for a given node j is
transformed in the wavenumber plane as: $f_j = \int F(k)e^{ikx_j}dk$. In a similar fashion,
we represent, $g_j = \int G(k')e^{ik'x_j}dk'$. Hence, for the jth node, the exact derivative
can be written as,

$$D_j = -\frac{1}{h^2}\int\int k_o^2 F(k)G(k')e^{i(k+k')x_j}dk\,dk'$$

(2.141)

where $k_o^2 = (kk' + k^2)h^2$. From Eq. (2.140) for the interior nodes subject to the
boundary closure schemes given by Eq. (2.139), D_j can be written in the matrix
form as,

$$[A]\{D\} = \frac{1}{h^2}([B_1]\{f_j g_{j+1/2}\} + [B_2]\{f_j g_{j-1/2}\})$$ (2.142)

It has been shown in [88], that the expression for equivalent wavenumber at jth node, $k_{eq}|_j$ obtained for the discretized form can be expressed as,

$$k_{eq}^2|_j = -\left[\sum_{l=1}^{N}(C_{1jl}P_{jl} + C_{2jl}Q_{jl})\right]$$ (2.143)

so that,

$$D_j = -\frac{1}{h^2}\int\int k_{eq}^2 F(k)G(k')e^{i(k+k')x_j}dk\,dk'$$ (2.144)

where $P_{jl} = e^{ik(x_j-x_l)}e^{ik'(x_j-x_l+\frac{h}{2})}$, $Q_{jl} = e^{ik(x_j-x_l)}e^{ik'(x_j-x_l-\frac{h}{2})}$, $[C_1] = [A^{-1}B_1]$ and $[C_2] = [A^{-1}B_2]$. The quantity (k_{eq}^2/k_o^2) is in general a complex quantity with the real part indicating resolution of the scheme and the imaginary part representing the dispersion error. In Fig. 2.13, the real part of (k_{eq}^2/k_o^2) is plotted in the $(k'h, kh)$-plane for different values of α_I for $j = 50$, with $N = 100$. The coefficient $\alpha_I = 0$, represents the explicit $CD2$ scheme given by Eq. (2.139). The imaginary part of (k_{eq}^2/k_o^2) is found to be very negligible for this case, even at near-boundary points [88].

Ideally it is desirable to have the real part of (k_{eq}^2/k_o^2) equal to one for all values of k and k' in the $(kh, k'h)$-plane. From Fig. 2.13, it is observed that at the Nyquist limit $(kh = \pi)$, the value of of (k_{eq}^2/k_o^2) is $4/\pi^2$, for the explicit scheme $(\alpha_I = 0)$. Clearly, there is an improvement in the resolution properties as α_I is increased. This is noted from the value of (k_{eq}^2/k_o^2) at the Nyquist limit that increases to 0.49 for $\alpha_I = 0.05$; 0.6086 for $\alpha_I = 0.1$; 0.68 for $\alpha_I = 0.125$; 0.74 for $\alpha_I = 0.15$ and 0.95 for $\alpha_I = 0.2$. It is noted that the compact scheme over-estimates the second derivative for some combinations of kh and $k'h$, in all the cases shown in Fig. 2.13. This is a desirable property for effective control of aliasing errors for DNS/ LES. For $\alpha_I = 0.1$, the resolution at Nyquist limit is about 50% higher than that is achieved by the explicit scheme. For the same α_I, the compact scheme over-estimates the derivative by only 10.0%- that too only for a very small range of values in $(kh, k'h)$-plane. However, too much overestimation of (k_{eq}^2/k_o^2) can also alter the physical dispersion property of the problem. In [88], the bi-directional wave equation given by

$$\frac{\partial^2 f}{\partial t^2} = c^2 \frac{\partial^2 f}{\partial x^2}$$ (2.145)

has been solved in the transformed ξ-coordinate. The transformation function is given by $x(\xi) = \left[1 - \frac{\tanh\{\bar{\beta}(1-2\xi)\}}{\tanh(\bar{\beta})}\right]$. In the transformed plane, Eq. (2.145) is transformed to

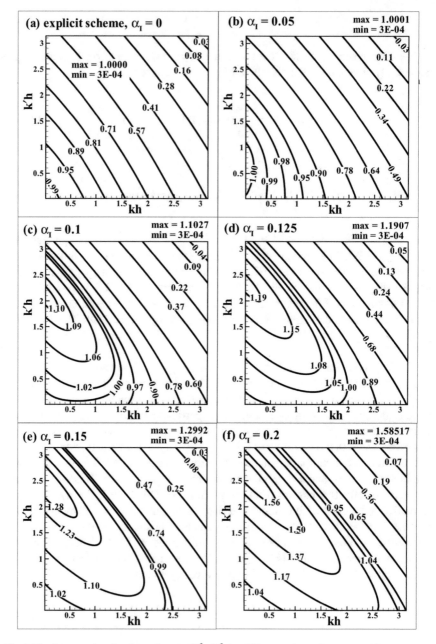

Fig. 2.13 Contour plots for the real part of k_{eq}^2 / k_o^2 for different values of α_I in $(kh, k'h)$-plane for 50th node, when total number points is $N = 100$ for the SACD scheme given by Eq. (2.140). The maximum and the minimum value for each cases have been indicated

$$\frac{\partial^2 f}{\partial t^2} = \frac{c^2}{x_\xi} \frac{\partial}{\partial \xi} \left(\frac{1}{x_\xi} \frac{\partial f}{\partial \xi} \right) \tag{2.146}$$

where $1/x_\xi$ represents the metric of the transformation. In [88], the squared integral error $SE = \sum_{j=1}^{N}(f_{exact} - f_{num})^2$ is plotted in $(\alpha_I, \bar{\beta})$-plane and it is found that least error is committed for $\alpha_I = 0.1$. This is proposed in [88] as the optimum value of α_I for the SACD scheme given by Eq. (2.140).

We have discussed optimized staggered compact scheme (OSCS) in Sect. 2.3.6. This is the optimized version of the staggered compact schemes proposed in [62]. One can also use these schemes twice to evaluate the self-adjoint diffusive terms. For the boundary closure, CD_2 scheme can be used. According to the proposed methodology, one has to first evaluate $f'_{j+1/2}$ (the first derivative of the function f at the $(j + 1/2)$th point) using the following stencil given as

$$\alpha_{II} f'_{j-1/2} + f'_{j+1/2} + \alpha_{II} f'_{j+3/2} = b_{II} \left(\frac{f_{j+2} - f_{j-1}}{3h} \right) + a_{II} \left(\frac{f_{j+1} - f_j}{h} \right) \tag{2.147}$$

subject to the following boundary stencils at $j = 3/2$ and $j = N - 1/2$ given as

$$f'_{3/2} = \frac{f_2 - f_1}{h}$$

$$f'_{N-1/2} = \frac{f_N - f_{N-1}}{h} \tag{2.148}$$

As mentioned in Sect. 2.3.6, Eq. (2.147) is a single parameter (α_{II}) family of fourth order schemes for $a_{II} = (9 - 6\alpha_{II})/8$ and $b_{II} = (-1 + 22\alpha_{II})/8$. After evaluating $f'_{j+1/2}$ from Eqs. (2.147) and (2.148) one proceeds to evaluate D_j using the stencil given as

$$\alpha_{II} D_{j-1} + D_j + \alpha_{II} D_{j+1} = b_{II} \left(\frac{\sigma_{j+3/2} - \sigma_{j-3/2}}{3h} \right)$$

$$+ a_{II} \left(\frac{\sigma_{j+1/2} - \sigma_{j-1/2}}{h} \right) \tag{2.149}$$

using the following boundary stencils at $j = 2$ and $j = N - 1$ given as

$$D_2 = \frac{\sigma_{5/2} - \sigma_{3/2}}{h}$$

$$D_{N-1} = \frac{\sigma_{N-1/2} - \sigma_{N-3/2}}{h} \tag{2.150}$$

where $\sigma_{j+1/2} = g_{j+1/2} f'_{j+1/2}$ and $g_{j+1/2} = (g_j + g_{j+1})/2$. The optimized value of $\alpha_{II} = 0.22$ is mentioned in Sect. 2.3.6, whereas $\alpha_{II} = 9/62$ makes Eqs. (2.147) and

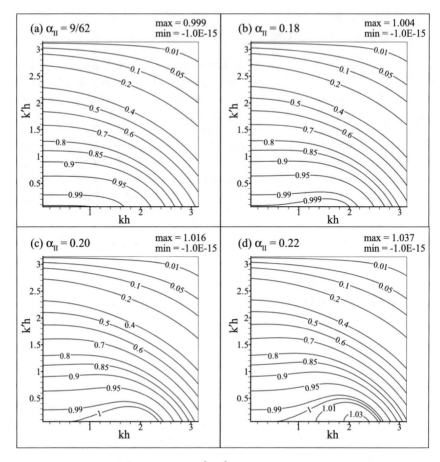

Fig. 2.14 Contour plots for the real part of $k_{eq}^2/k_\circ^2$ for different values of α_{II} in $(kh, k'h)$-plane for 50th node, when total number points is $N = 100$, using OSCS scheme twice to evaluate the second derivative, as given by Eqs. (2.147)–(2.150). The maximum and the minimum values for each cases have been indicated

(2.150) as 6th-order accurate [62]. A similar full-domain analysis can be performed for twice repeated OSCS scheme for evaluating self-adjoint terms. In Fig. 2.14, the real part of $(k_{eq}^2/k_\circ^2)$ is plotted for $j = 50$, with $N = 100$. It has been found that, $\alpha_{II} = 0.22$ provides the least error for the OSCS scheme. Although the proposed methodology requires the solution of tridiagonal matrix twice, as compared to only once for SACD scheme, it produces lesser error compared to the latter. This methodology is used for some 2D receptivity calculations and all 3D receptivity calculations, reported in later chapters.

2.3.8 Computing Methods for Unsteady Flows: Dispersion Relation Preserving (DRP) Methods

We have discussed the dispersion relation applicable to space-time dependent problems. In the context of transitional and turbulent flows, we have also noted the range of length and time scales in terms of spectra of such flows. It is noted that DNS would require preservation of the physical dispersion relation in the numerical sense. We have noted the role of group velocity in transporting energy of a dynamical system. One of the main aims of the present section is to relate group velocity with the dispersion property of the system numerically. However, apart from matching physical and numerical group velocity, one would also like to ensure the overall accuracy of the numerical solution and relate the accuracy with the properties of adopted numerical methods.

In this context, it has been noted that Eq. (2.43) serves well the purpose of a model equation to check the accuracy of any numerical method involving space and time discretization. We have already established that this is a non-dissipative and non-dispersive equation, which convects the initial solution unattenuated to the right at the speed c, representing both phase speed and group velocity.

This model equation is non-dispersive and provides a simple, yet a tough test for any combined space-time discretization method's dispersion property. We emphasize that the study of spatial and temporal discretizations separately is improper. Such studies can be found variously in [52, 111, 116]. Vichnevetsky and Bowles [120] have reported the dispersion property of a single finite difference and a finite element method by heuristic approaches. A proper estimate of the numerical dispersion property has been advanced in [92, 95, 101]. The explanation here is based on analysis of numerical schemes developed over the full computational domain in [96].

2.3.9 Numerical Amplification Factor for 1D Convection Equation

Analyses of numerical methods and associated error propagation study have been performed following different routes by many researchers. From the discretized governing equation, one can work backwards to obtain an equivalent differential equation, that has been effectively solved by the discrete equation, an approach advocated in [108]. According to this, if the truncation error terms are retained for space-time dependent problems, then the corresponding differential form is called the Γ-form. If the retained truncation error terms are for space derivatives alone, then the corresponding differential form is called the Π-form. While most analyses in the literature have used the Π-form, only in a few, the Γ-form has been utilized [93, 95]. The present approach follows the same route, where both space and time discretizations are considered simultaneously. The irrelevance of Π-form has been clearty explained in [95].

In the classical approach attributed to von Neumann, the evolving error of the discretized differential equation with linear constant coefficients is assumed to follow an identical discrete equation. In this approach, the difference between an exact and a computed solution arises due to round-off error and error in the initial data. For periodic problems, error is furthermore decomposed into Fourier series and the individual normal modes are investigated. This approach is also extended for nonlinear systems and linear systems with variable coefficients. For nonlinear systems, the principle of superposition does not hold and additionally problem of aliasing arises due to nonphysical transfer of energy to wrong wavenumbers and frequencies in the computational plane, whenever products are numerically evaluated with finite grid resolution.

There have been many efforts in analyzing error dynamics, using the method attributed to von Neumann, as in [16, 20]. The main assumption for linear problems that the error and the signal follow the same dynamics appears intuitively correct. It has been unambiguously shown in [95], that this assumption is flawed for any discrete computing and the difference is due to dispersion and phase errors, and when the numerical method causes physical amplification and/or attenuation.

We demonstrate the inadequacy of von Neumann analysis with the help of Eq. (2.43) in analyzing space-time discretization schemes. This equation helps in testing numerical methods for solution accuracy, error propagation and most importantly, the dispersion error — as has been variously attempted in [92, 95, 101, 120, 132].

We represent the unknown, by its Laplace transform at the jth node of a uniformly spaced discrete grid of spacing h as $u(x_j, t) = \int U(kh, t)\, e^{ikx_j}\, dk$. For the 1D convection equation, one can define the spectral amplification factor G as

$$G(\Delta t, kh) = \frac{U(kh, t^n + \Delta t)}{U(kh, t^n)} \tag{2.151}$$

Note that G is a complex quantity, as the Fourier–Laplace amplitude $U(kh, t)$ is complex. In the continuum limit of h, $\Delta t \to 0$, one must have

$$|G| \equiv 1 \tag{2.152}$$

However, real and imaginary parts of G depend on the differential equation solved and the discretization methods adopted for spatial and temporal derivatives. Other important numerical properties are obtained via the spectral representation of Eq. (2.43) and using Eq. (2.90) to represent the first derivative. This gives

$$\int \left[\frac{dU}{dt} + \frac{c}{h} \sum U C_{jl}\, e^{ik(x_l - x_j)} \right] e^{ikx_j}\, dk = 0 \tag{2.153}$$

Since the above equation is true for all wavenumbers, the integrand must be zero for any k. The implicit condition of Eq. (2.153) can be reinterpreted as

$$\frac{dU}{U} = -\left[\frac{cdt}{h}\right] \sum_{l=1}^{N} C_{jl}\, e^{ik(x_l - x_j)} \tag{2.154}$$

The first factor on the right-hand side of Eq. (2.154) is nothing but the CFL number (N_c). As the right-hand side of Eq. (2.154) is node dependent, we express the left-hand side by the nodal numerical amplification factor (G_j) given by

$$G_j = G|_{(x=x_j)} = 1 - N_c \sum_{l=1}^{N} C_{jl}\, e^{ik(x_l - x_j)} \tag{2.155}$$

for the Euler time discretization scheme. From Eq. (2.91), one can replace the summed up quantity on the right-hand side as

$$G_j = G|_{(x=x_j)} = 1 - i N_c \, [k_{eq}h]_j \tag{2.156}$$

This numerical property can not be expressed in terms of spatial discretization alone. Euler time integration is thus, not numerically stable, as $|G_j| > 1$. Two time level, multi-stage higher order methods are needed ideally for accuracy and stability - as explained with respect to the (RK_4) method [83, 95]. If one denotes the right-hand side of Eq. (2.43), by $L(u) = -c\frac{\partial u}{\partial x}$, then the steps used in (RK_4) are given by

$$\text{Step 1}: \quad u^{(1)} = u^{(n)} + \frac{\Delta t}{2} L[u^{(n)}]$$

$$\text{Step 2}: \quad u^{(2)} = u^{(n)} + \frac{\Delta t}{2} L[u^{(1)}]$$

$$\text{Step 3}: \quad u^{(3)} = u^{(n)} + \Delta t L[u^{(2)}]$$

$$\text{Step 4}: \quad u^{(n+1)} = u^{(n)} + \frac{\Delta t}{6}\{L[u^{(n)}] + 2L[u^{(1)}] + 2L[u^{(2)}] + L[u^{(3)}]\}$$

For the RK_4 time integration scheme, G_j is obtained as [95]

$$G_j = 1 - A_j + \frac{A_j^2}{2} - \frac{A_j^3}{6} + \frac{A_j^4}{24} \tag{2.157}$$

where

$$A_j = N_c \sum_{l=1}^{N} C_{jl}\, e^{ik(x_l - x_j)}$$

While G_j is a source of error, additional error arises due to dispersion, whose effects are subtle and often misunderstood.

2.3.10 Quantification of Dispersion Error and Error Propagation Equation

Using Fourier transform for the analysis of a numerical method, one obtains a numerical dispersion relation by expressing the equivalent differential equation in the wavenumber-circular frequency plane. This dispersion relation is different from the physical dispersion relation and the difference between the two gives rise to dispersion error, as explained with the 1D model convection equation for a typical space-time dependent system.

If we represent the initial condition for Eq. (2.43) as

$$u(x_j, t = 0) = u_j^0 = \int A_0(k)\, e^{ikx_j}\, dk \tag{2.158}$$

then the general solution at any arbitrary time can be obtained as

$$u_j^n = \int A_0(k)\, [|G_j|]^n\, e^{i(kx_j - n\beta_j)}\, dk \tag{2.159}$$

where $|G_j| = (G_{rj}^2 + G_{ij}^2)^{1/2}$ and $\tan(\beta_j) = -\frac{G_{ij}}{G_{rj}}$, with G_{rj} and G_{ij} as the real and imaginary parts of G_j, respectively. Thus, the phase of the solution is determined by $n\beta_j = kc_N t$, where c_N is the numerical phase speed. Although the physical phase speed is a constant for all k for the non-dispersive system, this analysis shows that c_N is, in general, k-dependent, i.e., the numerical solution is dispersive, in contrast to the non-dispersive nature of Eq. (2.43). The implications of this simple difference are profound, as demonstrated below.

The general numerical solution of Eq. (2.43) is denoted as

$$\bar{u}_N = \int A_0\, [|G|]^{t/\Delta t}\, e^{ik(x - c_N t)}\, dk \tag{2.160}$$

The numerical dispersion relation is now given as $\omega_N = c_N k$, instead of the physical dispersion relation, $\bar{\omega} = ck$. Non-dimensional phase speed and group velocity (from the general definition given in Eqs. (2.48) and (2.49)) at the jth node are expressed as

$$\left[\frac{c_N}{c}\right]_j = \frac{\beta_j}{\bar{\omega}\Delta t} \tag{2.161}$$

$$\left[\frac{V_{gN}}{c}\right]_j = \frac{1}{hN_c}\frac{d\beta_j}{dk} \tag{2.162}$$

If the computational error is defined as $e(x, t) = u(x, t) - \bar{u}_N$, then one obtains the governing equation for its dynamics in the following manner. Using Eq. (2.160) one obtains

$$\frac{\partial \bar{u}_N}{\partial x} = \int ik A_0 \, [|G|]^{t/\Delta t} \, e^{ik(x-c_N t)} \, dk \tag{2.163}$$

$$\frac{\partial \bar{u}_N}{\partial t} = - \int ik \, c_N \, A_0 \, [|G|]^{t/\Delta t} \, e^{ik(x-c_N t)} \, dk$$

$$+ \int \frac{Ln \, |G|}{\Delta t} A_0 \, [|G|]^{t/\Delta t} \, e^{ik(x-c_N t)} \, dk \tag{2.164}$$

Thus, the error propagation equation for Eq. (2.43) is given by [95],

$$\frac{\partial e}{\partial t} + c \frac{\partial e}{\partial x} = -c[1 - \frac{c_N}{c}] \frac{\partial \bar{u}_N}{\partial x} - \int \frac{dc_N}{dk} \left[\int ik' A_0 [|G|]^{t/\Delta t} e^{ik'(x-c_N t)} dk' \right] dk$$

$$- \int \frac{Ln \, |G|}{\Delta t} A_0 \, [|G|]^{t/\Delta t} \, e^{ik(x-c_N t)} \, dk \tag{2.165}$$

This is the correct error propagation equation, as opposed to that is obtained using the assumption made in von Neumann analysis, where the right-hand side is assumed to be identically equal to zero. This is on the premise that $c_N \cong c$, i.e., there is no dispersion error and the numerical method is perfectly neutral, so that the last term on the right-hand side of Eq. (2.165) is identically zero. Error can grow even faster when the numerical solution displays sharp spatial variation, due to the first term on the right-hand side of Eq. (2.165).

To understand the ramifications of Eq. (2.165) for the model equation, one needs to look at the numerical properties of a specific combination of spatial and temporal discretization methods. For this purpose, we show the properties of OUCS3 scheme [96]. The stencil of OUCS3 scheme is given in Eq. (2.83). This scheme is for a non-periodic problem and here we show some typical results for a central and another node close to the inflow boundary, when RK_4 time integration strategy is used to solve Eq. (2.43). In Fig. 2.15, $|G|$, V_{gN}/c and $(1 - c_N/c)$ are plotted as contours in the indicated ranges of kh and N_c for a node adjacent to the boundary ($j = 2$) and one in the center of the domain ($j = 51$).

Results show that the numerical properties of $j = 2$ are significantly different from those for $j = 51$. It is noted from the top frame for numerical amplification contours that the scheme is stable for $N_c \leq 1.301$, i.e., $|G| \leq 1$ for interior nodes. The scheme is neutrally stable for very small values of N_c and a limited range of kh – a property absolutely essential for DNS, as one notes that the last term on the right-hand side of Eq. (2.165) vanishes for the neutrally stable case. In the middle frame of Fig. 2.15, V_{gN}/c contours display significant dispersion effects for high ks, which would invalidate long time integration results, even when $|G_j| = 1$ is ensured by computing with vanishingly small N_c. In fact, above $kh \geq 2.4$ the numerical solution will travel in the wrong direction, as $V_{gN} \leq 0$ for $N_c \cong 0$. Such spurious numerical waves are termed q-waves, which propagate upstream, in contrast to the physical or p-waves

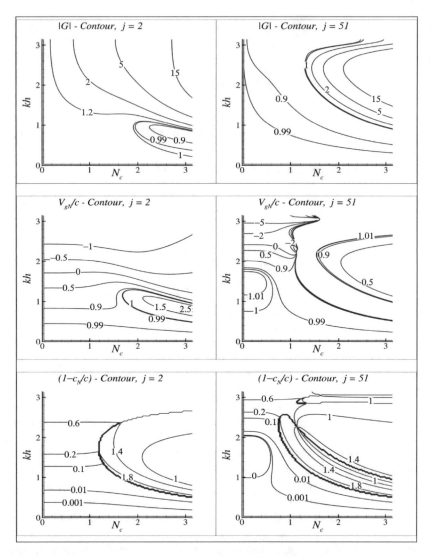

Fig. 2.15 Comparison of numerical amplification factor ($|G|$), normalized numerical group velocity (V_{gN}/c) and phase error ($1 - c_N/c$) contours for the near-boundary node $j = 2$ (left column) and the central node $j = 51$ (right column) for the 1D convection equation solved by the OUCS3-RK_4 method

traveling downstream for Eq. (2.43) [91]. In Eq. (2.165), we note that the first term on the right hand side affects error evolution via the numerical property $(1 - \frac{c_N}{c})$. In the bottom frame of Fig. 2.15, contours of this quantity are shown. The bottom two frames of Fig. 2.15 indicate effects of dispersion error, which cannot be simply eliminated or reduced by grid refinement alone.

One can show the connection between q-waves as the extreme form of dispersion error for different numerical methods used in CFD. Even for a large range of k, corresponding p-waves travel at incorrect group velocity, constituting dispersion error. The q-wave is entirely due to dispersion in many discrete computation methods [91].

Flow transition was thought often to occur for external flows by TS wave packets created as physical instability to small disturbances inside a shear layer. However, many flows do not show TS wave and are collectively stated to have suffered bypass transition. In [82], one particular form of bypass transition is explained theoretically and experimentally, when the flow is excited by a disturbance source remaining always outside the shear layer. In this scenario, the resultant disturbance occurring inside the shear layer is seen to propagate upstream, with respect to the source of disturbance convecting outside the shear layer. Thus, there are flow transition scenarios where physical disturbances travel upstream and it is necessary to compute these, without the interference of q-waves created numerically.

For attached flows, q-waves have been reported in [68] as very high k events. Thus, for DNS of bypass transitional flows, one would like to capture p-waves associated with bypass transition, while avoiding q-waves which arise solely due to dispersion error at high wavenumbers. For the zero pressure gradient boundary layer, upstream propagating damped waves were detected theoretically in [99]. We emphasize that in many flows, disturbances appear to move downstream with respect to an inertial frame — but an observer moving with the source at the convection speed will see the response inside the shear layer move upstream.

2.3.11 Dispersion Relation Preserving Schemes

Significant progress has been made in developing numerical methods for solving space-time dependent problems with schemes, which preserve physical dispersion relation over a wide range of parameters. For discrete methods, closeness between the physical and numerical dispersion relation can be realized only for limited ranges of space and time steps. In [96], a spectral analysis was developed to characterize numerical schemes in the full domain, with numerical group velocity used as a measure to quantify dispersion error. Space-time discretization was treated independently in many earlier works. This has been corrected in [95, 100] and with the derivation of Eq. (2.165), a DRP scheme was identified as essential for DNS and acoustics problems. These results have been used in [71, 100] to optimize time discretization schemes for high accuracy spatial discretization schemes, by minimizing numerical error in the spectral plane.

Here, we have chosen representative methods from finite difference, finite volume and finite element approaches to explain q-waves and quantifying DRP properties. For the finite difference method, OUCS3 scheme [96] is chosen. In solving Eq. (2.43), the first spatial derivative indicated by a prime is obtained from the general representation

$$[A] \{u'\} = \frac{1}{h}[B] \{u\} \tag{2.166}$$

For the explicit discretization method, $[A]$ is the identity matrix and the $[C]$ matrix in Eqs. (2.90) and (2.91) is identical to the $[B]$ matrix. For implicit schemes, $[C]$ is equal to $[A]^{-1} [B]$ matrix. Having obtained $[C]$ for spatial discretization, one evaluates the complex amplification factor G_j for any time integration schemes. In particular, for the RK_4 method one can obtain G_j from Eq. (2.157) with

$$A_j = N_c \sum_{l=1}^{N} C_{jl} \, e^{ik(x_l - x_j)}$$

This approach is used to obtain G_j for explicit CD_2 and the OUCS3 schemes for spatial discretization (Eq. (2.83)) with RK_4 time integration scheme.

For a representative finite volume scheme, the QUICK scheme [54] is considered, which uses flux vector splitting of the convection term as [83],

$$\int \left[\frac{\partial u}{\partial t}\right] dx + c \, (u_{j+1/2}^+ - u_{j-1/2}^+) = 0 \tag{2.167}$$

In the above, RK_4 time integration is used for the jth cell and the superscript sign for the second set of terms indicates right moving quantities showing the balance of incoming and outgoing fluxes through the cell interfaces. One of the representative flux quantities is given by (see [42, 83] for details)

$$u_{j-1/2}^+ = u_{j-1} + \frac{1}{4}[(1 - \kappa)(u_{j-1} - u_{j-2}) + (1 + \kappa)(u_j - u_{j-1})] \tag{2.168}$$

where $\kappa = 1/2$ for the QUICK scheme. Other flux terms $u_{j+1/2}^+$ can be similarly obtained. These expressions help one obtain the $[C]$ matrix in Eq. (2.90), which in turn yields G_j. Readers are referred to [97] for a similar analysis for a compact scheme based flux vector splitting finite volume method.

For the finite element method, the streamwise upwind Petrov–Galerkin (SUPG) method described in [36] is analyzed. Among finite element formulations, Galerkin methods belong to the class of solutions for PDEs in which solution residue is minimized, giving rise to the well-known weak formulation of problems. In this approach, dependent variable $u(x, t)$ for a 1D space-time dependent problem is expressed in the form

$$u(x, t) = \sum_{j=1}^{N} \phi_j(x) \, u_j(t) \tag{2.169}$$

where $\phi_1, \ldots, \phi_N$ are the chosen low order polynomials as the basis functions, localized about elements. Note that the weak formulation considers space-time dependent terms together, giving rise to better DRP properties for Galerkin formulations.

However, it has been identified in [83, 91] that the Galerkin methods display instability near the *inflow* when linear basis functions are used (G1FEM) for Eq. (2.43).
Similarly, observations on the amplification factor holds for solving the 1D convection equation by quadratic basis functions by G2FEM.

One method of removing the problems of Galerkin FEM is to make the discrete
equation dissipative. Authors in [24, 123, 124] have proposed dissipative Galerkin
procedures. This procedure was furthermore adopted in [75], who advocated an optimal procedure following an approximate form of phase error proposed in [111] for 1D
convection equation. Brooks and Hughes [14] have adopted the same methodology
in the Petrov–Galerkin formulation and called it the "Streamline Upwind/Petrov–
Galerkin (SUPG)" formulation. Raymond and Garder [75] have also referred to
"ghost waves" following [22]. A complete analysis of the "ghost" or q-waves is presented here for the SUPG formulation, following the work on q-wave [83, 91] for
different discrete computing methods.

The discretized form of Eq. (2.43) for G1FEM is obtained using linear basis
functions on a uniform grid as

$$\frac{h}{6}\left(\frac{du_{j+1}}{dt} + 4\frac{du_j}{dt} + \frac{du_{j-1}}{dt}\right) + \frac{c}{2}(u_{j+1} - u_{j-1}) = 0 \qquad (2.170)$$

for any interior jth-node. For boundary nodes ($j = 1, N$) one obtains,

$$j = 1: \qquad \frac{h}{3}\left(2\frac{du_1}{dt} + \frac{du_2}{dt}\right) + c(u_2 - u_1) = 0 \qquad (2.171)$$

$$j = N: \qquad \frac{h}{3}\left(2\frac{du_N}{dt} + \frac{du_{N-1}}{dt}\right) + c(u_N - u_{N-1}) = 0 \qquad (2.172)$$

The derivations are as given in [83, 91].

Three aspects are evident from the above discrete equations for G1FEM: (i) the
non-dissipative nature of the discrete equation for the interior nodes; (ii) instability at
$j = 1$ due to the one-sided nature of the stencil, with the information propagating to
the boundaries from the interior, which is contrary to the physical description given
by Eq. (2.43) and (iii) the overly dissipative nature of the discrete equation at $j = N$.

Using the hybrid Fourier–Laplace representation of u in Eqs. (2.170)–(2.172),
one obtains the effectiveness of the derivative discretization $k_{eq}^{(1)}$ for the G1FEM as
[81, 83]

$$j = 1: \qquad k_{eq}^{(1)} = \frac{3}{ih}\left(\frac{e^{ikh} - 1}{e^{ikh} + 2}\right) \qquad (2.173)$$

$$2 \le j \le N - 1: \qquad k_{eq}^{(1)} = \frac{3\sin kh}{[h(2 + \cos kh)]} \qquad (2.174)$$

$$j = N: \qquad k_{eq}^{(1)} = \frac{3}{ih}\left(\frac{1 - e^{-ikh}}{2 + e^{-ikh}}\right) \qquad (2.175)$$

If one adopts Euler time stepping for discretized Eqs. (2.170)–(2.172), the numerical amplification factor for G1FEM is given, as in [81, 83]

$$j = 1: \qquad G^{(1)} = 1 - 3N_c \left[\frac{e^{ikh} - 1}{e^{ikh} + 2} \right] \qquad (2.176)$$

$$2 \le j \le N - 1: \qquad G^{(1)} = 1 - 3i N_c \left[\frac{\sin kh}{2 + \cos kh} \right] \qquad (2.177)$$

$$j = N: \qquad G^{(1)} = 1 - 3N_c \left[\frac{1 - e^{-ikh}}{2 + e^{-ikh}} \right] \qquad (2.178)$$

If instead one uses quadratic basis functions (G2FEM) for a uniformly spaced grid, then the discrete equations for the Galerkin approximation of Eq. (2.43) are as given in [81, 83]. Here, we report only the equation for the interior nodes as

$$3 \le j \le N - 2: \qquad \frac{h}{15} \left(-\frac{du_{l+2}}{dt} + 4\frac{du_{l+1}}{dt} + 24\frac{du_l}{dt} + 4\frac{du_{l-1}}{dt} - \frac{du_{l-2}}{dt} \right)$$

$$+ \frac{c}{6} \left(u_{l-2} - 8u_{l-1} + 8u_{l+1} - u_{l+2} \right) = 0 \qquad (2.179)$$

Using hybrid Fourier–Laplace transform of the unknown u in the above, one obtains the effectiveness of the derivative discretization $k_{eq}^{(2)}$ for G2FEM for an interior node as

$$3 \le j \le N - 2: \qquad k_{eq}^{(2)} = \frac{5 \sin kh(4 - \cos kh)}{h(12 + 4 \cos kh - \cos 2kh)} \qquad (2.180)$$

If one adopts Euler time stepping for the discretized equations for G2FEM, one obtains the numerical amplification factor for G2FEM as

$$3 \le j \le N - 2: \qquad G^{(2)} = 1 - 5i N_c \frac{\sin kh(4 - \cos kh)}{(12 + 4 \cos kh - \cos 2kh)} \qquad (2.181)$$

It is seen that G1FEM suffers from numerical instability at $j = 1$ and G2FEM suffers the same for $j = 1$ and 2. Also, it is shown in [81, 83] that there is overstability at $j = N$ for G1FEM and at $j = (N - 1)$ and N for G2FEM.

To apply Galerkin methods for wave propagation problems, numerical instabilities occurring at and near boundaries must be cured — as mentioned in [24, 75, 123, 124]. Brooks and Hughes [14] adopted the prescription in [75] and renamed it as SUPG method. In Grescho and Sani [36], the discrete equation obtained by the SUPG method at an interior node ($2 \le j \le (N - 1)$) for Eq. (2.43) is given as

$$\frac{h}{6}\left[\left(1+\frac{\beta}{2}\right)\frac{du_{j-1}}{dt}+4\frac{du_j}{dt}+\left(1-\frac{\beta}{2}\right)\frac{du_{j+1}}{dt}\right]+\frac{c}{2}\left(u_{j+1}-u_{j-1}\right)$$

$$=\beta c\left(u_{j+1}-2u_j+u_{j-1}\right) \tag{2.182}$$

where β is the stream-wise diffusion parameter. An optimal value of β is given as $1/\sqrt{15}$ in [14] following the work in [75]. In [75], the analytical solution of Eq. (2.182) was taken from [52, 111] in arriving at the optimal value. However, the analytical solution was derived on the basis of a semi-discrete analysis, considering no error being committed in time discretization. Hence, this value is not universal and it varies from one time discretization method to another, a fact often ignored among practitioners. The right-hand side of Eq. (2.182) represents the lowest order dissipation term. The k_{eq}^{SUPG} of the SUPG method for discrete Eq. (2.182) can be obtained as

$$k_{eq}^{SUPG}=\left(\frac{6}{h}\right)\left(\frac{\sin kh-2i\beta(1-\cos kh)}{4+2\cos kh-i\beta\sin kh}\right) \tag{2.183}$$

Adopting Euler time stepping for Eq. (2.182), the numerical amplification factor for the SUPG method can be obtained as

$$G^{SUPG}=1-6\beta N_c\left[\frac{4(1-\cos kh)(2+\cos kh)-\sin^2 kh}{4(2+\cos kh)^2+\beta^2\sin^2 kh}\right]$$

$$-6i N_c\left[\frac{(4+2\cos kh)\sin kh+2\beta^2(1-\cos kh)\sin kh}{4(2+\cos kh)^2+\beta^2\sin^2 kh}\right] \tag{2.184}$$

In Figs. 2.16 and 2.17, we specifically show $G_j(N_c, kh)$ and V_{gN}/c, respectively, for the chosen four methods: (a) RK_4-OUCS3, (b) RK_4-CD_2, (c) RK_4-QUICK and (d) Euler-SUPG obtained using Eqs. (2.155), (2.157) and (2.162) [83, 91]. In these figures, properties of these methods for only the interior nodes are shown. From the contour plots in Fig. 2.16, one notes a narrow range of N_c available for which the OUCS3 and CD_2 schemes have neutral stability for a full range of available kh. The region in the (N_c, kh)-plane where this is feasible is marked by hatched lines in Fig. 2.16. In comparison, neither the QUICK nor the SUPG scheme has a neutrally stable region and would not qualify for DNS.

The main interest here is to compare the dispersion property of these methods in terms of numerical group velocity and also find the reason for which q-waves are created numerically in solving Eq. (2.43). The V_{gN}/c contours shown in Fig. 2.17, indicate that all these methods produce dispersion error, for the computational parameters, when the kh and N_c combination takes values away from the origin. The V_{gN}/c contours show for these four methods, presence of line(s) along which V_{gN} is zero. For the CD_2 and OUCS3 schemes, this is a $kh = constant$ line, parallel to the N_c-axis, while for the other two methods, these lines are curved. From Eq. (2.162), the condition of zero group velocity corresponds to

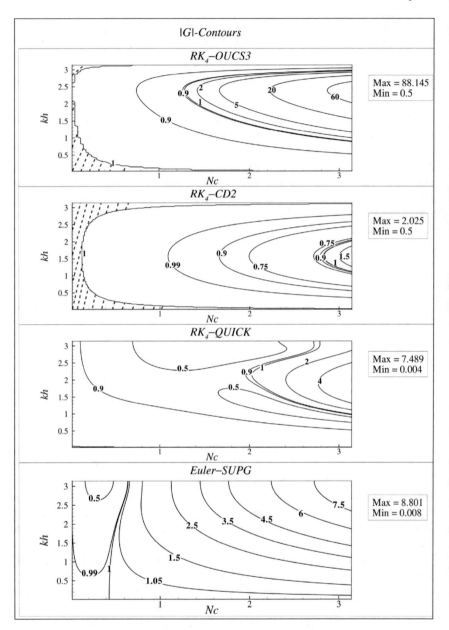

Fig. 2.16 Comparison of the numerical amplification factor ($|G|$) for interior nodes by indicated space-time discretization schemes

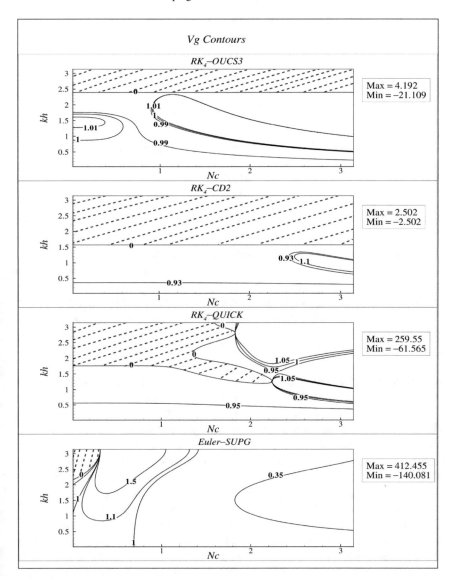

Fig. 2.17 Comparison of normalized numerical group velocity (V_{gN}/c) for interior nodes for the indicated combinations of space-time discretizations

$$\frac{d}{dk}\left[\tan^{-1}\left(-\frac{G_{ij}}{G_{rj}}\right)\right] = 0$$

This can be further simplified to

$$G_{rj}\frac{\partial G_{ij}}{\partial k} = G_{ij}\frac{\partial G_{rj}}{\partial k}$$

For the RK_4 time integration and central spatial schemes, this condition further simplifies to

$$\frac{dk_{eq}}{dk} = 0$$

as $A_j = -ik_{eq}N_c$. Above this line for $CD_2 - RK_4$ and $OUCS3 - RK_4$ methods, the numerical group velocity of the methods for the solution of Eq. (2.43) is negative, i.e., the numerical waves would propagate upstream, despite the physical require-ment of downstream movement. The region in the (N_c, kh)-plane where q-waves are created is marked by hatched lines in Fig. 2.17. Thus, in solving Eq. (2.43) by the RK_4-CD_2 method, q-waves are created for $kh > \pi/2$, and for the RK_4-OUCS3 method, q-waves are created for $kh > 2.391$, for any choice of time steps. This criti-cal value, $(kh)_{cr}$, for the OUCS3 method is significantly higher in comparison to the other three methods under consideration. In Fig. 2.17 for the CD_2 method, V_{gN}/c contours are symmetric about $(kh)_{cr}$ and also due to the symmetry of the stencil, minimum and maximum values of V_{gN}/c are same. Maximum and minimum val-ues of numerical group velocities are lowest for the CD_2 method, followed by the OUCS3 method. In comparison, the QUICK and SUPG methods have a significantly higher magnitude of maximum and minimum group velocity − essentially due to the addition of excessive numerical diffusion. It is often wrongly presumed that the addition of diffusion does not alter the dispersion property of numerical methods. The presence of q-waves indicates effects similar to what is noted as unsteady flow separation, which is often the harbinger of bypass transition, as shown in [82].

The presence or absence of q-waves depends not only upon the existence of a region with $V_{gN} < 0$, but also on the bandwidth of the implicit filter (as given by the real part of k_{eq}/k) and added diffusion (as given by the imaginary part of k_{eq}/k) of the basic numerical method. It is easy to reason that excessive filtering and damping removes q-waves. The appearance of q-waves is more likely for the OUCS3 method as compared to other methods, due to its lower filtering and less added diffusion. For the CD_2 method, q-waves will appear for $kh > \pi/2$, where the low-pass property of the method will filter the signal that is 38% at each time step, and in contrast, for the OUCS3 method at $(kh)_{cr} = 2.391$, corresponding filtering is equal to only 7%, and there is an additional small attenuation due to low numerical diffusion. It is seen that at $kh = 2.4$, the CD_2 method filters the signal (while spatially discretizing) by more than 70%. In comparison to these two methods, the QUICK and SUPG methods add excessive numerical diffusion, which prevents the appearance of q-waves at the cost of numerical accuracy. For the SUPG method, q-waves are created for very small values of N_c and very high values of kh. However, at these high values of kh, the SUPG method introduces a very large amount of numerical diffusion, due to which q-waves are not seen, once again.

The above discussion shows the relative merits of different methods with respect to dispersion properties, including creation of spurious upstream propagating waves, as example of an extreme form of dispersion error. It is shown that high accuracy compact schemes used with the RK_4 scheme provide the best option in terms of the DRP property. There are many other subtle issues of DRP schemes.

2.4 Design of DRP Schemes for DNS/LES

There has been considerable effort to improve the accuracy of numerical methods in solving space-time-dependent problems by developing DRP schemes. Earlier attempts are reported in [39, 112]. In many earlier attempts, DRP methods were attempted to be obtained by considering the spatial discretization alone, mainly by minimizing the truncation error [12, 112]. In [5, 45, 67, 72], space-time discretizations have been considered together to minimize the error between the numerical amplification factor and the true amplification factor. Based on this approach, Hu et al. [45], proposed a class of low-dissipation and low-dispersion Runge–Kutta (LDDRK) schemes. Correct numerical dispersion relation for combined space-time discretization schemes was identified for convection equation in [25, 83, 93], where the use of DRP schemes for DNS and acoustics problems (see also [106]) was emphasized. To obtain a scheme which would provide ideal DRP property in [71, 100], space-time discretization has been considered together to minimize dispersion and phase error, while keeping the scheme neutrally stable for 1D convection equation. Such a scheme should be used for DNS and acoustics.

Following Eq. (2.157), numerical amplification factor G_{num} for p-stage explicit Runge–Kutta scheme can be expressed as

$$g_{num} = 1 + \sum_{1}^{p} (-1)^j a_j A^j \tag{2.185}$$

One notes that for 1D convection equation, exact amplification factor is given as $G_{exact} = e^{iN_ckh}$, Eq. (2.185) is a polynomial approximation of G_{exact} and $a_j = 1/j!$ minimizes the corresponding truncation error in temporal space. Hu et al. [45] considered the 1D convection equation to minimize the difference between G_{num} and G_{exact}. Similarly in [5, 67], an optimization was performed by minimizing $|G_{num} - e^{iN_ckh}| \, ||u_0||_2$, where $n = t/\Delta t$ and $||u_0||_2$ defines the L_2-norm of the initial condition $u_0(x)$. In [71, 100] following objective functions were chosen to obtain the DRP scheme:

$$F(a_j, N_c) = \int_0^\gamma |G_{num} - G_{exact}|^2 d(kh) \tag{2.186}$$

where $\gamma = \alpha\pi$ with $\alpha = [0, 1]$. This condition is identical to what was used in [45]. However, in addition to minimizing F, additional explicit constraints were also imposed in [71, 100], to ensure neutral stability, minimum phase and dispersion errors. Following the source terms in the correct error propagation equation in Eq. (2.165). These constraints are given as [71, 100],

$$F_1(a_j, N_c) = \int_0^{\gamma_1} |1 - |G_{num}|| d(kh) \le \varepsilon_1 \tag{2.187}$$

$$F_2(a_j, N_c) = \int_0^{\gamma_2} |1 - \frac{c_N}{c}| d(kh) \le \varepsilon_2 \tag{2.188}$$

$$F_3(a_j, N_c) = \int_0^{\gamma_3} |1 - \frac{V_{gN}}{c}| d(kh) \le \varepsilon_3 \tag{2.189}$$

where ε_i and γ_i are constants chosen to satisfy the numerical properties of the basic method. Although, a small tolerance in ε_i was prescribed, main focus was to look for numerical parameters, which ensure neutral stability for large range of kh and N_c. Grid-search technique was employed to solve the constrained optimization problem and to locate the feasible values of parameters a_j for a fixed value of N_c. While optimized a_j's are functions of N_c, a near optimal values of these parameters were searched that could be used over a longer range of N_c.

2.5   Numerical Filtering: Error Control via Stabilization and Dealiasing

Computing of transitional or turbulent flows is very challenging. If not treated carefully, the high wavenumber components (components whose wavenumbers are near to $kh = \pi$) causes numerical instability due to aliasing error. Hence, the use of filters are quite essential to periodically remove these critical high wavenumber fluctuations, without affecting the actual dispersion property of the solution. Controlled amount of explicit fourth-order diffusion is already added to the first derivative, but their excessive use is not recommended, as this might alter the numerical dispersion relation. Instead the use of filters, that attenuates the high wavenumber components are prescribed in Bhumkar [10], without altering the physical dispersion relation, and is recommended.

2.5.1   Use of Filters for DNS and LES

Filtering is always implicitly present in all discrete computations [53, 81, 83]. In numerical computations, the nature of filter used has to be low-pass, which should leave smaller wavenumber components of the variable unaffected, while filtering the higher wavenumber components, which can create numerical problems as explained above. In traditional LES, the governing equation is analytically filtered before discretization. In [34, 48, 59, 61, 80, 110, 119] the effects of different types of filters used in traditional LES are illustrated. The filtering operation in traditional LES requires adding additional empirical stresses (termed as SGS or Leonard stresses) to be incorporated into the governing equation at the formulation stage itself. Such an operation can lead to numerical instabilities due to aliasing operation as traditional LES requires the variables to be multiplied with a spatial kernel as

$$\tilde{f}(\overrightarrow{X}, t) = \int G(\overrightarrow{X} - \overrightarrow{X'}) \, f(\overrightarrow{X'}, t) d^3 X' \qquad (2.190)$$

In contrast, the spatial filters [10, 28–30, 77, 78, 89, 90, 122] are applied at the end of time advancement of the governing variables, without the need for any artificial modeling of SGS stresses. Though, no strict rules exist on the frequency and order of filtering, significant advantages can be achieved, if LES is performed using spatial filters, where one can completely dispense with empirical SGS model. Moreover, it can also help avoiding numerical instabilities due to aliasing operation involved in traditional LES, as explained above.

Explicit Padè type filters have been proposed in [10, 28–30, 83, 89, 90, 122] to control instabilities arising from mesh non-uniformities and the application of numerical boundary conditions and advanced the use of spatial filters as a tool for LES. These spatial filters are central in nature for the interior nodes, while one-sided stencil have been proposed for non-periodic problems [29, 121] near the boundary. However, it has been shown in [90] by analysis, using matrix-spectral theory that one-sided stencils at the boundary can be destabilizing near the inflow of a computational domain. Such numerical instability can affect also interior points.

One dimensional filters have been proposed and developed in [122]. The general interior stencil of these 1D filters are given as,

$$\alpha_f \hat{u}_{j-1} + \hat{u}_j + \alpha_f \hat{u}_{j-1} = \sum_{n=0}^{N} \frac{a_n}{2} (u_{j+n} + u_{j-n}) \qquad (2.191)$$

which is defined as a $(2N)$th-order filter. Here, $\hat{u}_j$ is the filtered value of the variable u_j at jth grid-point It is to be emphasized that the definition of the order of filter is meaningless activity, as in Eq. (2.191). one does not even match the filtered and unfiltered variable itself. Original definition in [29, 30, 122] have been inspired by explicit derivative evaluation, even though one is performing an implicit filtering process. Writing, u_j and $\hat{u}_j$ in terms of Fourier–Laplace transforms as $u_j = \int U(k)e^{ikx_j} dk$ and $\hat{u}_j = \int \hat{U}(k)e^{ikx_j} dk$, one can define the corresponding transfer function for the periodic filtering stencil as $TF(k) = \frac{\hat{U}(k)}{U(k)}$. For the filter stencil given in Eq. (2.191), the transfer function for the corresponding periodic problem is obtained as

$$TF(k) = \frac{a_0 + \sum_{n=1}^{N} a_n \cos(nkh)}{1 + 2\alpha_f \cos(kh)} \tag{2.192}$$

For a $(2N)$th-order 1D filter, one matches up to the coefficients of u_j^{2N-2} and $\hat{u}_j^{2N-2}$ from the Taylor series expansion of Eq. (2.191). This gives N equations for the $(N + 1)$ coefficients a_0 to a_N if α_f is treated as a parameter. The additional equation required to determine the values of a_0 to a_N in terms of the filtering coefficients α_f is found by imposing the condition that $TF(k) = 0$ at $kh = \pi$. The values of the coefficients a_0 to a_N in terms of the filter coefficient α_f for 2nd- to 10th-order filters are tabulated in Table IV of [122]. It should be noted that, the value of α_f should range from $+0.5$ to -0.5, with $\alpha_f = 0.5$ indicates no filtering [83]. Generally positive values of α_f is considered for computation purpose. In Fig. 2.18, $TF(k)$ is shown plotted as a function of kh for 2nd-, 4th- and 6th-order filter at indicated values of α_f. One finds from Fig. 2.18, that the 2nd-order filter not only affects the high-kh components, but also attenuates the mid wavenumber components for α_f below 0.47.

To implement 4th and higher order filters for non-periodic problems, authors in [122] proposed the use of one-sided boundary filters, along with the central interior filters. For example, at a near boundary point j, they have proposed filter stencils of the following type to retain the tridiagonal form the resultant matrix

$$\alpha_{f1}\hat{u}_{j+1} + \hat{u}_j + \alpha_{f1}\hat{u}_{j-1} = \sum_{n=1}^{11} a_{n,i} u_n, \, i\varepsilon 2, \ldots, 5$$

$$\alpha_{f1}\hat{u}_{j+1} + \hat{u}_j + \alpha_{f1}\hat{u}_{j-1} = \sum_{n=0}^{10} a_{N-n,i} u_{N-n}, \, i\varepsilon(N-4), \ldots, (N-1) \tag{2.193}$$

For a 6th-order boundary stencil at $j = 3$, these coefficients are given as [122]: $a_{1,3} = -1/64 + \alpha_f/32$, $a_{2,3} = 3/32 + 13\alpha_f/16$, $a_{3,3} = 49/64 + 15\alpha_f/32$, $a_{4,3} = 5/16 + 3\alpha_f/8$, $a_{5,3} = -15/64 + 15\alpha_f/32$, $a_{6,3} = 3/32 - 3\alpha_f/16$ and $a_{7,3} = -1/64 + \alpha_f/32$. For such non-periodic filtering stencils, one can express the filtering operation as

$$[A]\{\hat{u}_j\} = [B]\{u_j\} \tag{2.194}$$

where the special boundary closure filters given in Eq. (2.193) define the first and last few rows of $[A]$ and $[B]$ matrices. Expressing, $\hat{u}_j$ and u_j by corresponding Fourier–Laplace transforms, and equating the contributions at the jth-node, one obtains

$$\sum_{l=1}^{N} a_{jl} e^{ik(x_l-x_j)} \hat{U}(k) = \sum_{l=1}^{N} b_{jl} e^{ik(x_l-x_j)} U(k) \tag{2.195}$$

where $\hat{U}$ and U represent the Fourier–Laplace transform of $\hat{u}_j$ and u_j, respectively. Thus, a transfer function for the jth-node is defined in the spectral plane as,

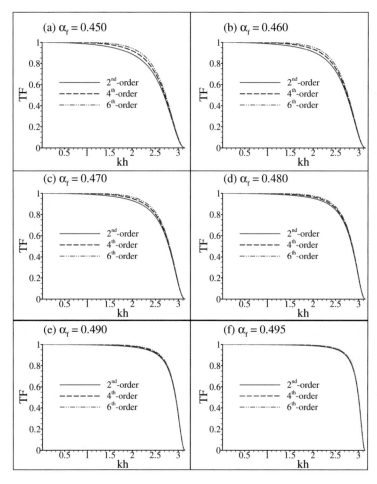

Fig. 2.18 Transfer function (TF) of the periodic 1D filtering stencil plotted as a function of kh for 2nd-, 4 and 6th-order filters for indicated values of α_f

$$TF_j(k) = \frac{\sum_{l=1}^{N} a_{jl}e^{ik(x_l - x_j)}}{\sum_{l=1}^{N} b_{jl}e^{ik(x_l - x_j)}} \qquad (2.196)$$

Let, $u_j^n = u(x_j, t^n)$ be the solution of 1D convection equation at the jth-node and nth time step. When a filter is applied after obtaining u_j^{n+1} to obtain the filtered variable $\hat{u}_j^{n+1}$, then the full-step would have an equivalent amplification factor given as

$$\hat{G}_j(kh, N_c) = TF_j(kh)G_j(kh, N_c) \qquad (2.197)$$

In Fig. 2.19a, b, real and imaginary parts of TF_j are plotted, respectively for the sixth order filter with $\alpha_f = 0.45$. The boundary closure stencils are given by Eq. (2.193).

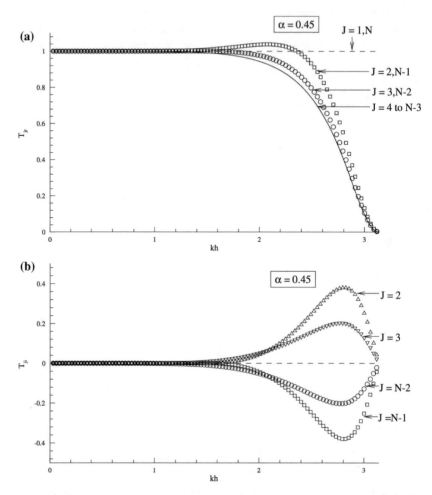

Fig. 2.19 **a** Real and **b** imaginary parts of transfer function TF_j for non-periodic problems with indicated filter coefficient α for the indicated nodes for the 6th-order interior filter with near-boundary filters of [122]

In Fig. 2.19a, one notices higher bandwidth of filters for the near-boundary points and overshoot of TF_j for $j = 2$ and $N - 1$. The imaginary part of TF_j shown in Fig. 2.19b, indicates numerical instability for $j = 2$ and 3 and excessive attenuation at $j = N - 1$ and $N - 2$. One also notes that the stability property near the boundaries is accentuated for higher order filters, with more number of points affected near the boundaries. At interior points, imaginary part of TF_j is identically zero. The imaginary part of TF_j is associated with dispersive effect. To study dispersive effects of a filter, one needs to look at specific space-time dependent differential equations like 1D convection equation. Similar studies have been carried out in [83, 90].

Because of the displayed property of filter transfer function in Fig. 2.19, G_j (kh, N_c) will be attenuated strongly near the Nyquist limit $(kh = \pi)$, where by design, TF_j is forced to be equal to zero. For the near-boundary points of a non-periodic problem, $TF_j(kh)$ is a complex number and therefore, it would alter $G_j(kh, N_c)$ leading to attenuation/amplification and dispersion. Following Eq. (2.163), the filtered solution of 1D convection equation can be written as

$$\hat{u}_N = \int A_0 \, [|\hat{G}|]^{t/\Delta t} \, e^{ik(x-c_N t)} \, dk \qquad (2.198)$$

and corresponding numerical and physical group velocities can be obtained as

$$\left(\frac{c_N}{c}\right)_j = \frac{\hat{\beta}_j}{N_c kh} \qquad (2.199)$$

$$\left(\frac{V_{gN}}{c}\right)_j = \frac{1}{N_c} \frac{\partial \hat{\beta}_j}{\partial (kh)} \qquad (2.200)$$

where $\beta_j = -tan^{-1}\left([\hat{G}_j]_{imag}/[\hat{G}_j]_{real}\right)$. For periodic problems dispersion properties are not affected, as TF_j is purely a real quantity (See Eq. (2.191)) and therefore, $[\hat{G}_j]_{imag}/[\hat{G}_j]_{real} = [G_j]_{imag}/[G_j]_{real}$. This is also true for interior points of a non-periodic filter, irrespective of the choice of α_f, the filter coefficient. For these cases, central filtering does not alter the numerical dispersion properties of the computed solution and this can be considered a favorable feature.

For non-periodic stencils, as the order of the filter is increased, additional numerical problems are noted at the inflow boundary points, while points near the outflow boundary display massive dissipation. To circumvent these issues, application of 1D upwinded filter was proposed in [90]. The general stencil at the interior points for the proposed upwinded filter is given as [90]

$$\alpha_f \hat{u}_{j-1} + \hat{u}_j + \alpha_f \hat{u}_{j-1} = \eta f_{j+4} + (a_3/2 - 5\eta) f_{j+3} + (a_2/2 + 10\eta) f_{j+2}$$
$$+ (a_1/2 - 10\eta) f_{j+1} + (a_0 + 5\eta) f_j + (a_1/2 - \eta) f_{j-1} + (a_2/2) f_{j-2}$$
$$+ (a_3/2) f_{j-3} \quad (2.201)$$

This is formally a 5th-order upwinded filter whose coefficients are obtained in terms of filter coefficient α_f and the upwinding constant η as: $a_0 = (22 + 20\alpha_f)/32 - 10\eta$, $a_1 = (15 + 34\alpha_f)/32 + 15\eta$, $a_2 = (-6 + 12\alpha_f)/32 - 6\eta$ and $a_3 = (1 - 2\alpha_f)/32 + \eta$. For $\eta = 0$, one recovers the 6th-order central filter, while for $\eta \neq 0$, a complex transfer function is obtained which would alter the dispersion relation.

In Fig. 2.20, real and imaginary parts of the transfer function are plotted for a 5th-order upwinded filter [90]. In applying this upwinded filter, central filters have been used at near-boundary nodes, where Eq. (2.201) is not applicable. The above filter were used from $j = 5$ to $j = (N - 4)$; a second order central filter was employed at

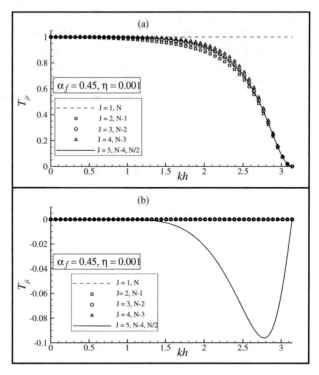

Fig. 2.20 **a** Real and **b** imaginary part of $T F_j$ plotted for a 5th-order upwinded filter with indicated α_f and η. The figure is taken from [90]

$j = 2$ and $(N - 1)$; a fourth order central filter at $j = 3$ and $(N - 2)$ and sixth order central filter at $j = 4$ and $(N - 3)$ were used. Such an upwinded filter completely eliminates the problem of instability near the inflow and excessive damping near the outflow of the type shown in Fig. 2.19 for 6th-order central filter. It is readily seen that there is hardly any difference among $T F_{jr}$ values for different nodes in Fig. 2.20a, while for the near-boundary points, $T F_{ji}$ is equal to zero in Fig. 2.20b - due to the use of central filters at these nodes. In Fig. 2.20b for all the other nodes from $j = 5$ to $(N - 4)$, the transfer function has the same imaginary value shown by the solid line. Further details on the use of upwinded filter is provided in [90]. Specifically, the authors have shown that:

1. The upwind filter allows one to add controlled amount of dissipation in the interior of the domain to remove instability and excessive damping.
2. Absolute control over the imaginary part of the transfer function allows one to mimic the hyper-viscosity that the SGS models often need for traditional LES.
3. The upwind filters through the presence of the imaginary part of the transfer function, alters the amplification property differentially for the real and imaginary part, which in turn changes the numerical group velocity for the upwind filters.

4. Use of upwinded filter either stabilize the computation or allows one to adopt larger time-step without sacrificing the accuracy of computation for flow problems, where bypass transition takes place. These types of computations are very difficult to handle in traditional computations without the use of filters, due to the sensitive dependence on the numerical dispersion properties at high wavenumbers. In this sense, such an approach bears resemblance with process adopted in DES [4, 117] without requiring the use of different equations in different part of the domain.

2.5.2 Use of 1D Filters at the Outflow

Here, we show another application of 1D filter to aid in specifying boundary conditions at the outflow, particularly for turbulent flow-field. Consider a situations, where for wall-bounded turbulent flow-field, massive unsteady separation bubbles have formed on the wall. To aid these separation bubbles to smoothly pass through the outflow boundary, one has to resort to special treatment. For incompressible turbulent flows, the outflow boundary conditions are generally specified in terms of Neumann type boundary conditions or Sommerfeld type convective boundary conditions at the outflow. The Sommerfeld type convective boundary condition on the primary variable vorticity in derived variable formulation is given as

$$\frac{\partial \omega}{\partial t} + U_c \frac{\partial \omega}{\partial x} = 0 \qquad (2.202)$$

where U_c is the artificial convective velocity. When the outflow boundary is located in a turbulent zone where massive vortices pass through this boundary, such boundary conditions are not adequate for these. For such scenarios, one can use 1D filter at the outflow to allow the vortices to smoothly pass out of the domain, by damping their amplitude, without affecting the solution much, in the interior of the computational zone. This sort of application of the 1D filter is equivalent to the buffer-domain technique used in the literature [49, 60], where the flow Reynolds number is varied at very thin layer close to the outflow. A specific example is presented here, where 1D filter was used to smoothly convect the massive vortices out of the computational domain, without affecting the solution at the interior.

In one of the computations of transition of zero pressure gradient boundary layer (ZPGBL), wall excitation is provided by SBS strip, exciting the flow harmonically with a non-dimensional frequency of $F_f = 1.0 \times 10^{-4}$ and amplitude of excitation is 1% of the free-stream speed U_∞. The details of the flow transition for such situations are provided in details in later chapters. Such an excitation generates STWF at later stages, which induces massive separation bubbles on the wall. The domain of computation is taken to be from $-0.05 \leq x \leq 120$ and $0 \leq y \leq 1.5$ (in nondimensional units) in the streamwise and wall-normal directions, respectively. First, the computations are performed without applying the 1D filter at the outflow boundary.

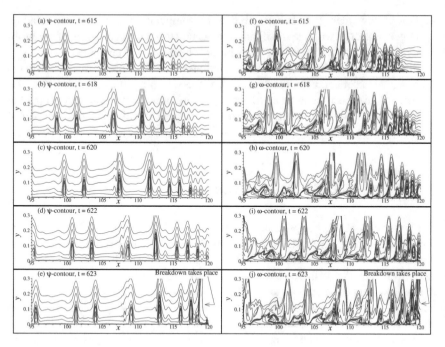

Fig. 2.21 Stream-function ψ- (frames **a–e**) and vorticity ω-contours (frames **f–j**) shown at indicated times for $95 \leq x \leq 120$ before applying 1D 2nd-order filter at the outflow

In Fig. 2.21, the ψ- (frames (a)–(e)) and ω-contours (frames (f)–(j)) are plotted for this case at $t = 615$, 618, 620, 622 and 623 before the application of the 1D, 2nd-order, filter at the outflow, as described next. One-dimensional 2nd-order filter at the outflow is used only for the last 10 streamwise grid-points, whenever its application becomes necessary. The filter coefficient in this zone is smoothly varied from 0.5 at the onset of the filtered zone to 0.495 at the outflow boundary as

$$\alpha_f(i) = 0.5 - 0.05 \left(\frac{x_i - x_{st}}{x_{out} - x_{st}} \right)^2 \tag{2.203}$$

where x_{st} is the x-location of the beginning of the filtered zone and i is the streamwise index.

Only a part of the computational domain ($95 \leq x \leq 120$) near the outflow boundary at $x = 120$ is shown in Fig. 2.21. One notices several separation bubbles on the wall in Fig. 2.21. The tallest of these separation bubbles have a height of $y = 0.2$ which translates into roughly 3.5 times the local displacement thickness δ^* at its location. Hence, these bubbles pierce through the boundary layer and is responsible for the generation of turbulence. The leading separation bubble, though is quite smaller than the tallest one, convects downstream and hits the outflow boundary at $t = 620$ as shown in Fig. 2.21d and (i). A separation indicate a massive concentration of vortical

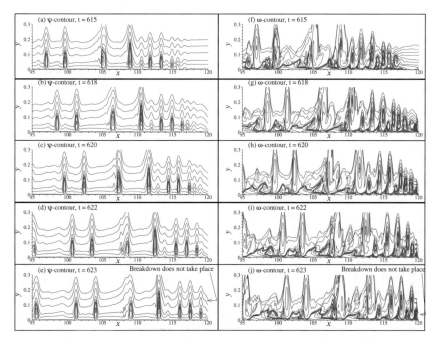

Fig. 2.22 Stream-function ψ- (frames **a–e**) and vorticity ω-contours (frames **f–j**) shown at indicated times for $95 \le x \le 120$ after applying 1D 2nd-order filter at the outflow from $t = 15$

structures and the radiative outflow boundary condition prescribed in Eq. (2.202) is simply not sufficient enough to handle such large vortices and convect these out of the outflow boundary, even if the artificial convective velocity U_c is equal to the free-stream speed, U_∞. The inability of the radiative boundary condition (Eq. (2.202)) to convect the vortices smoothly out of the domain causes accumulation of vorticity in a narrow strip near the outflow boundary, and this causes the simulation to breakdown just after $t = 623$. The ω-contours shown in Fig. 2.21j at $t = 623$ shows substantial accumulation of vorticity at the outflow boundary, which manifests itself as a massive separation bubble in the respective ψ-contour shown in Fig. 2.21e at the outflow. Now the 1D, 2nd-order filter is applied at the outflow from $t = 615$. The unfiltered solution at $t = 615$ is prescribed as the initial solution for this modified approach. The results are shown in Fig. 2.22. The corresponding ψ- (frames (a) to (e)) and ω-contours (frames (f) to (j)) obtained after applying the 1D, 2nd-order filter at the last ten streamwise grid-points near outflow are shown in Fig. 2.22 at $t = 615, 618, 620, 622$ and 623. One notes that by doing so, the previous problem of accumulation of vorticity at the outflow is solved. Now, the radiative boundary condition (Eq. (2.202)) handles this attenuated level of vorticity at the outflow properly and the simulation does not breakdown, as it happened in the previous case shown in Fig. 2.21. One also observes that, the application of the 1D filter does not alter the flow in any way for $x \le 115$, as all the minute details shown in Figs. 2.21 and 2.22 are identical for both ψ and ω for $x \le 115$. After this application of the filter at the outflow, it was possible to continue this simulation even for much longer time up to $t = 655$ till the time the solution breaks down at an interior point due to aliasing. This problem of aliasing is cured by the application of 2D adaptive filter first proposed in Bhumkar and Sengupta [11], which is described next.

2.5.3   Two-Dimensional Higher Order Filters

Despite displaying several advantages, 1D filters also show strong directionality in the computed solution, for 2D fluid flow problems. In [90], it was shown that filtering in azimuthal direction for flows past an airfoil or a circular cylinder performing rotational oscillations, causes unphysical smearing of vortical structures in that direction. Researchers have identified aliasing error as a potent source of error occurring during discrete computations. The origin of the aliasing problem, while solving the NSE, occurs whenever one encounters a term which is a product of more than one quantities. Two such terms in NSE are due to nonlinear convection terms and the linear self-adjoint diffusion terms, that appear while solving the NSE in transformed plane. In [89], a new 2D filter is proposed, which is capable of removing the aliasing problem more effectively than 1D filters, without introducing any directionality to the solution. These 2D filters are significantly different from the conventional 1D Padè type filters described in Sect. 2.5.1.

First, we describe the origin of the aliasing error and why it occurs in any discrete computations. As stated, aliasing errors occur whenever the product terms are evaluated numerically. Consider the product of two terms in the physical space as

$$w(x_j) = u(x_j)\, v(x_j) \tag{2.204}$$

Now, each of these individual terms can be represented by respective bilateral Fourier–Laplace transform as

$$w(x_j) = \int W(k)e^{ikx_j}\,dk$$

$$u(x_j) = \int U(k_1)e^{ik_1x_j}\,dk_1$$

$$v(x_j) = \int V(k_2)e^{ik_2x_j}\,dk_2 \tag{2.205}$$

From Eqs. (2.204) and (2.205), one notes that

$$w(x_j) = \int_{-k_m}^{k_m}\int_{-k_m}^{k_m} U(k_1)V(k_2)e^{i(k_1+k_2)x_j}\,dk_1dk_2 \tag{2.206}$$

Fig. 2.23 Spectral domain
of interest for finite grid
computations. The shaded
areas (*EBF* and *HDG*)
contribute to aliasing in
evaluating product terms to
zones *HOG* and *EOF*

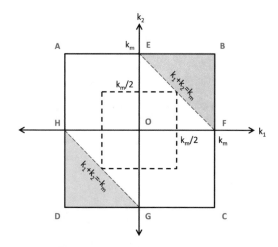

For any discrete computation, the maximum resolvable wavenumber k_m is given by
the Nyquist limit, where $k_m = \pi/h$, with h representing the uniform grid-spacing.
Therefore, the limit of integration for k_1 and k_2 in Eq. (2.206) varies from $-k_m$ to
k_m for both k_1 and k_2. This zone is shown in Fig. 2.23 by the solid box (*ABCD*).
One notes from Eq. (2.206) that wavenumber component of $w(x_j)$ is represented
by $k = k_1 + k_2$ and as both k_1 and k_2 varies from $-k_m$ to k_m, k varies from $2k_m$ to
$-2k_m$. For example, in the triangular zone *EBF* in the top right corner of Fig. 2.23,
$k_m \leq k \leq 2k_m$. Similarly, inside the triangular zone *HDG* (in the bottom left corner
of Fig. 2.23), $-2k_m \leq k \leq -k_m$. However, for any discrete computation, one can
resolve scales only up to the Nyquist limit, i.e., $-k_m \leq k \leq k_m$. Therefore, the region
where $(k_1 + k_2) \geq |k_m|$, should fold back inside the Nyquist limit, while maintain
the phase relation in either of this zones. Defining a modified wavenumber $\bar{k} =
(k_1 + k_2) - 2k_m = k - 2k_m$, one notes that

$$e^{ikx_j} = e^{i\bar{k}x_j}$$

Moreover, when $k_m \leq k \leq 2k_m$, $-k_m \leq \bar{k} \leq 0$. This implies that such a transformation in wavenumber retains the phase relationship, while maintaining the modified wavenumber $\bar{k}$ to lie within the Nyquist limit. In essence, the region *EBF*
would be aliased to the region *HOG*. Similar arguments can also be provides for
the zone *HDG*, which would be aliased to the region *EOF*. Mapping the integral
in Eq. (2.206) from (k_1, k_2)-plane to (k, k_2)-plane, one obtains

$$w(x_j) = \int_{-k_m}^{0} \int_{-k_m}^{k+k_m} U(k - k_2)V(k_2)e^{ikx_j} dk_2 dk$$
$$+ \int_{0}^{k_m} \int_{k-k_m}^{k_m} U(k - k_2)V(k_2)e^{ikx_j} dk_2 dk$$

$$+ \int_{-2k_m}^{-k_m} \int_{-k_m}^{k+k_m} U(k - k_2)V(k_2)e^{ikx_j}\,dk_2 dk$$

$$+ \int_{k_m}^{2k_m} \int_{k-k_m}^{k_m} U(k - k_2)V(k_2)e^{ikx_j}\,dk_2 dk \qquad (2.207)$$

Derivation of Eq. (2.207) is left as an exercise to the reader. One notes from Eq. (2.207) that the first and the last term falls within the Nyquist limit for wavenumber component k, whereas the last two integrals lie outside of it. Following the transformations $\bar{k} = k + 2k_m$ and $\bar{k} = k - 2k_m$ in the third and fourth integrals, respectively, one rewrites Eq. (2.207) as

$$w(x_j) = \int_{-k_m}^{0} \int_{-k_m}^{k+k_m} U(k - k_2)V(k_2)e^{ikx_j}\,dk_2 dk$$

$$+ \int_{0}^{k_m} \int_{k-k_m}^{k_m} U(k - k_2)V(k_2)e^{ikx_j}\,dk_2 dk$$

$$+ \int_{0}^{k_m} \int_{-k_m}^{k-k_m} U(k - k_2 - 2k_m)V(k_2)e^{ikx_j}\,dk_2 dk$$

$$+ \int_{-k_m}^{0} \int_{k+k_m}^{k_m} U(k - k_2 + 2k_m)V(k_2)e^{ikx_j}\,dk_2 dk \qquad (2.208)$$

In Eq. (2.208), $\bar{k}$ has been replaced by k for uniformity. One clearly notes the effect of aliasing in the last two integrals of Eq. (2.208). Expressing $w(x_j)$ by its Fourier–Laplace transform within the Nyquist limit and equating it with the right hand side of Eq. (2.208), the spectral amplitude of it is given as

$$W(k) = \int_{-k_m}^{k+k_m} U(k - k_2)V(k_2)dk_2 + \int_{k-k_m}^{k_m}$$
$$U(k - k_2)V(k_2)dk_2 + W_{alias}(k) \qquad (2.209)$$

where the aliased component $W_{alias}(k)$ is given as

$$W_{alias}(k) = \int_{-k_m}^{k-k_m} U(k - k_2 - 2k_m)V(k_2)dk_2$$

$$+ \int_{k+k_m}^{k_m} U(k - k_2 + 2k_m)V(k_2)dk_2 \qquad (2.210)$$

Equation (2.210) shows that aliasing causes the shifting of the resultant spectral amplitude to a different wavenumber than the combined wavenumber of constituent terms. This problem also arises while solving NSE in uniform grid because of the presence of the nonlinear convection term. A more comprehensive discussion on the aliasing error and its effects are given in [83] and are not repeated here for brevity.

Two-dimensional filter, only removes the high wavenumber components that are responsible for aliasing. General form of the proposed 2D filters in [89] is given by,

$$
\hat{u}_{i,j} + \alpha_{2f}(\hat{u}_{i-1,j} + \hat{u}_{i+1,j} + \hat{u}_{i,j-1} + \hat{u}_{i,j+1})
$$

$$
= \sum_{n=0}^{\hat{M}} \frac{a_n}{2}(u_{i\pm n,j} + u_{i,j\pm n}) \tag{2.211}
$$

In Eq. (2.211), $\hat{u}_{i,j}$ is the filtered variable, while u_{ij} is the unfiltered counterpart, which is being filtered here. The variable $\hat{M}$ in Eq. (2.211) represents the order of the 2D-filter, which is equal to one for a 2D, 2nd-order filter and every increase of $\hat{M}$ by one increases the order of the filter by two. Let, N_i and N_j be the total number of grid-points along i- and j-directions, respectively. Then the indices i and j varies from $i = 2$ to $(N_i - 1)$ and $j = 2$ to $(N_j - 1)$, for the 2nd-order, 2D-filter, respectively. Similarly, for 4th-order, 2D-filter, Eq. (2.211) is applicable from $i = 3$ to $(N_i - 2)$ and $j = 3$ to $(N_j - 2)$. Similarly, higher order filters can also be obtained. Equation (2.211) implies that application of 2D-filter amounts to solving a linear algebraic equation with the left-hand side giving rise to a pentadiagonal matrix. Any iterative method employed to solve this linear algebraic equation, requires the diagonal dominance of the resultant matrix. This can be ensured by prescribing $|\alpha_{2f}| \leq 1/4$ [89].

Expressing the filtered ($\hat{u}$) and unfiltered (u) variables by respective bi-directional Fourier–Laplace transform as

$$
u_{j,l} = \int\int U(k_x, k_y)e^{i(k_x x_j + k_y y_l)} \, dk_x \, dk_y
$$

$$
\hat{u}_{j,l} = \int\int \hat{U}(k_x, k_y)e^{i(k_x x_j + k_y y_l)} \, dk_x \, dk_y
$$

one obtains the transfer function of the 2D-filter given by Eq. (2.211) as

$$
TF(k_x h_x, k_y h_y) = \frac{a_0 + \sum_{n=1}^{\hat{M}} a_n[\cos(nk_x h_x) + \cos(nk_y h_y)]}{1 + 2\alpha_{2f}[\cos(k_x h_x) + \cos(k_y h_y)]} \tag{2.212}
$$

where k_x and k_y are the wavenumbers and h_x and h_y are grid-spacing along i and j-directions, respectively. The choice of $\alpha_{2f} = 1/4$ indicates no filtering of the variable. The coefficients a_k of the 2D-filter can be obtained as a function of the filter coefficient α_{2f} by expressing the right-hand and left-hand sides of Eq. (2.211) by Taylor series expansion and equating coefficients of appropriate terms on either sides of Eq. (2.211). For example, for 2nd-order filter, one has to match the coefficients of unfiltered and filtered quantities at the (i, j)th node, for consistency of the variable (consistency condition), which gives the following condition

$$
a_0 + 2a_1 = 1 + 4\alpha_{2f} \tag{2.213}
$$

for obtaining a_0 and a_1 for 2nd-order, 2D-filter, the above equation is supplemented by an additional condition which requires that the transfer function should be zero when $k_x h_x = k_y h_y = \pi$, i.e, at point B in Fig. 2.23. This condition is imposed, so that the 2D-filter can effectively control the aliasing error without distorting or unphysically attenuating the physically relevant components. This condition imposes the relation

$$a_0 = 2a_1 \qquad (2.214)$$

Following Eqs. (2.213) and (2.214), one obtains

$$a_0 = \frac{1}{2} + 2\alpha_{2f}$$
$$a_1 = \frac{1}{4} + \alpha_{2f} \qquad (2.215)$$

For 4th-order, 2D-filter, the consistency condition is given as

$$a_0 + 2a_1 + 2a_2 = 1 + 4\alpha_{2f} \qquad (2.216)$$

For this filter, one also has to equate the coefficients of 2nd-order partial derivatives (including the mixed derivatives) at (i, j)th-node, which gives the following relation

$$a_1 + 4a_2 = 2\alpha_{2f} \qquad (2.217)$$

Equations (2.216) and (2.217) are to be augmented by vanishing conditions of the transfer function at point B (see Fig. 2.23), which gives

$$a_0 - 2a_1 + 2a_2 = 0 \qquad (2.218)$$

Equations (2.216)–(2.218) gives the coefficients (a_0, a_1, a_2) for the 4th-order 2D-filter as

$$a_0 = \frac{(5 + 12\alpha_{2f})}{8}, \quad a_1 = \frac{(1 + 4\alpha_{2f})}{4}, \quad a_2 = \frac{(-1 + 4\alpha_{2f})}{16} \qquad (2.219)$$

Similar conditions can also be derived for the 6th-order 2D-filter and the corresponding coefficients are obtained as [89],

$$a_0 = \frac{(11 + 20\alpha_{2f})}{16}, \quad a_1 = \frac{(15 + 68\alpha_{2f})}{64}$$
$$a_2 = \frac{(-3 + 12\alpha_{2f})}{32}, \quad a_3 = \frac{(1 - 4\alpha_{2f})}{64} \qquad (2.220)$$

In Fig. 2.24, the transfer function of a 2nd-order, 2D filter is shown, for indicated values of α_{2f} as a function of nondimensional wavenumbers $k_x h_x$ and $k_y h_y$. As noted earlier, the region corresponding to $(k_x h_x + k_y h_y) > \pi$ contributes to aliasing

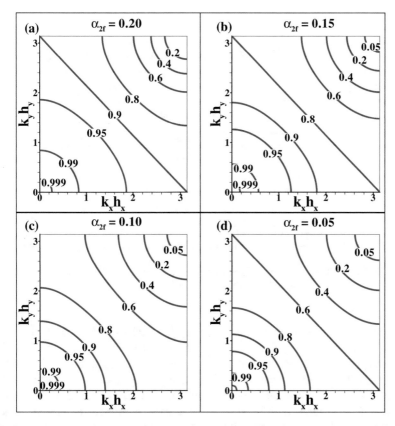

Fig. 2.24 The transfer functions of the 2nd-order 2D filter are shown in frames **a** and **b** for indicated α_{2f}

error. It is clear from Fig. 2.24, that the transfer function of the 2D-filter effectively removes the aliasing error without unduly attenuating the mid-wavenumber region in each direction [89]. It has also been shown in Bhumkar and Sengupta [11] that such 2D filters can be used in an adaptive manner on a small patch of the grid to only affect the zone, where some high wavenumber fluctuations of the solution have been developed. Such adaptive filters have been used while performing receptivity calculations in [6, 85, 86] for effective de-aliasing. It is impossible to simulate transitional or turbulent flows correctly without effective de-aliasing of the solution. In the next subsection, a brief description of the application procedure of such filters are provided.

2.5.4 Adaptive 2D Filter

Adaptive 2D filters have been first proposed in [11], which can be effectively used over small localized patches of computational domain to avoid local numerical insta-

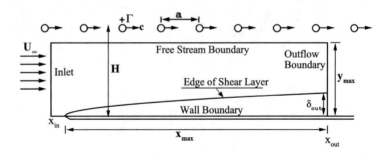

Fig. 2.25 Schematic of the problem solved for free stream excitation by an infinite array of convecting vortices over a semi-infinite flat plate

bilities occurring due to aliasing error during physical secondary and higher order instabilities. For this purpose, the filter operation given by Eq. (2.212) is used in adaptive manner at some selective points, based on a predetermined criterion. The resultant linear algebraic equations can be solved iteratively using Bi-CGSTAB method [118].

Here, one particular transitional and turbulent flow problem is described, where application of adaptive 2D-filter is essential to carry out the numerical calculations accurately. The problem is regarding the receptivity of the boundary layer to the free-stream convecting vortices. The schematic diagram of the problem is shown in Fig. 2.25. Here, an infinite row of convecting vortices migrate at a fixed height, with a fixed speed. The imposition of infinite overhead vortex system requires an associated image vortex system of opposite sign (see Fig. 2.25), so that the wall-normal velocity component at the wall is zero. The separation between the successive vortices (a), the strength of each vortex Γ, convection speed c, and its height over the flat plate H, are the parameters of the problem. Here a case is described, where $c = 0.2$, $a = 10$, $H = 2.0$ and $\Gamma = 0.25$. To compute the receptivity of the shear-layer to the free-stream vortices, 2D incompressible NSE given by Eqs. (2.31) and (2.32) are solved using OUCS3 scheme for spatial discretization scheme for convective terms and optimized 3-stage Runge–Kutta scheme ORK3 [100] is used for time-integration.

The overhead convecting vortices causes vortical eruptions inside the boundary layer, which are directly responsible for transition to turbulence. While calculating this case, in a computational domain with 4500×400 points spanning from -0.05 to 90 in the streamwise direction and 1.5 in the wall-normal direction, the problem of aliasing was faced. This resulted in the complete breakdown of the simulation at $t = 113$. In Fig. 2.26, the vorticity contours are plotted at $t = 111$, 112 and 113 in frames (a), (b) and (c), respectively. One notices the unsteady vortical eruptions

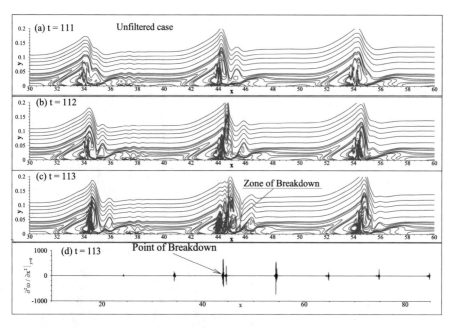

Fig. 2.26 Vorticity contours of the unfiltered case plotted at indicated times for the free stream convecting vortex problem with $c = 0.2$, $a = 10$, $H = 2.0$ at indicated times in frames **a, b** and **c**. **d** $\frac{\partial^2 \omega_w}{\partial x^2}$ plotted of the filtered case at $t = 113$

and associated induction of high wavenumber components nearby, due to aliasing. In Fig. 2.26d, the quantity $\partial^2 \omega_w / \partial x^2$ is plotted as a function of x. One observes high wavenumber oscillations at some selective locations underneath the overhead vortex. The point where the actual breakdown of the solution has taken place at $t = 113$ is also marked in Fig. 2.26c, d.

To rectify these issues of aliasing error causing breakdown of the numerical solution, 2D adaptive filter is used very close to the wall. One identifies a rectangular/square sub-domain in the transformed (ξ, η)-plane where filtering is performed, with variation of α_{2f} with respect to the indices i and j is given as

$$\alpha_{2f} = l_1 - \frac{(l_1 - l_0)}{2} \left[1 + \cos\left(\pi \frac{i_{max} - 2i + i_{min}}{i_{max} - i_{min}} \right) \right]$$
$$\cos\left(\frac{\pi}{2} \frac{j - j_{min}}{j_{max} - j_{min}} \right) \qquad (2.221)$$

where, $l_1 = 0.25, l_2 = 0.05$ and i_{max}; i_{min}; j_{max}; j_{min} are the upper and lower limits of i and j indices, respectively i.e., $i_{min} \le i \le i_{max}$ and $j_{min} \le j \le j_{max}$. In Fig. 2.27a a typical variation of α_{2f} is plotted over a 9×9 filter-tent, where $i_{max} = 9, i_{min} = 1$, $j_{max} = 9$ and $j_{min} = 1$. The corresponding 3D perspective plot of the variation of α_{2f} is shown in Fig. 2.27b. Maximum filtering is performed at $(i = (i_{max} + i_{min})/2$,

Fig. 2.27 **a** A typical variation of α_{2f} is shown over a 9×9 filter tent in the transformed plane. The corresponding 3D perspective plot of the variation of α_{2f} in the filter tent is shown in frame **b**

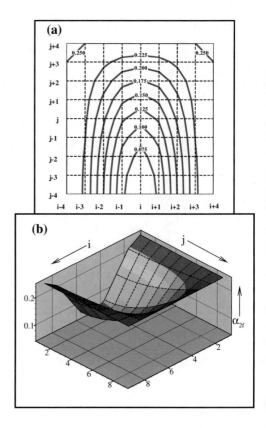

$j = j_{min}$)th point of the tent, which progressively reduces to no filtering at the edges of it, where $\alpha_{2f} = 0.25$.

In applying this filter, the streamwise second derivative of the wall-vorticity ($\partial^2 \omega_w / \partial x^2$, where ω_w is the vorticity at the wall) is calculated and if this value exceed a certain predefined value, then the 2D adaptive filter is activated at that particular x-location. There are several reasons to chose such a criterion. First, it was noted that breakdown of simulations happened due to the occurrence of the high-wavenumber fluctuations in the vorticity and for wall-bounded boundary layers, maximum vorticity generally occurs on the wall itself. Secondly, as the differentiation operation magnifies the high wavenumber amplitudes, the chosen criterion would exactly highlight the points on the wall, where high wavenumber oscillations appear in the vorticity. These are the reasons behind choosing the quantity $\partial^2 \omega_w / \partial x^2$ as a test-parameter. The specification of the cutoff value for the test-parameter was performed, based on experience and there is no definitive rule to prescribe it.

The same case was run with applying 2D adaptive filter at the selected locations as described in the last paragraph, whenever $\partial^2 \omega_w / \partial x^2$ exceeded 100. This case was run from results at $t = 100$, of earlier unfiltered simulation. The results of the filtered case is shown in Fig. 2.28. One notices that the application of adaptive filter has been able to

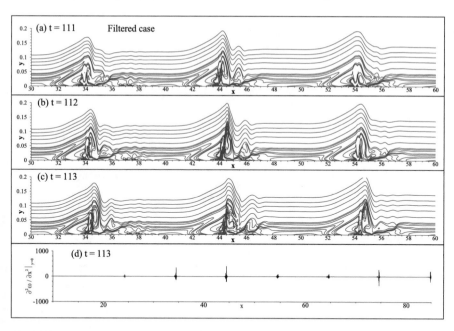

Fig. 2.28 Vorticity contours plotted at indicated times for the free stream convecting vortex problem with $c = 0.2$, $a = 10$, $H = 2.0$ times in frames **a**, **b** and **c** after applying adaptive 2D filter from $t = 100$. $\frac{\partial^2 \omega_w}{\partial x^2}$ plotted of the filtered case at $t = 113$ in frame **d**

significantly control the high wavenumber oscillations, while completely preserving flow features at lower wavenumber. Fig. 2.28d reveals that the amplitude of the high frequency oscillations are also substantially reduced due to the application of this filter. To further reveal the effectiveness of the 2D adaptive filter, the wall vorticity and its FFT at $t = 113$ is plotted in Fig. 2.29 for both unfiltered (frames (a) and (b)) and filtered case (frames (c) and (d)). One finds in Fig. 2.29b, that the aliasing effectively increasing the high wavenumber components as indicated. Fig. 2.29d also testifies to the fact that 2D adaptive filtering not only suppresses the amplitude of the high wavenumber components by an order of magnitude, but also keep the low wavenumber perturbation amplitudes intact, which only determines the nature of transition. Similar application of 2D adaptive filter to control aliasing errors during the computations of flows past circular cylinder performing rotary oscillation and flows past a natural laminar aerofoil at a very high Reynolds number have been reported in [11].

2.6   Role of Solenoidality Errors in Velocity-Vorticity Formulations

One of the major issues of high accuracy finite difference computing of NSE using different formulations is the identification of an error metric and different investigators have tried to identify different derived quantities as a marker for the

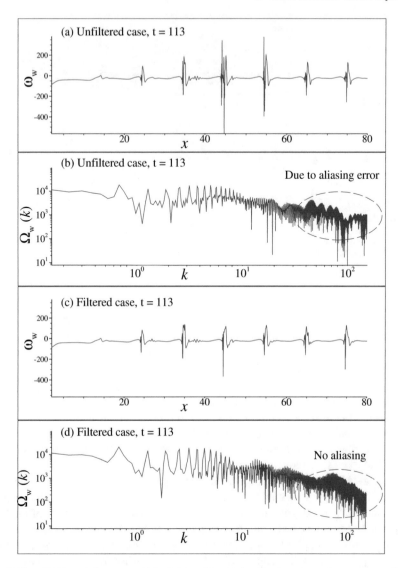

Fig. 2.29 a Wall vorticity at $t = 113$ of the unfiltered free stream convecting vortex problem shown in Fig. 2.27 and **b** its Fourier transform plotted as a function of streamwise distance x and wavenumber k, respectively. **c** Wall vorticity at $t = 13$ of the case where 2D adaptive filtered is used for the free stream convecting vortex problem shown in Fig. 2.28 and **d** its Fourier transform plotted as a function of streamwise distance x and wavenumber k, respectively

accuracy of the numerical solution. Here, the solenoidality of velocity and vorticity as the error metrics are focused upon, and the relevance of these for some benchmark 2D and 3D flow problems are presented. In order to develop an accurate method using $(\overrightarrow{V}, \overrightarrow{\omega})$-formulations for simulating transitional flows, the effects of solenoidality of velocity on flow evolution has to be assessed and for this purpose, one must focus on the flow regimes, when instabilities play a dominant role for both internal and external flows. For 3D flows, solenoidality of vorticity vector also has to be considered. This has prompted the development of the conservative rotational formulation of VTE, which has been given previously in Eq. (2.24a). Here, application of this formulation is described along with the development of new optimum versions of derivative evaluation and interpolation of unknowns.

2.6.1 Computation of Physically Unstable Flows in Square Lid Driven Cavity

Vorticity is generated at no-slip walls or due to flow instabilities in free shear layers, for incompressible flows. For turbulent flows, vorticity and other derived variables play a central role in understanding some of the physical mechanisms of creating turbulence. Hence, VTE is found very relevant for the analysis and solution of viscous incompressible flows. Attendant velocity field can be obtained from the solution of the Poisson equation for stream-function (for 2D case) or by solving Poisson equations for the velocity components, as described previously in Sects. 2.1.5 and 2.1.6.

As an illustrative example, first the flow inside a 2D square lid-driven cavity (LDC) is considered. For this problem, each side of the square and the lid-velocity is used as the reference length and velocity scale to nondimensionalize the governing equations. This determines the Reynolds number for the flow. A staggered variable arrangement, as used in Guj and Stella [37] and Napolitano and Pascazio [63] has been adopted. The staggered arrangement of variables are shown in Fig. 2.10a, where the velocity components are defined at the center of each sides and the vorticity components are defined at each integral grid-nodes. It has been shown that numerical error is smaller in staggered grid than non-staggered (or collocated) grid using $(\overrightarrow{V}, \omega)$-formulation [44]. It is known that unlike (ψ, ω)-formulation, velocity-vorticity formulation does not identically ensure satisfaction of solenoidality of the velocity field everywhere inside the flow domain. The governing equations are 2D incompressible NSE in physical x- and y-domain with uniform distribution of grid-points. Therefore, $h_1 = h_2 = 1$ in Eqs. (2.36)–(2.39) and these are given as

$$\frac{\partial \omega}{\partial t} + \frac{\partial}{\partial x}(u\omega) + \frac{\partial}{\partial y}(v\omega) = \frac{1}{Re}\left(\frac{\partial^2 \omega}{\partial x^2} + \frac{\partial^2 \omega}{\partial y^2}\right) \tag{2.222}$$

$$\frac{\partial^2 u}{\partial x^2} + \frac{\partial^2 u}{\partial y^2} = -\frac{\partial \omega}{\partial y} \qquad (2.223)$$

$$\frac{\partial^2 v}{\partial x^2} + \frac{\partial^2 v}{\partial y^2} = \frac{\partial \omega}{\partial x} \qquad (2.224)$$

where the velocity $\vec{V}\,[=(u, v)]$ is expected to satisfy the divergence-free condition given as

$$D_v = \frac{\partial u}{\partial x} + \frac{\partial v}{\partial y} = 0 \qquad (2.225)$$

As the purpose is to ascertain the role of non-solenoidality of velocity field in velocity-vorticity formulation, an alternate equation derived by differentiating Eq. (2.225) with respect to y is used, which is used to obtain one of velocity components. This ensures solenoidality of velocity identically everywhere inside the flow domain. In the present case, Eq. (2.224) is replaced by the equation given by

$$\frac{\partial^2 v}{\partial y^2} = -\frac{\partial^2 u}{\partial x \partial y} \qquad (2.226)$$

In evaluating the role of solenoidality of velocity field, the LDC problem is solved employing two schemes as described next.

Scheme-A: In this approach, Eq. (2.222)–(2.224) are solved. Here first Eq. (2.222) is solved for time advancing the vorticity field, followed by Eqs. (2.223) and (2.224) for the velocity components. Thus, the solution obtained by this scheme will not identically satisfy solenoidality of the velocity field.

Scheme-B: Here, the vorticity field is time advanced by solving Eq. (2.222) first. The velocity components are obtained by solving Eqs. (2.223) and (2.226). Solving Eq. (2.226) instead of Eq. (2.224), ensures solenoidality of the velocity field by this scheme.

In both the schemes, time advancement of vorticity is carried out using 4th-order, four-stage Runge–Kutta (RK_4) scheme, while convective derivatives, i. e., $\frac{\partial}{\partial x}(u\omega)$ and $\frac{\partial}{\partial y}(v\omega)$ terms are discretized using OUCS3 scheme, and second derivative terms are discretized using second-order central difference schemes. To evaluate convective derivatives in the staggered grid, it is required to interpolate u- and v-components at the location of the vorticities; which is carried out by the optimized compact mid-point interpolation scheme of [8]. This is described in Sect. 2.3.2. To obtain u- and v-velocity components, either scheme A or scheme B is employed by using central difference explicit schemes for all pure and mixed second derivatives.

Results are presented here for the 2D LDC problem, for $Re = 9000$ and $Re = 10,000$. Both these Reynolds numbers belong to post-critical range for the 2D flow [105]. Results obtained in [105] using (ψ, ω)-formulation are accurate for this post-critical Reynolds number and is used here for validation and comparison purposes. The Poisson equations are solved using Bi-CGSTAB [118] iterative scheme.

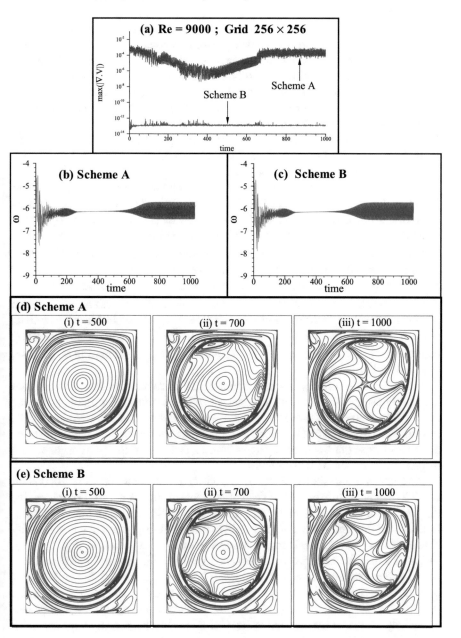

Fig. 2.30 Flow in a lid-driven cavity for $Re = 9000$. **a** Maximum absolute divergence error plotted as a function of time for schemes A and B. **b, c** Vorticity time-history at $x = 0.95$, $y = 0.95$ plotted for schemes A and B, respectively. **d, e** Vorticity contours at indicated times shown for schemes A and B, respectively

A pre-determined tolerance value for the solution error (ε) is used in the Bi-CGSTAB solver. Maximum solenoidality error occurring inside the domain for the scheme A, is directly dependent on the level of ε chosen. In computing both cases, a grid with (257×257) uniformly distributed points have been used. After every 100 time-steps, D_v is computed using 2nd-order central difference scheme and maximum absolute value of D_v is stored. In Fig. 2.30a, time history of D_v are compared for $Re = 9000$ for schemes-A and -B, for $\varepsilon = 10^{-6}$ used with the Bi-CGSTAB solver. In Fig. 2.30b, c, vorticity value at a point with co-ordinates $x = 0.95$ and $y = 0.95$ (with the origin at the lower left corner) is plotted as a function of time. Even though the levels of D_v shown in Fig. 2.30a, are orders of magnitude different, this does not affect either the onset time of instability or the amplitude of vorticity fluctuations in Fig. 2.30b, c. As noted in [105], this flow for $Re = 9000$ also shows nonlinear saturation of disturbance amplitude. Despite the difference in D_v, these two schemes show transient variation that also looks almost similar in Fig. 2.30b, c. This seems to counter the observation of [17, 128] for ($p, \overrightarrow{V}$)-formulation; who have talked about numerical instabilities due to non-satisfaction of solenoidality for velocity vector during transient and later stages of flow evolution computed by NSE. Recently it has also been pointed out in [125] that using ($p, \overrightarrow{V}$)-formulation with pseudo-spectral DNS slowly evolving spurious numerical turbulence is induced inside the flow-field, when the effect of machine round-off on the divergence-free condition is not carefully controlled. In contrast, for the ($\overrightarrow{V}, \omega$)-formulation, this has not seen to be the case, as shown in Fig. 2.30.

In Fig. 2.30d, e, vorticity contours at indicated times plotted for these two schemes are also indistinguishable. Both schemes show the presence of a triangular vortex at the center of the cavity, at $t = 700$, as reported earlier for 2D supercritical LDC flow in [98, 105, 107]. But one notices a small time lag among the satellite vortices, as shown in the third frames of Fig. 2.30d, e. This is due to different phase and dispersion errors for the two schemes.

Similar results are shown in Fig. 2.31a, b for the LDC flow for $Re = 10,000$. In Fig. 2.31a, time history of D_v has been plotted for schemes A and B. For scheme-A, different tolerance limit given by ε for the solution of Poisson equation are used. It is noted that tightening the tolerance limit by orders of magnitude, D_v decreases by similar orders, yet scheme-B shows negligibly small levels of D_v in Fig. 2.31a(ii). In Fig. 2.31b, time histories of disturbance vorticity are shown for different values of ε. For scheme-A three different levels of tolerance, $\varepsilon = 10^{-4}$, $\varepsilon = 10^{-6}$ and $\varepsilon = 10^{-9}$ are considered with Bi-CGSTAB solver. In using scheme-B, a tolerance value of $\varepsilon = 10^{-6}$ has been used in solving the Poisson equation for u-component of velocity. Note that the solution of Eq. (2.226) is obtained directly via exact solution of tridiagonal matrix system. From the displayed figures, one again notices that maximum solenoidality error is markedly different for all the four cases, even though the vorticity time history at $x = 0.95$ and $y = 0.95$ are almost similar.

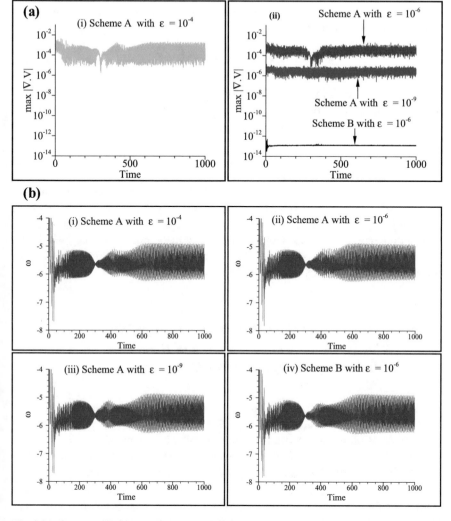

Fig. 2.31 Flow in a lid-driven cavity for $Re = 10000$. **a** Maximum absolute divergence error time-history plotted for the indicated schemes with different ε. **b** Vorticity time-history at $x = 0.95$ and $y = 0.95$ shown plotted for (i) scheme A with $\varepsilon = 10^{-4}$, (ii) scheme A with $\varepsilon = 10^{-6}$, (iii) scheme A with $\varepsilon = 10^{-9}$ and (iv) scheme B with $\varepsilon = 10^{-6}$

Presented results in Figs. 2.30 and 2.31, using velocity-vorticity formulations (which does not satisfy solenoidality of velocity vector to machine zero) show identical results for a long time without showing any tendency for numerical instability for a flow, which demonstrates physical instability and its nonlinear saturation. This shows that the velocity-vorticity formulation is insensitive to divergence error beyond a certain level. It is because of the following reasons.

In velocity-vorticity formulation of scheme-A, these two quantities are related by the vector Poisson equation as,

$$\nabla^2 \vec{V} = -\nabla \times \vec{\omega} \qquad (2.227)$$

Irrespective of the fact whether D_v is zero or not, this equation uses solenoidality in its derivation. As a consequence, taking divergence of Eq. (2.227) and using the vector identity $\nabla \cdot (\nabla \times \vec{G}) = 0$ for any arbitrary vector $\vec{G}$ one obtains

$$\nabla^2 D_v = 0 \qquad (2.228)$$

Thus, if one chooses D_v as the error metric in velocity-vorticity formulation, it becomes time-independent and its property is given by the minmax theorem of the governing Laplace equation. If any time dependence or instability has to occur in this formulation, then it must come from the kinematic boundary condition. In the absence of any such time-dependent boundary forcing, the governing equation for this metric is perfectly stable, as noted in Figs. 2.30 and 2.31.

2.6.2 Solution of 3D Cubic Lid Driven Cavity Using Velocity-Vorticity Formulation and Role of Solenoidality Error for Vorticity

In solving 2D flows using (ψ, ω)- or $(\vec{V}, \omega)$-formulations, one notes that the non-zero component of the vorticity lies in a plane perpendicular to the flow. Thus, the vorticity in 2D flows always satisfies solenoidality condition $(D_\omega = \nabla \cdot \vec{\omega} = 0)$. However for 3D flows this is not automatic, i.e., $D_\omega \neq 0$. Thus, one would like to investigate the role of D_ω for 3D flows. This is described in Sect. 2.1.5. Here, the role of D_ω in $(\vec{V}, \vec{\omega})$-formulations for 3D flows are estimated, by showing results for flows inside 3D cubic lid driven cavity. In Sect. 2.1.5, three different formulations of VTE is provided, namely, (a) the non-conservative form, Eq. (2.22), (b) the Laplacian form, Eq. (2.23), and (c) the rotational form, Eq. (2.24b), and also the corresponding evolution equation for divergence of vorticity (D_ω) is illustrated through Eqs. (2.25a)–(2.25c). It was noted that for the rotational form of VTE, one only requires D_ω to be zero, while prescribing the initial conditions. In contrast, for the Laplacian form of VTE, the evolution of D_ω follows a 3D-unsteady diffusion equation, Eq. (2.25b), and therefore, to maintain the solenoidality of the vorticity field, not only it has to be zero inside the computational domain, but one must also has to ensure that no spurious generation of it takes place from the boundary. The non-conservative formulation of VTE is most erroneous in terms of preserving the divergence of vorticity D_ω, as explained with the help of Eq. (2.25c). To maintain $D_\omega = 0$ identically at all time-instants, it not only requires all the constraints on D_ω, but also requires that $D_v = 0$, which acts as a source term for the evolution of D_ω (see Eq. (2.25c)).

The Laplacian and rotational variants of ($\vec{V}$, $\vec{\omega}$)-formulation of NSE is solved for flow inside the cubic LDC and results compared for sub-critical and post-critical Reynolds numbers. For this purpose, the staggered grid used in [131] is employed, as shown here in Fig. 2.10b on an elementary cell. The velocity components are defined at the center of the plane, on which the component is perpendicular; while the vorticity components are placed at mid-location on the edges of the cube which are parallel to the direction of the vorticity components. Such grid staggering for 3D flows require additional evaluation of derivatives of vorticity components (see Eqs. (2.23) and (2.24a)). For example, the x-component of VTE would require evaluation of derivatives of ω_y and ω_z at the location of ω_x. This derivative evaluation has been performed using optimized staggered compact scheme, Eq. (2.135), as described in Sect. 2.3.2.

For the Laplacian variant of ($\vec{V}$, $\vec{\omega}$)-formulation, the diffusion terms, i.e., terms appearing in $\nabla^2 \vec{\omega}$ are numerically evaluated using second order central difference scheme. In contrast, in the rotational variant of ($\vec{V}$, $\vec{\omega}$)-formulation, one requires to evaluate the terms appearing in $\nabla \times (\nabla \times \vec{\omega})$, which is performed by applying Eq. (2.135) twice in each of the x-, y- and z-directions. Boundary conditions used for this cubic cavity case appear directly from the zero-normal and no-slip conditions at the wall, and is discussed elaborately in the literature [31, 63, 131]. A total of $81 \times 81 \times 81$ uniformly distributed grid-points have been chosen with a time step of $\Delta t = 10^{-3}$. As was reported for the previous $2D$-simulations, here also, RK_4 scheme is used for time integration.

First, a set of results for the sub-critical Reynolds number of $Re = 1000$ (based on the side of the cube and lid velocity) is computed using Laplacian and rotational forms of ($\vec{V}$, $\vec{\omega}$)-formulation. Time histories of maximum value of D_ω and D_v in the full domain, are shown in Fig. 2.32. It is noted that the values do not change beyond $t = 70$ and hence the displayed histories are for the transient state only. Displayed values of D_ω are obtained using the compact scheme and it is clearly noted that the rotational form is superior over the Laplacian form of the NSE in ($\vec{V}$, $\vec{\omega}$)-formulation. Although, one notes almost similar time variation of D_v for both the forms (evaluated using simple CD_2 scheme). It is to be noted that in literature [37, 63, 131] results have been reported for the cubic LDC using Laplacian form. In [2, 55, 56], this flow is computed using primitive variable formulation by spectral element method with Gauss–Lobatto–Chebyshev collocation points.

In Fig. 2.33, u- and v-velocity components obtained at the mid-plane are shown using these two variants of ($\vec{V}$, $\vec{\omega}$)-formulation. The rotational form results match identically with the Laplacian form for this sub-critical Reynolds number, despite the large difference in D_ω noted in Fig. 2.32. Also, the results obtained using Laplacian form are seen to match with the results of [2], obtained using primitive variable by spectral element method. The stream-traces and vorticity contours are shown in Figs. 2.34 and 2.35 for different components, obtained using Laplacian and rotational forms, and one cannot see visually any differences. Whatever differences are noted

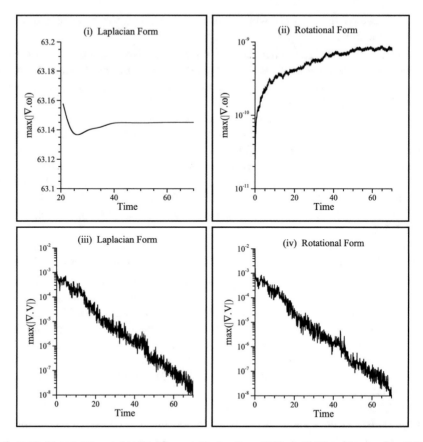

Fig. 2.32 Flow inside a cubic lid-driven cavity for $Re = 1000$. (i, ii) Time-history of vorticity divergence error plotted for computations using Laplacian and conservative rotational forms of VTE, respectively. (iii, iv) Time-history of velocity divergence error plotted for Laplacian and conservative rotational forms of VTE, respectively

for D_ω, these occur at the top left corner only of the displayed (x, y)-plane. This will become more evident when post-critical Reynolds number flows are studied. Figs. 2.34 and 2.35 also show that the flow is symmetric about the mid (x, y)-plane. Symmetry of the flow about the mid (x, y)-plane is retained, even during the transient stage of flow development to steady state for this Re. At this plane of symmetry, the cross-flow velocity w is found to be exactly zero. The strong similarity for the results of the simulations using Laplacian and rotational forms of VTE are even noted for all the intermediate times despite large difference in the vorticity solenoidality error levels in Fig. 2.32. In Fig. 2.36, maximum values of D_ω and D_v in the computational domain, are compared between the rotational and the Laplacian forms of $(\vec{V}, \vec{\omega})$-formulation for $Re = 3200$, which is a post-critical Reynolds number flow, as it does not reach a steady state even after $t = 2000$. Also, the experimental results of

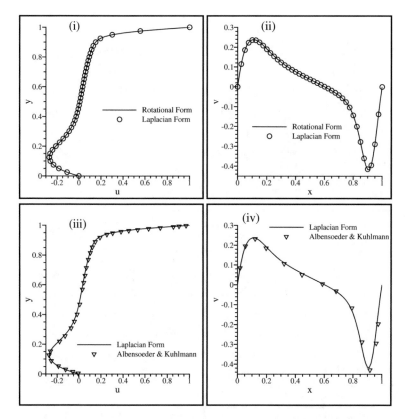

Fig. 2.33 (i, ii) *u*- and *v*-velocity components at the mid-plane plotted for Laplacian and rotational forms of VTE and (iii, iv) compared with the results of Albensoeder and Kuhlmann [2] for flow inside a cubic lid-driven cavity at $Re = 1000$

[51] for a rectangular parallelepiped (with square cross section) of aspect ratio of $L/D = 3 : 1$, has shown this flow for $Re = 3200$ to be time-dependent. The present computational results with the experimental results of [51] at the mid-plane are compared. For this Re, both forms of $(\vec{V}, \vec{\omega})$-formulation show identical level of D_v, although this does not decrease monotonically as was noted for $Re = 1000$ in Fig. 2.32. For D_ω, levels of error remain the same for these two forms, with rotational form showing billion times smaller value of D_ω.

In Fig. 2.37, computed velocity components in the mid (x, y)-plane are compared with the results reported in [51] for the same (x, y)-plane. For comparison purpose, computed results are time averaged between $t = 1900$ and $t = 2000$. The match appears quite well, considering the fact that the presented computed results are for a cubic cavity, while the experimental results are for a rectangular cavity with square cross-section whose length is three-times longer than its width. Results are also compared between rotational and Laplacian forms at $t = 250$ and $t = 2000$. Though the u-velocity profiles do not show any distinguishable differences at $t = 250$ and $t =$

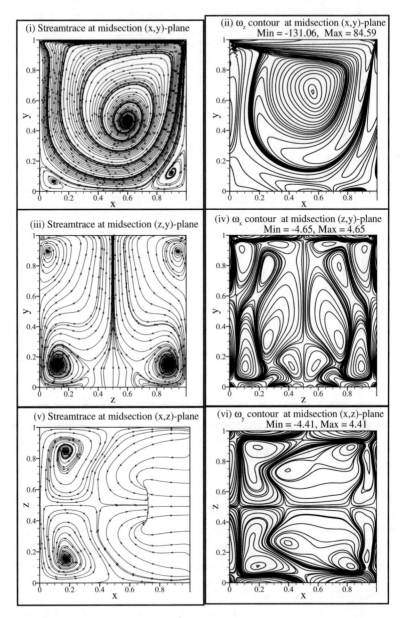

Fig. 2.34 Stream-traces and contours of normal component of vorticity are shown plotted at mid-(x, y) plane (frames (i) and (ii)), mid-(y, z) plane (frames (iii) and (iv)) and mid-(z, x) plane (frames (v) and (vi)) for $Re = 1000$ after steady state is reached. All the plots correspond to the simulations done using Laplacian form of VTE

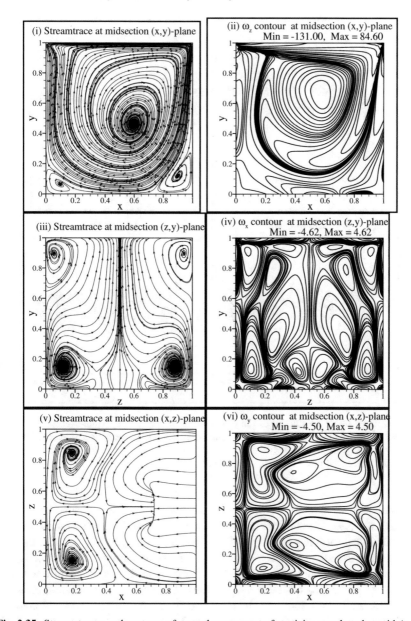

Fig. 2.35   Stream-traces and contours of normal component of vorticity are plotted at mid-(x, y) plane (frames (i) and (ii)), mid-(z, y) plane (frames (iii) and (iv)) and mid-(z, x) plane (frames (v) and (vi)) for $Re = 1000$ after steady state is reached. All the plots correspond to the simulations done using conservative rotational form of VTE

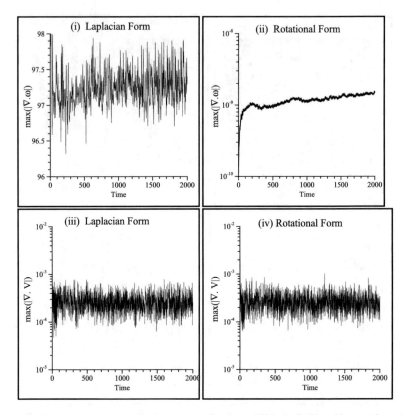

Fig. 2.36 Flow inside a cubic lid-driven cavity for $Re = 3200$. (i, ii) Time-history of vorticity divergence error plotted for computations using Laplacian and conservative rotational forms of VTE, respectively. (iii, iv) Time-history of velocity divergence error plotted for Laplacian and conservative rotational forms of VTE, respectively

2000, but there are differences for the v-velocity profiles at $t = 250$, specifically near the rear wall. Also, one notices that in frame (vi), v-velocity profile corresponding to the rotational form has slightly higher maximum and slightly lower minimum.

To further investigate the role played by the solenoidality error in vorticity, the stream-traces and contours of normal components of vorticity at three different mid-sections are plotted, as has been done in Figs. 2.34 and 2.35, at different times, in Figs. 2.38, 2.39, 2.40 and 2.41. In Fig. 2.38 stream-traces are shown plotted for (a) Laplacian and (b) rotational forms at $t = 250$ and $t = 350$, while in Fig. 2.40, these correspond to $t = 1800$ and $t = 2000$. In Figs. 2.39 and 2.41, contours of normal components of vorticity are shown plotted at three different mid-sections at $t = 250, \ 350$ and at $t = 1800, \ 2000$, respectively, for these forms. It can be noticed from Figs. 2.38, 2.39, 2.40 and 2.41 for this post-critical Reynolds number, that the plotted contours and stream-traces show significant differences. These differences are more visibly apparent in the mid (y, z)- or (z, x)-planes. The main differences in

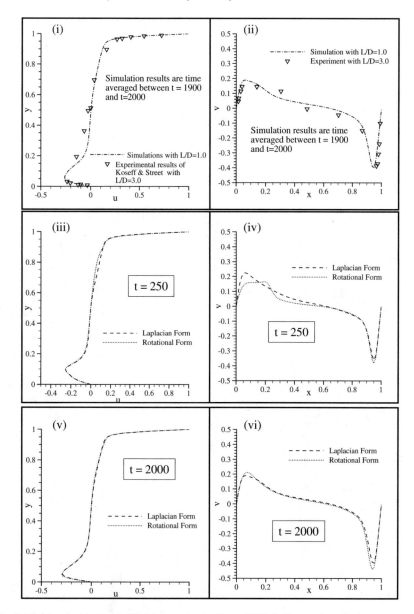

Fig. 2.37 Flow inside a cubic lid-driven cavity for $Re = 3200$. (i, ii) u- and v-velocity components at the mid-plane and time averaged between $t = 1900$ and 2000 by using Laplacian form are compared with the experimental time averaged results of [51]. (iii, v) u- and (vi, vi) v-velocity components at mid-plane obtained by using rotational form and Laplacian form are compared at $t = 250$ and $t = 2000$ as indicated in the frames

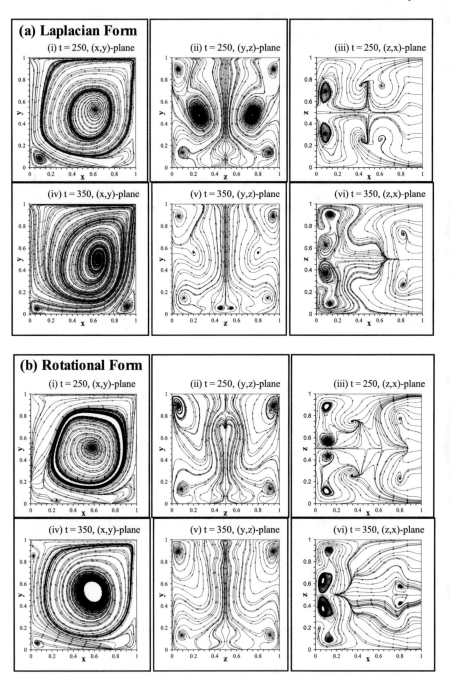

Fig. 2.38 Stream-traces shown plotted for **a** Laplacian form of VTE and **b** rotational form of VTE used, at $t = 250$ and 350 on mid-planes

(a) Laplacian Form

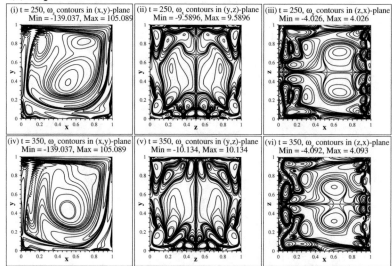

(b) Rotational Form

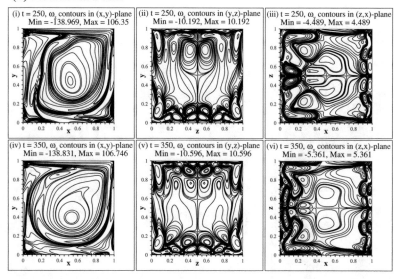

Fig. 2.39 Contours of vorticity component normal to the indicated mid-planes shown plotted **a** Laplacian form of VTE and **b** rotational form of VTE are used at $t = 250$ and $t = 350$

(a) Laplacian Form

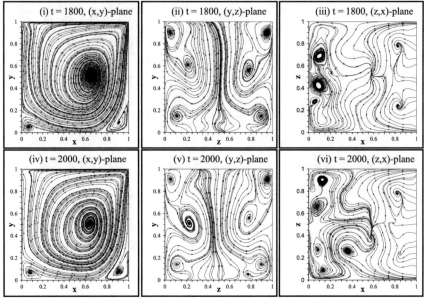

(b) Rotational Form

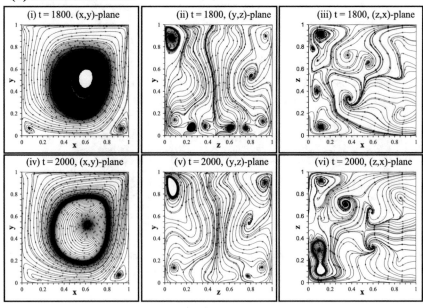

Fig. 2.40 Stream-traces shown plotted for **a** Laplacian form of VTE and **b** rotational form of VTE used at $t = 1800$ and 2000 on mid-planes

(a) Laplacian Form

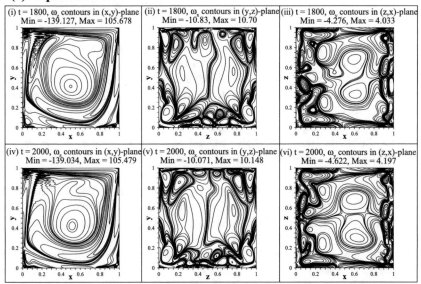

(b) Rotational Form

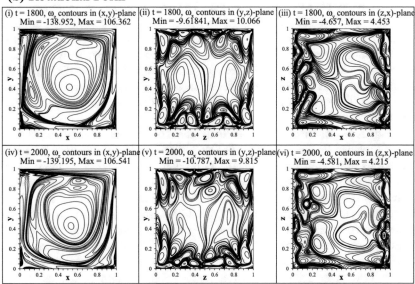

Fig. 2.41 Contours of vorticity component normal to the indicated mid-planes shown plotted for **a** Laplacian form of VTE and **b** rotational form of VTE used at $t = 1800$ and 2000

the flow topology is the location and the nature of the wall stagnation points, as seen in Fig. 2.38. As noted in Fig. 2.38a, right and left wall stagnation points are of saddle type, for both frames (i) and (iv). The nature of singularity on the right wall is same for both the forms in Fig. 2.38a, b. However, there is qualitative difference in the nature of singularity noted on the left wall between these two forms. For the rotational form, it is node at $t = 250$, that changes into a saddle at $t = 350$. For the Laplacian form, no such time variation is noted during this time interval. The singular points on the left wall changes more rapidly for the rotational form, from being a node to a saddle point. For example, it changes again to a node at $t = 450$. This phenomenon happens alternately without a single time period. Similar alteration of the singular point on the left wall takes place with the Laplacian form more infrequently. During the initial stages, the rotational form also captures a coherent vortical structure at the top left corner of the cavity (as can be seen in the frame (iv) of Fig. 2.38b), which is absent in the Laplacian form. Hence, it can be concluded that the rotational form is more receptive in capturing flow instabilities, as compared to the Laplacian form. Undoubtedly, this points to the higher accuracy of the rotational form over the other forms, using velocity and vorticity. This is also noted by the fact that the rotational form looses symmetry earlier (at around $t = 550$), as compared to the Laplacian form, which looses symmetry at around $t = 650$. A possible explanation for this is that the rotational form is more receptive to the numerical errors and hence it looses symmetry at an earlier time instant. The maximum and the minimum values of w-velocity at $t = 250$, in the mid (x, y)-plane for the Laplacian form are noted as, 3.5×10^{-7} and -1.25×10^{-7}, respectively. In contrast, for the rotational form, corresponding values are 1.23×10^{-6} and -2.45×10^{-6}, respectively. Thus, apart from the earlier onset time of instability, the rotational form also displays higher growth rate of disturbances by an order of magnitude. As noted earlier for the flow at $Re = 1000$, w-velocity is exactly zero at the mid (x, y)-plane of symmetry, for both the formulations.

In Fig. 2.39, contours of normal components of vorticity are shown plotted in three mid-planes. Frames (i) and (iv) in Fig. 2.39a, show higher amount of numerical fluctuations at top left corners for the Laplacian form. Such fluctuations at grid scale level are due to aliasing error and in the conservative rotational form, the quantum of it is significantly lower in the frames (i) and (iv) of Fig. 2.39b. Similar observations are noted in Fig. 2.41 for $Re = 3200$ case, where normal component of vorticity contours are plotted at the later time, $t = 1800$ and $t = 2000$. In Fig. 2.40, corresponding stream-traces at three different indicated planes are shown. In this figure, one notices significant lack of symmetry in the mid (y, z)- and (z, x)-planes. But, this asymmetry is more pronounced for the rotational form. As a result of such a high level of asymmetry, one can clearly observe the presence of multiple limit cycles in the stream-trace plots in the mid (x, y)-plane of Fig. 2.40b. Comparing frames (iii) and (vi) of Fig. 2.40a, with corresponding frames in Fig. 2.40b, higher number of vortices along y-axis can be observed near the rear wall for the rotational form. From these observations of the simulated flow inside a cubic cavity for $Re = 3200$, one notes the rotational form to capture details of unsteady flows relatively more accurately, than the Laplacian form.

References

1. Adams, Y. (1977). Highly accurate compact implicit method and boundary conditions. *Computer Physics, 24*, 10–22.
2. Albensoeder, S., & Kuhlmann, H. C. (2006). Nonlinear three-dimensional flow in the lid-driven square cavity. *Journal of Fluid Mechanics, 569*, 465–480.
3. Auld, B. (1973). *Acoustic fields and waves in solids*. New York: Wiley-Interscience.
4. Barone, M. F., & Roy, C. J. (2006). Evaluation of detached eddy simulation for turbulent wake applications. *AIAA Journal, 44*(12), 3062–3071.
5. Bernardini, M., & Pirozzoli, S. (2009). A general strategy for the optimization of Runge-Kutta schemes for wave propagation phenomena. *Journal of Computational Physics, 228*, 4182–4199.
6. Bhaumik, S. (2013). *Direct numerical simulation of inhomogeneous transitional and turbulent flows*. Ph.D. Thesis, I. I. T. Kanpur.
7. Bhaumik, S., Sengupta, T. K. (2011). On the divergence-free condition of velocity in two-dimensional velocity-vorticity formulation of incompressible Navier–Stokes equation. In AIAA-2011-3238, AIAA CFD Conference, Honolulu, Hawaii, USA.
8. Bhaumik, S., & Sengupta, T. K. (2015). A new velocity-vorticity formulation for direct numerical simulation of 3D transitional and turbulent flows. *Journal of Computational Physics, 284*, 230–260.
9. Bhaumik, S., & Sengupta, T. K. (2014). Precursor of transition to turbulence: Spatiotemporal wave front. *Physical Review E, 89*(4), 043018.
10. Bhumkar, Y. G. (2011). *High performance computing of bypass transition*. Ph.D. Thesis, I. I. T. Kanpur.
11. Bhumkar, Y. G., & Sengupta, T. K. (2011). Adaptive multi-dimensional filters. *Computers Fluids, 49*(1), 128–140.
12. Bogey, C., & Bailly, C. (2004). A family of low dispersive and low dissipative explicit schemes for flow and noise computations. *Journal of Computational Physics, 194*, 194–214.
13. Brillouin, L. (1960). *Wave propagation and group velocity*. New York: Academic Press.
14. Brooks, A. N., & Hughes, T. J. R. (1982). Streamline upwind/Petrov-Galerkin formulation for convection dominated flows with particular emphasis on the incompressible Navier–Stokes equations. *Computer Methods in Applied Mechanics and Engineering, 32*, 199–259.
15. Buresti, G. (2015). A note on Stokes hypothesis. *Acta Mechanica, 226*, 3555–9.
16. Charney, J. G., Fjørtoft, R., & Von Neumann, J. (1950). Numerical integration of the barotropic vorticity equation. *Tellus, 2*(4), 237–254.
17. Chorin, A. J. (1968). Numerical solution of the Navier–Stokes equation. *Mathematics of Computation, 22*, 745–762.
18. Chu, Peter C., & Fan, Chenwu. (1998). A three-point combined compact difference scheme. *Journal of Computational Physics, 140*(2), 370–399.
19. Cramer, M. S. (2012). Numerical estimates for the bulk viscosity of ideal gases. *Physics of Fluids, 24*(066102), 1–23.
20. Crank, J., & Nicolson, P. (1947). A practical method for numerical evaluation of solutions of partial differential equations of the heat conduction type. *Proceedings of Cambridge Philosophical Society, 43*(50), 50–67.
21. Crouch, J. D., Garbaruk, A., Magidov, D., & Travin, A. (2009). Origin of transonic buffet on aerofoils. *Journal of Fluid Mechanics, 628*, 357–369.
22. Cullen, M. J. P. (1974). A finite-element method for a non-linear initial value problem. *J. Int. Math. Appl., 31*, 233–247.
23. Davidson, P. A. (2004). *Turbulence: An introduction for scientists and engineers*. Oxford, UK: Oxford University Press.
24. Dendy, F. E. (1974). Sediment trap efficiency of small reservoirs. *Transaction of the American Society of Agricultural Engineers, 17*(5), 898–908.
25. Dipankar, A., & Sengupta, T. K. (2006). Symmetrized compact scheme for receptivity study of 2D transitional channel flow. *Journal of Computational Physics, 215*(1), 245–253.

26. Emanuel, G. (1990). Bulk viscosity of a dilute polyatomic gas. *Physics of Fluids A*, *2*, 2252–2254.
27. Gad-el-Hak, M. (1995). Questions in fluid mechanics. Stokes' hypothesis for a Newtonian, isotrspic fluid. *Journal of Fluids Engineering*, *117*, 3–5. (Technical Forum).
28. Gaitonde, D. V., Shang, J. S., & Young, J. L. (1999). Practical aspects of higher-order numerical schemes for wave propagation phenomena. *International Journal for Numerical Methods in Engineering*, *45*, 1849.
29. Gaitonde, D. V. & Visbal, M. R. (1999). Further development of a Navier-Stokes solution procedure based on higher order formulas. In *37th Aerospace Sciences Meeting and Exhibition, AIAA 99–0557, Reno, NV*.
30. Gaitonde, D. V., & Visbal, M. R. (2000). Padé– type higher-order boundary filters for the Navier–Stokes equations. *AIAA Journal*, *38*(11), 2103.
31. Gatski, T. B. (1991). Review of incompressible fluid flow computations using the vorticity-velocity formulation. *Applied Numerical Mathematics*, *7*, 227–239.
32. Gatski, T. B., Grosch, C. E., & Rose, M. E. (1982). A numerical study of the two-dimensional Navier-Stokes equations in vorticity-velocity variables. *Journal of Computational Physics*, *48*, 1–22.
33. Gatski, T. B., Grosch, C. E., & Rose, M. E. (1989). A numerical solution of the Navier–Stokes equations for three-dimensional, unsteady, incompressible flows by compact schemes. *Journal of Computational Physics*, *82*, 298–329.
34. Ghosal, S., & Moin, P. (1995). The basic equations for the large eddy simulation of turbulent flows in complex geometry. *Journal of Computational Physics*, *118*, 24.
35. Graves, R. E., & Argrow, B. M. (1999). Bulk viscosity: Past to present. *Journal of Thermophysics and Heat Transfer*, *13*(3), 337–342.
36. Grescho, P. M., & Sani, R. L. (1998). *Incompressible flow and the finite element method*. Chichester, UK: Wiley.
37. Guj, G., & Stella, F. (1993). A vorticity-velocity method for the numerical solution of 3D incompressible flows. *Journal of Computational Physics*, *106*, 286–298.
38. Hamilton, W. R. (1839). *The collected mathematical papers* (Vol. 4). Cambridge: Cambridge University Press.
39. Haras, Z., & Ta'asan, S. (1994). Finite difference scheme for long time integration. *Journal of Computational Physics*, *14*, 265–279.
40. Hirasaki, G. J., & Hellums, J. D. (1968). A general formulation of the boundary conditions on the vector potential in three-dimensional hydrodynamics. *Quarterly of Applied Mathematics*, *26*(3), 331–342.
41. Hirasaki, G. J., & Hellums, J. D. (1970). Boundary conditions on the vector and scalar potentials in viscous three-dimensional hydrodynamics. *Quarterly of Applied Mathematics*, *28*(2), 293–296.
42. Hirsch, C. (1990). *Numerical computation of internal and external flows* (Vol. I and II)., Computational methods for inviscid and viscous flows. Chichester, UK: Wiley.
43. Holdeman, J. T. (2012). A velocity-stream function method for three-dimensional incompressible fluid flow. *Computer Methods in Applied Mechanics and Engineering*, *209*, 66–73.
44. Huang, H., & Li, M. (1997). Finite-difference approximation for the velocity-vorticity formulation on staggered and non-staggered grids. *Computer & Fluids*, *26*(1), 59–82.
45. Hu, F. Q., Hussani, M. Y., & Manthey, J. L. (1996). Low-dissipation and low-dispersion Runge-Kutta schemes for computational acoustics. *Journal of Computational Physics*, *124*, 177–191.
46. Jenkins, F., & White, H. (1973). *Fundamentals of physical optics*. New York: McGraw-Hill.
47. Karim, S. M., & Rosenhead, L. (1952). *Review of Modern Phys.*, *24*, 108–16.
48. Kennedy, C. A., & Carpenter, M. H. (1994). Several new numerical methods for compressible shear-layer simulations. *Applied Numerical Mathematics*, *14*, 397.
49. Kloker, M., Konzelmann, U., & Fasel, H. (1993). Outflow boundary conditions for spatial Navier–Stokes simulations of transitional boundary layers. *AIAA Journal*, *31*, 620.
50. Kopal, Z. (1966). *Numerical analysis*. New York, USA: Springer.

51. Koseff, J. R., & Street, R. L. (1984). On end wall effects in a lid-driven cavity flow. *Journal of Fluids Engineering, 106*, 385–398.
52. Kreiss, H., & Oliger, J. (1972). Comparison of accurate methods for the integration of hyperbolic equations. *Tellus, 24*, 199–215.
53. Lele, S. K. (1992). Compact finite difference schemes with spectral-like resolution. *Journal of Computational Physics, 103*(1), 16–42.
54. Leonard, B. P., Leschziner, M. A. & McGuirk, J. (1978). Third order finite-difference method for steady two-dimensional convection. In C. Taylor, K. Morgan, & C. A. Brebbia (Eds.) *Numerical Methods in Laminar and Turbulent Flows* (pp. 807–819). London: Pentech Press.
55. Leriche, E. (2006). Direct numerical simulation in a lid-driven cubical cavity at high Reynolds number by a Chebyshev spectral method. *Journal of Scientific Computing, 27*, 335–345.
56. Leriche, E., & Gavrilakis, S. (2000). Direct numerical simulation of the flow in a lid-driven cubical cavity. *Physics of Fluids, 12*(6), 1363–1376.
57. Liebermann, L. N. (1949). The second viscosity of liquids. *Physical Review, 75*(9), 1415.
58. Lighthill, M. J. (1978). *Fourier analysis and generalized functions.* Cambridge, UK: Cambridge University Press.
59. Mathew, J., Lechner, R., Foysi, H., Sesterhenn, J., & Friedrich, R. (2003). An explicit filtering method for large eddy simulation of compressible flows. *Journal of Computational Physics, 15*(8), 2279.
60. Meitz, H. L., & Fasel, H. F. (2000). A compact-difference scheme for the Navier–Stokes equations in vorticity-velocity formulation. *Journal of Computational Physics, 157*, 371–403.
61. Najjar, F. M., & Tafti, D. K. (1996). Study of discrete test filters and finite difference approximations for the dynamic subgrid-scale stress model. *Journal Computational Physics, 8*(4), 1076.
62. Nagarajan, S., Lele, S. K., & Ferziger, J. H. (2003). A robust high-order compact method for large eddy simulation. *Journal of Computational Physics, 19*, 392–419.
63. Napolitano, M., & Pascazio, G. (1991). A numerical method for the vorticity-velocity Navier-Stokes equations in two and three dimensions. *Computer & Fluids, 19*, 489–495.
64. Ortega-Casanova, Joaqun, & Fernandez-Feria, Ramn. (2008). A numerical method for the study of nonlinear stability of axisymmetric flows based on the vector potential. *Journal of Computational Physics, 227*(6), 3307–3321.
65. Oswald, G. A., Ghia, K. N. & Ghia, U. (1988). *Direct solution methodologies for the unsteady dynamics of an incompressible fluid.* In S. N. Atluri, G. Yagawa (Eds.) International Conference on Computer Engineering Science, vol. 2 Atlanta. Berlin: Springer.
66. Oberleithner, K., Sieber, M., Nayeri, C. N., Paschereit, C. O., Petz, C., Hege, H.-C., et al. (2011). Three-dimensional coherent structures in a swirling jet undergoing vortex breakdown: stability analysis and empirical mode construction. *Journal of Fluid Mechanics, 679*, 383–414.
67. Pirozzoli, S. (2007). Performance analysis and optimization of finite-difference schemes for wave propagation problems. *Journal of Computational Physics, 222*, 809–831.
68. Poinsot, T., & Veynante, D. (2005). *Theoretical and numerical combustion* (2nd ed.). PA: Edwards.
69. Rajagopal, K. R. (2013). A new development and interpretation of the Navier-Stokes fluid which reveals why the "Stokes assumption" is inapt. *International Journal of Non-Linear Mechanics, 50*, 141–151.
70. Rajpoot, M. K., Bhaumik, S., & Sengupta, T. K. (2012). Solution of linearized rotating shallow water equations by compact schemes with different grid-staggering strategies. *Journal of Computational Physics, 231*, 2300–2327.
71. Rajpoot, M. K., Sengupta, T. K., & Dutt, P. K. (2010). Optimal time advancing dispersion relation preserving schemes. *Journal of Computational Physics, 229*(10), 3623–3651.
72. Ramboer, J., Broeckhoven, T., Smirnov, S., & Lacor, C. (2006). Optimization of time integration schemes coupled with spatial discretization for use in CAA applications. *Journal of Computational Physics, 213*, 777–802.
73. Rayleigh, L. (1889). *Scientific papers* (Vol. 1). Cambridge: Cambridge University Press.
74. Rayleigh, L. (1890). *Scientific papers* (Vol. 2). Cambridge: Cambridge University Press.

75. Raymond, W. H., & Garder, A. (1976). Selective damping in a Galerkin method for solving wave problems with variable grids. *Monthly Weather Review, 104*, 1583–1590.
76. Reynolds, W. C., & Hussain, A. K. M. F. (1972). The mechanics of an organized wave in turbulent shear flow. Part 3. Theoretical models and comparisons with experiments. *Journal of Fluid Mechanics, 54*(2), 263–288.
77. Rizzetta, D. P., Visbal, M. R., & Blaisddell, G. A. (2003). A time-implicit high-order compact differencing and filtering scheme for large-eddy simulation. *International Journal for Numerical Methods in Fluids, 42*, 655.
78. Rizzetta, D. P., Visbal, M. R. & Morgan, P. E. (2008). *A high-order compact finite-difference scheme for large-eddy simulation of active flow control.* In 46th aerospace sciences meeting and exhibition, AIAA 2008-526, Reno, NV.
79. Rosenhead, L. (1954). *Proceedings of Royal Society London A, 226*, 1–6.
80. Sagaut, P. (2002). *Large eddy simulation for incompressible flows.* Berlin: Springer.
81. Sengupta, T. K. (2004). *Fundamentals of computational fluid dynamics.* Hyderabad (India): Universities Press.
82. Sengupta, T. K. (2012). *Instabilities of flows and transition to turbulence.* Florida, USA: CRC Press, Taylor & Francis Group.
83. Sengupta, T. K. (2013). *High accuracy computing methods: fluid flows and wave phenomenon.* New York, USA: Cambridge University Press.
84. Sengupta, T. K., Ballav, M., & Nijhawan, S. (1994). Generation of Tollmien-Schlichting waves by harmonic excitation. *Physics of Fluids, 6*(3), 1213–1222.
85. Sengupta, T. K., & Bhaumik, S. (2011). Onset of turbulence from the receptivity stage of fluid flows. *Physics Review Letter, 154501*, 1–5.
86. Sengupta, T. K., Bhaumik, S., & Bhumkar, Y. G. (2012). Direct numerical simulation of two-dimensional wall-bounded turbulent flows from receptivity stage. *Physical Review E, 85*(2), 026308.
87. Sengupta, T. K., Bhaumik, S., & Bose, R. (2013). Direct numerical simulation of transitional mixed convection flows: Viscous and inviscid instability mechanisms. *Physics of Fluids, 25*, 094102.
88. Sengupta, T. K., Bhaumik, S., & Usman, S. (2011). A new compact difference scheme for second derivative in non-uniform grid expressed in self-adjoint form. *Journal of Computational Physics, 230*(5), 1822–1848.
89. Sengupta, T. K., & Bhumkar, Y. G. (2010). New explicit two-dimensional higher order filters. *Computers & Fluids, 39*, 1848–1863.
90. Sengupta, T. K., Bhumkar, Y., & Lakshmanan, V. (2009). Design and analysis of a new filter for LES and DES. *Computers & Structures, 87*, 735–750.
91. Sengupta, T. K., Bhumkar, Y., Rajpoot, M. K., Suman, V. K., & Saurabh, S. (2012). Spurious waves in discrete computation of wave phenomena and flow problems. *Applied Mathematics and Computation, 218*, 9035–9065.
92. Sengupta, T. K., & Dey, S. (2004). Proper orthogonal decomposition of direct numerical simulation data of by-pass transition. *Computers & Structures, 82*, 2693–2703.
93. Sengupta, T. K., & Dipankar, A. (2004). A comparative study of time advancement methods for solving Navier–Stokes equations. *Journal of Scientific Computing, 21*(2), 225–250.
94. Sengupta, T. K., & Dipankar, A. (2005). Subcritical instability on the attachment-line of an infinite swept wing. *Journal of Fluid Mechanics, 529*, 147–171.
95. Sengupta, T. K., Dipankar, A., & Sagaut, P. (2007). Error dynamics: Beyond von Neumann analysis. *Journal of Computational Physics, 226*, 1211–1218.
96. Sengupta, T. K., Ganeriwal, G., & De, S. (2003). Analysis of central and upwind compact schemes. *Journal of Computational Physics, 192*, 677–694.
97. Sengupta, T. K., Jain, R., & Dipankar, A. (2005). A new flux-vector splitting compact finite volume scheme. *Journal of Computational Physics, 207*, 261–281.
98. Sengupta, T. K., Lakshmanan, V., & Vijay, V. V. S. N. (2009). A new combined stable and dispersion relation preserving compact scheme for non-periodic problems. *Journal of Computational Physics, 228*(8), 3048–3071.

99. Sengupta, T. K. & Nair, M. T. (1997). A new class of wave Blasius boundary layer. In Proceedings 7th Asian Congress of Fluid Mechanics
100. Sengupta, T. K., Rajpoot, M. K., & Bhumkar, Y. G. (2011). Space-time discretizing optimal DRP schemes for flow and wave propagation problems. *Computers & Fluids*, *47*(1), 144–154.
101. Sengupta, T. K., Rao, A. K., & Venkatasubbaiah, K. (2006). Spatio-temporal growing wave fronts in spatially stable boundary layers. *Physical Review Letters*, *96*(22), 224504.
102. Sengupta, T. K., Rao, A. K., & Venkatasubbaiah, K. (2006). Spatio-temporal growth of disturbances in a boundary layer and energy based receptivity analysis. *Physics of Fluids*, *18*, 094101.
103. Sengupta, T. K., Sengupta, A., Sengupta, S., Bhole, A. & Shruti, K. S. (2016). Non-equilibrium thermodynamics of Rayleigh–Taylor instability. *International Journal Thermophysics*, *37*(4), 1–2. https://doi.org/10.1007/s10765-016-2045-1
104. Sengupta, T. K., Sengupta, A., Sharma, N., Sengupta, S., Bhole, A., & Shruti, K. S. (2016). Roles of bulk viscosity on Rayleigh-Taylor instability: Non-equilibrium thermodynamics due to spatio-temporal pressure fronts. *Physics of Fluids*, *28*, 094102.
105. Sengupta, T. K., Singh, N., & Vijay, V. V. S. N. (2011). Universal instability modes in internal and external flows. *Computers & Fluids*, *40*, 221–235.
106. Sengupta, T. K., Sircar, S. K., & Dipankar, A. (2006). High accuracy compact schemes for DNS and acoustics. *Journal of Scientific Computing*, *26*(2), 151–193.
107. Sengupta, T. K., Vijay, V. V. S. N., & Bhaumilk, S. (2009). Further improvement and analysis of CCD scheme: Dissipation discretization and de-aliasing properties. *Journal of Computational Physics*, *228*(17), 6150–6168.
108. Shokin, Y. I. (1983). *The method of differential approximation*. Berlin: Springer.
109. Stokes, G. G. (1845). On the theories of inertial friction of fluids in motion. *Transaction of Cambridge Philosphical Society*, *8*, 287–305.
110. Stoltz, S., Adams, N. A., & Kleiser, L. (2001). An approximate deconvolution model for large-eddy simulation with application to incompressible wall-bounded flows. *Physics of Fluids*, *13*(4), 997.
111. Swartz, B., & Wendroff, B. (1974). The relation between the Galerkin and collocation methods using smooth splines. *SIAM Journal on Numerical Analysis*, *11*(5), 994–996.
112. Tam, C. K. W. (1971). Directional acoustic radiation from a supersonic jet generated by shear layer instability. *Journal of Fluid Mechanics*, *46*, 757–768.
113. Tam, C. K. W., Viswanathan, K., Ahuja, K. K. & Panda, J. (2008). The sources of jet noise: experimental evidence. *Journal of Fluid Mechanics*, *615*, 253–292.
114. Tennekes, H., & Lumley, J. L. (1971). *First course in turbulence*. Cambridge, MA: MIT Press.
115. Touber, E. & Sandham, N. D. (2009). Large-eddy simulation of low-frequency unsteadiness in a turbulent shock-induced separation bubble. *Theoretical and Computational Fluid Dynamics*, *23*(2), 79–107.
116. Trefethen, L. N. (1982). Group velocity in finite difference schemes. *SIAM Review*, *24*(2), 113–136.
117. Tucker, P. G. (2003). Differential equation-based wall distance computation for DES and RANS. *Journal of Computational Physics*, *190*, 229–248.
118. Van der Vorst, H. A. (1992). Bi-CGSTAB: A fast and smoothly converging variant of Bi-CG for the solution of non-symmetric linear systems. *SIAM Journal on Scientifc and Statistical Computing*, *12*, 631–644.
119. Vasilyev, O. V., Lund, T. S., & Moin, P. (1998). A general class of commutative filters for LES in complex geometries. *Journal of Computational Physics*, *146*, 82.
120. Vichnevetsky, R., & Bowles, J. B. (1982). *Fourier analysis of numerical approximations of hyperbolic equations.*, SIAM studies of applied mathematics Philadelphia, USA.
121. Visbal, M. R., & Gaitonde, D. V. (1999). High-order-accurate methods for complex unsteady subsonic flows. *AIAA Journal*, *37*(10), 1231.
122. Visbal, M. R., & Gaitonde, D. V. (2002). On the use of higher-order finite-difference schemes on curvilinear and deforming meshes. *Journal of Computational Physics*, *181*, 155.

123. Wahlbin, L. B. (1974). A dissipative numerical method for the numerical solution of first order hyperbolic equations. In C. de Boor (Ed.) *Mathematical aspects of finite elements in partial deferential equations* (pp. 147–170). New York: Academic Press.
124. Wahlbin, L. B. (1975). A modified Galerkin procedure with cubics for hyperbolic problems. *Mathematics of Computation, 29,* 978–984.
125. Wang, L. P., & Rosa, B. (2009). A spurious evolution of turbulence originated from round-off error in pseudo-spectral simulation. *Computers & Fluids, 38,* 1943–1949.
126. Weinan, E., & Liu, Jian-Guo. (1996). Vorticity boundary condition and related issues for finite difference schemes. *Journal of Computational Physics, 124*(2), 368–382.
127. Whitham, G. B. (1974). *Linear and nonlinear waves.* New York: Wiley-Intescience.
128. Williams, G. P. (1969). Numerical integration of the three dimensional Navier-Stokes equation for incompressible flow. *Journal of Fluid Mechanics, 37,* 727–750.
129. Wong, A. K., & Reizes, J. A. (1984). An effective vorticity-vector potential formulation for the numerical solution of three-dimensional duct flow problems. *Journal of Computational Physics, 55*(1), 98–114.
130. Wong, A. K., & Reizes, J. A. (1986). The vector potential in the numerical solution of three-dimensional fluid dynamics problems in multiply connected regions. *Jorunal of Computational Physics, 62*(1), 124–142.
131. Wu, X. H., Wu, J. Z., & Wu, J. M. (1995). Effective vorticity-velocity formulations for 3D incompressible viscous flows. *Journal of Computational Physics, 122,* 68–82.
132. Zingg, D. W. (2000). Comparison of high-accuracy finite-difference schemes for linear wave propagation. *SIAM Journal on Scientific Computing, 22*(2), 476–502.

Chapter 3
Receptivity and Instability

3.1 Linear Stability/Receptivity Theories: Classical Approaches and Signal Problem

In this chapter, linear stability and receptivity analysis of the zero-pressure gradient (ZPG) boundary layer, under the parallel flow assumption, is discussed. This assumption implies that the equilibrium flow quantities do not grow in the streamwise direction and requires solving the Orr-Sommerfeld equation (OSE) to study evolution of disturbance field in a linearized analysis. The concept of the spatio-temporal wave-front (STWF) originates from the receptivity analysis with the OSE solved for the response field. First, the simplified description of equilibrium flow in terms of a similarity solution for ZPG boundary layer is presented. Following which the OSE is derived for boundary layers, making use of the parallel flow approximation [19, 53]. This equation have been solved for the ZPG boundary layer using analytical approaches in [28, 48, 71]. We instead introduce the compound matrix method, a robust method for stiff differential equation useful for the OSE. Finally, the receptivity analysis of the ZPG boundary layer flow is provided, with results taken from [61, 62]. The unique feature of the materials in this chapter is the topic of instability of mixed convection flows for which two theorems are enunciated for an inviscid linear mechanism, based on materials extensively taken from Sengupta et al., Physics of Fluids, 25, 094102 (2013).

The unique feature of the materials in this chapter is the topic of instability of mixed convection flows for which two theorems are enunciated for an inviscid linear mechanism, based on materials extensively taken from a publication.[1]

---

[1][Reproduced from *Direct numerical simulation of transitional mixed convection flows: Viscous and inviscid instability mechanism.* Sengupta et al., *Physics of Fluids*, **25**, 094102 (2013), with the permission of AIP Publishing.]

© Springer Nature Singapore Pte Ltd. 2019
T. K. Sengupta and S. Bhaumik, *DNS of Wall-Bounded Turbulent Flows*,
https://doi.org/10.1007/978-981-13-0038-7_3

3.1.1 The Equilibrium Flow Equation

For flow past a flat plate the effects of shear stress due to fluid viscosity is limited to a very thin layer, commonly known as the boundary layer (or alternately wall-bounded shear layer), whose thickness increases as one moves downstream along the streamwise direction. This the boundary layer assumption introduced by Prandtl, which revolutionized our understanding of fluid mechanics and is the bedrock of perturbation theory used in many branches of mathematics and physics. For thin shear layers, the governing NSE can be simplified based on thin shear layer assumption, culminating in the boundary layer equation [14]. Furthermore, if the edge velocity at the free-stream U_e varies as x^m, where x is the streamwise distance from the plate or wedge leading edge, the boundary layer equations can be further simplified by what is known as similarity transformation [14], that reduces this to a nonlinear ordinary differential equations, called the Falkner–Skan equation and is given as [14, 53],

$$\frac{d^3 f}{d\bar{\eta}^3} + m\left[1 - \left(\frac{df}{d\bar{\eta}}\right)^2\right] + \frac{m+1}{2} f \frac{d^2 f}{d\bar{\eta}^2} = 0 \tag{3.1}$$

where, $\bar{\eta} = y\sqrt{U_e/\nu x}$, $\psi = (U_e \nu x)^{1/2} f(\bar{\eta})$ and $u(x, y)/U_e(x) = df/d\bar{\eta}$. Here, U_e, y, ψ, u and v are the boundary layer edge velocity, wall-normal co-ordinate, stream function, the streamwise velocity component and the kinematic viscosity, respectively. To obtain the similarity profile, one has to solve Eq. (3.1) subject to the no-slip and zero-normal velocity at the wall. The boundary condition at the free-stream is obtained as $u(x, y) \rightarrow U_e(x)$ as $y \rightarrow \infty$. In terms of f, these conditions translate to

$$f(0) = f'(0) = 0$$
$$f' \rightarrow 1 \quad \text{as} \quad \bar{\eta} \rightarrow \infty \tag{3.2}$$

where, $'$ the prime refers to first derivative with respect to the similarity variable, $\bar{\eta}$. For ZPG boundary layer $U_e(x) = U_\infty = $ Const.. This simplifies Eq. (3.1) to the the well known Blasius equation [14] by noting that $m = \frac{x}{U_e}\frac{dU_e}{dx} = 0$ for ZPG condition and the governing equation is given as,

$$f''' + \frac{1}{2} f f'' = 0 \tag{3.3}$$

In Fig. 3.1, f', f'' and f''' obtained from solving Eq. (3.3) are plotted as a function of $\bar{\eta}$. The governing equation is solved using uniformly distributed points up to a limit of $\bar{\eta}_{max} = 12$ by fourth order, RK4 method [30]. In terms of physical significance, f', f'' and f''' are related to the streamwise component of velocity, its first and second derivatives with respect to $\bar{\eta}$, respectively. One notes from Fig. 3.1c that $f'''(0) = 0$, which signifies that the inflection-point (i.e., the point where $d^2u/dy^2 = 0$) for the ZPG boundary layer is exactly at the wall. Hence, the ZPG Blasius profile does

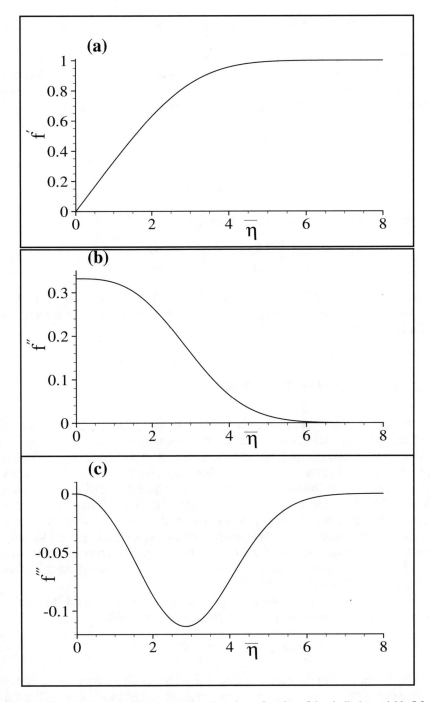

Fig. 3.1 The solution of the Blasius equation plotted as a function of the similarity variable $\bar{\eta}$ for **a** f', **b** f'' and **c** f'''

not strictly satisfy the Rayleigh's inflection-point theorem for inviscid instability [19, 38].

The velocity profile of the mean flow at any streamwise location is characterized by, (i) the edge velocity U_e, (ii) the displacement thickness δ^* and (iii) the momentum thickness θ, at that particular location (see [14, 53] for details). The displacement thickness and the momentum thickness for any velocity profile is defined as

$$\delta^* = \int_0^\infty \left(1 - \frac{u}{U_e}\right) dy$$

$$\theta = \int_0^\infty \frac{u}{U_e}\left(1 - \frac{u}{U_e}\right) dy \tag{3.4}$$

For Blasius profile given by Eq. (3.3), it can be shown that [14]

$$\delta^* \simeq \frac{1.72x}{\sqrt{Re_x}} \quad \text{and} \quad \theta \simeq \frac{0.664x}{\sqrt{Re_x}} \tag{3.5}$$

where, $Re_x = U_\infty x/\nu$ is the local Reynolds number. It can also be shown that for Blasius profile, boundary layer thickness $\bar{\delta} \simeq 3\delta^*$, where $\bar{\delta}$ is defined such that $u(x, \bar{\delta}) \simeq 0.99U_\infty$.

3.2 Linear Stability Equation

In classical instability theory, the governing equation is linearized, as one would be looking for growth of infinitesimally small background disturbances. Thus, one linearizes the governing NSE. However, solving a set of 3D linearized perturbation equation is as difficult as solving the nonlinear equation itself. To circumvent this problem, early researchers made additional assumption of a parallel flow, where the growth of underlying shear layer is neglected. Defining the equilibrium flow, only in terms of wall-normal co-ordinate (y), one can introduce Fourier-Laplace transform for the perturbation quantities. This was also the basis of deriving Rayleigh's stability equation, as mentioned in Chap. 1. However, in this chapter and subsequently, we will only be discussing about disturbance growth for viscous flows, and in the following we describe the formulation of the OSE.

If U, V and W are the streamwise (x), spanwise (z) and wall–normal (y) components of an equilibrium flow, then under parallel flow approximation,

$$U = U(y); \quad V = 0; \quad W = W(y)$$

which will be used in deriving the OSE in this section. The governing equations for the evolution of the small perturbations in the shear layer are obtained from the linearized NSE [53], with parallel flow approximation given in component form, along with equation of continuity by,

$$\frac{\partial u'}{\partial t} + U\frac{\partial u'}{\partial x} + W\frac{\partial u'}{\partial z} + v'\frac{dU}{dy} = -\frac{\partial p'}{\partial x} + \nu\nabla^2 u' \tag{3.6}$$

$$\frac{\partial v'}{\partial t} + U\frac{\partial v'}{\partial x} + W\frac{\partial v'}{\partial z} = -\frac{\partial p'}{\partial y} + \nu\nabla^2 v' \tag{3.7}$$

$$\frac{\partial w'}{\partial t} + U\frac{\partial w'}{\partial x} + W\frac{\partial w'}{\partial z} + v'\frac{dW}{dy} = -\frac{\partial p'}{\partial z} + \nu\nabla^2 w' \tag{3.8}$$

$$\frac{\partial u'}{\partial x} + \frac{\partial v'}{\partial y} + \frac{\partial w'}{\partial z} = 0 \tag{3.9}$$

where, u', v', w' and p' are the perturbation components of velocity and pressure, respectively. This implies that the instantaneous quantity (q) is split as,

$$q(x, y, z, t) = Q(x, y, z) + \varepsilon q'(x, y, z, t)$$

with ε as a small parameter representing the quantum of imposed disturbance. In a linearized study, the response scales directly with ε, and is not considered important to specify it explicitly. However, for receptivity analysis this has to be specified to obtain the response of the dynamical system for input disturbance amplitude given by ε.

Since, U and W components of the mean flow are invariant with either streamwise or spanwise coordinate, it is preferred to take the streamwise edge velocity U_e and the displacement thickness δ^* as the velocity and length scales, respectively, in non-dimensionalizing the perturbation equations. Furthermore, one can express the velocity and pressure disturbances in terms of Fourier-Laplace transform due to the adoption of parallel flow approximation, with x and z being treated as the homogeneous directions. Thus,

$$[u', v', w', p']^T = [f_o(\tilde{y}), \phi(\tilde{y}), h_o(\tilde{y}), \pi_o(\tilde{y})]^T e^{i(\alpha\tilde{x} + \beta\tilde{z} - \omega_0 t)}$$

where, $\tilde{x} = x/\delta^*$, $\tilde{y} = y/\delta^*$ and $\tilde{z} = z/\delta^*$. Substituting these in the linearized NSE (Eqs. (3.6)–(3.9)) and further simplification, one obtains the OSE given by [28, 48, 53, 71],

$$(D^2 - \gamma^2)^2\phi = i\tilde{R}e\left[\{\alpha U(\tilde{y}) + \beta W(\tilde{y}) - \omega_0\}(D^2 - \gamma^2)\phi\right.$$

$$\left. -\alpha D^2 U(\tilde{y}) + \beta D^2 W(\tilde{y})\phi\right] \tag{3.10}$$

where, $D = \frac{d}{d\tilde{y}}$, $\tilde{R}e = U_e\delta^*/\nu$, and $\gamma^2 = \alpha^2 + \beta^2$. To study stability of the equilibrium flow, homogeneous boundary conditions at the wall and free-stream for the OSE have to be satisfied. For example, at the wall, one uses the no-slip boundary conditions and at the free-stream (i.e., as $\tilde{y} \to \infty$), one requires the disturbance quantities to decay to zero. The implication of this set of homogeneous boundary conditions is that the instability is triggered by disturbances within the shear layer, which is progressively decaying with increase in height. Physically, this implies that the imposed disturbances within the shear layer has finite energy and hence the disturbances progressively decay with height. Thus, a boundary layer destabilized from free stream would require different approach of study and would be explained in greater details while discussing about receptivity of boundary layer. The following homogeneous boundary conditions are used for the eigenvalue problem associated with disturbance imposed from within shear layer (as performed experimentally by near-wall excitation in [50]),

$$f_o(0) = \phi(0) = h_o(0) = 0 \tag{3.11}$$

$$f_o(\tilde{y}), \phi(\tilde{y}), h_o(\tilde{y}) \to 0 \qquad \text{as } \tilde{y} \to \infty \tag{3.12}$$

The non-trivial solutions of Eq. (3.10), subject to homogeneous boundary conditions given by Eqs. (3.11) and (3.12), exist only for particular combinations of the parameters α, β, ω_0 and $\tilde{Re}$. This defines the dispersion relation of the resultant eigenvalue problem to be satisfied at the wall as,

$$\hat{D}(\alpha, \beta, \omega_0, \tilde{Re}) = 0 \tag{3.13}$$

While discussing about receptivity of a boundary layer to wall excitation, we will note that the dispersion relation originates, so as to satisfy the no-slip boundary conditions at the wall. The common meaning of dispersion relation is to relate spatial and temporal scales of the disturbance field of an equilibrium flow and can arise from the governing differential equation of the disturbance field and/or the associated boundary conditions. It is to be noted that the equilibrium flow can be time-independent and/ or its space dependence be given by similarity solution. But, no such assumption is made on the disturbance field. As noted here, the governing equation for disturbance field in classical instability theory is derived from the NSE with the small perturbation assumption. While the equilibrium boundary layer can be obtained via many simplifications, even though the equation is nonlinear.

Historically, first few attempts to solve the above mentioned eigenvalue problem were made by Heisenberg [28], Tollmien [71] and Schlichting [48]. In the dispersion relation given by Eq. (3.13), α, β and ω_0 can thus be complex quantities, but traditionally only two approaches have been adopted: (i) temporal amplification theory, which considers complex ω_0 with α and β as real and (ii) spatial amplification theory that considers α and β to be complex, while keeping ω_0 as real. A more elaborate and detailed discussions on the finer aspects of both these theories and their mutual relationship can be found in [53]. In the subsequent discussions, only 2D mean flow subjected to 2D disturbances will be considered for which the governing OSE given by Eq. (3.10) reduces to,

$$(D^2 - \alpha^2)^2 \phi = i\tilde{Re}\left[(\alpha U(\tilde{y}) - \omega_0)(D^2 - \alpha^2)\phi - \alpha(D^2 U)\phi\right] \quad (3.14)$$

For the solution of Eq. (3.14) satisfying the homogeneous boundary conditions given by,

$$\text{at} \quad \tilde{y} = 0: \quad \phi, \phi' = 0 \quad (3.15)$$

$$\text{and as} \quad \tilde{y} \to \infty: \quad \phi, \phi' \to 0 \quad (3.16)$$

the associated dispersion relation for the 2D disturbance field is,

$$\hat{D}(\alpha, \omega_0, \tilde{Re}) = 0 \quad (3.17)$$

The solution for the 4th-order ODE given by Eq. (3.14) for general excitation can be represented in terms of four fundamental solutions $(\phi_1, \phi_2, \phi_3, \phi_4)$ as,

$$\phi = a_1\phi_1 + a_2\phi_2 + a_3\phi_3 + a_4\phi_4 \quad (3.18)$$

It will be shown shortly that out of these four modes, two grow with height (say, ϕ_2 and ϕ_4) and the other two decay with height, which we call as ϕ_1 and ϕ_3. If we consider wall excitation only, then it is implied that the disturbance field decay in the free stream ($\tilde{y} \to \infty$). At the free stream, $U = 1$ and all mean flow derivatives with respect to $\tilde{y}$ are zero, simplifying Eq. (3.14) to,

$$(D^2 - \alpha^2)^2 \phi = i\tilde{Re}(\alpha - \omega_0)(D^2 - \alpha^2)\phi \quad (3.19)$$

The special solution of Eq. (3.19) for $\tilde{y} \to \infty$, after removing exponentially growing terms (by enforcing $a_2 = a_4 = 0$) is given as,

$$\phi_\infty = a_1 e^{-\alpha\tilde{y}} + a_3 e^{-Q\tilde{y}} \quad (3.20)$$

where $Q = \sqrt{\alpha^2 + i\tilde{Re}(\alpha - \omega_0)}$, considering the cases where real parts of α and Q are positive. It is to be emphasized that the complex wavenumbers, α (inviscid mode due to independence of $\tilde{Re}$) and Q (viscous mode, that depends on $\tilde{Re}$) can have both positive and negative real part. Apart from [53], other sources in literature do not explain this aspect of differentiating between wall and free-stream excitations clearly. As we go along, we will emphasize this aspect of instability and receptivity. Hence, the general solution of the OSE, subject to the prescribed homogeneous boundary conditions at the free stream given by Eq. (3.16) can only be in terms of ϕ_1 and ϕ_3 given by,

$$\phi = a_1\phi_1 + a_3\phi_3 \quad (3.21)$$

From Eq. (3.19), one can easily show the retained fundamental modes obey, $\phi_{1\infty} = e^{-\alpha \tilde{y}}$ and $\phi_{3\infty} = e^{-Q\tilde{y}}$. It is noted that these two retained fundamental solutions decay exponentially at two different widely disparate rates in the free stream, which indicates that Eq. (3.14) constitute a stiff ODE. Solution of Eq. (3.14) can not be obtained in a straight-forward manner due to this stiffness problem. This problem is avoided by using compound matrix method (CMM), as described in [1, 43, 53] for external problems.

3.3   Linear Receptivity Analysis of Parallel Boundary Layer

Here, a brief description of receptivity of the parallel shear layer is presented, while more details of the method are reported in [57, 61, 62]. Schubauer and Skramstad [50] in their experiment excited the boundary-layer by a vibrating ribbon near the wall. Such excitation can be idealized by a localized harmonic exciter at the wall, with perturbation streamwise and wall-normal velocity components given by,

$$u'(\tilde{x}, 0, t) = 0 \tag{3.22}$$

$$v'(\tilde{x}, 0, t) = \delta(\tilde{x}) \, e^{i\bar{\omega}_0 t} \, H(t) \tag{3.23}$$

where $\delta(\tilde{x})$ is the Dirac-delta function in space, to show the localized nature of excitation, and $H(t)$ is the Heaviside function, that indicates the finite impulsive start-up time of the harmonic excitation at the circular frequency $\bar{\omega}_0$, which is a real quantity. In Fig. 3.2, an equivalent parallel boundary layer at the location of a harmonic exciter is shown by dotted line. Whereas the actual boundary layer is shown by the solid line, that starts with a zero thickness at the leading edge of the plate. In the receptivity theory, two distinct approaches have been followed: (i) *Signal problem* is the one where one assumes, a priori, that the response of the system will be exactly at the frequency of excitation ($\bar{\omega}_0$) and this approach was introduced in [52] with the use of a Bromwich contour in the α-plane, and (ii) *spatio-temporal receptivity approach* in which one does not fix the response at the forcing frequency ($\bar{\omega}_0$) and the spatio-temporal transfer function determines the response of the system at the complex frequency ω_0, as was pioneered in [57] for the receptivity of Blasius boundary layer. Both these approaches are also called the Bromwich contour integral method (BCIM). Based on further results in [61, 62], the second approach is definitely more meaningful for fluid-dynamical system prone to instability. All of these results are described together in [53].

The boundary conditions at the wall, for the spatio-temporal approach are given by Eqs. (3.22) and (3.23); these fix the wall-boundary conditions on ϕ as,

$$\phi(0) = \frac{i}{(\bar{\omega}_0 - \omega_0)} + \pi \delta(\omega_0 - \bar{\omega}) \tag{3.24}$$

$$\phi'(0) = 0 \tag{3.25}$$

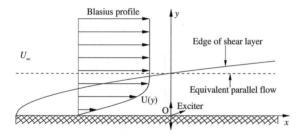

Fig. 3.2 Harmonic excitation of a parallel boundary layer corresponding to the location of the exciter

Since the perturbations must decay at the free stream, Eq. (3.21) is applicable and one can fix the values of the unknown constants, a_1 and a_3, from Eqs. (3.24) and (3.25) as

$$a_1\phi_1(0) + a_3\phi_3(0) = \frac{i}{(\bar\omega_0 - \omega_0)} \tag{3.26}$$

$$a_1\phi_1'(0) + a_3\phi_3'(0) = 0 \tag{3.27}$$

From Eqs. (3.26) and (3.27), one obtains the general expression of $\phi(\tilde{y})$ from Eq. (3.21) as

$$\phi(\tilde{y}) = \left[\frac{i}{(\bar\omega_0 - \omega_0)}\right]\left[\frac{\phi_3(0)\phi_1(\tilde{y}) - \phi_1(0)\phi_3(\tilde{y})}{\phi_1(0)\phi_3'(0) - \phi_3(0)\phi_1'(0)}\right] \tag{3.28}$$

After obtaining $\phi(\tilde{y})$ from Eq. (3.28), one can get back the perturbation velocity components by inverse Fourier-Laplace transform as

$$u'(\tilde{x}, \tilde{y}, t) = \left(\frac{1}{2\pi}\right)^2 \int_{\alpha_{Br}} \int_{\omega_{0Br}} -i\alpha\phi'(\tilde{y})e^{i(\alpha\tilde{x}-\omega_0 t)} \, d\alpha \, d\omega_0 \tag{3.29}$$

$$v'(\tilde{x}, \tilde{y}, t) = \left(\frac{1}{2\pi}\right)^2 \int_{\alpha_{Br}} \int_{\omega_{0Br}} \phi(\tilde{y})e^{i(\alpha\tilde{x}-\omega_0 t)} \, d\alpha \, d\omega_0 \tag{3.30}$$

In Eqs. (3.29) and (3.30), the integration is performed along the Bromwich contours in complex α- and ω_0-planes [53, 74], as shown in Fig. 3.3. The denominator in the expression for $\phi(\tilde{y})$ in Eq. (3.28) is basically the dispersion relation obtained from the satisfaction of wall boundary condition, $\hat{D}(\alpha, \omega_0, \tilde{Re})$, which is also known as the characteristic determinant of the eigenvalue problem given in Eqs. (3.14)–(3.17). This establishes the connection between the receptivity and the corresponding stability problem. Hence, the eigenvalues of the OSE given by Eq. (3.14), which are the zeros of the characteristic determinant $\hat{D}$ given by Eq. (3.17), seen in the denominator of Eq. (3.28). Thus, the eigenvalues are nothing but the poles of the integrand of ϕ defining the receptivity problem.

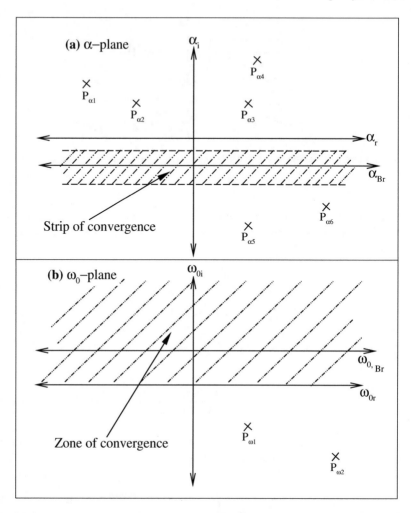

Fig. 3.3 Bromwich contour along with the region of convergence shown in **a** α- and **b** ω_0-plane

Now, since the Laplace transform with respect to $\tilde{x}$ is basically a bilateral Laplace transform (as $\tilde{x}$ varies from $-\infty$ to $+\infty$), and therefore a strip of convergence for the Bromwich contour must exist in the α-plane, as shown in Fig. 3.3a [53, 74]. All the poles lie outside this strip of convergence, and the ones above this strip correspond to downstream propagating disturbances, while those below the strip of convergence are for which the disturbances propagate upstream of the exciter.

In contrast, the Laplace transform with respect to t is called the unilateral Laplace transform, as the integration in t is performed from 0 to $+\infty$. Hence for the Bromwich contour in ω_0-plane, the convergence of Bromwich contour integral would be achieved, only if it is placed above all eigenvalues. This is done to satisfy the causality principle [74]. It has been noted that in α-plane, all the eigenvalues lying above the Bromwich contour signify downstream propagating modes (with positive group velocity, $V_g \geq 0$), while the modes lying beneath it would imply upstream propagating ones (with $V_g \leq 0$). These are the criteria followed in [57, 61, 62] to fix the Bromwich contour in the α-plane, while calculating the receptivity of a parallel boundary layer in linearized framework. In general a Bromwich contour in α-plane, can be taken parallel and below the α_r-axis. There is also the added advantage for this choice of contour in applying discrete fast Fourier transform (DFFT) along this contour, as numerically evaluating the integral given in Eqs. (3.29) and (3.30).

3.3.1 Receptivity of Blasius Boundary Layer

Traditionally, stability analysis is carried out using parallel flow approximation of the mean flow at any streamwise location of interest, as shown in Fig. 3.2. Parallel flow approximation for stability analysis of the shear layers is valid only away from the leading edge. Since at the leading order, boundary layer grows slowly with respect to $\tilde{x}$ (as at the leading order $\bar{\delta}$, δ^* and θ varies with streamwise distance from the leading edge as $\tilde{x}^{1/2}$), where similarity transform applies. Without going through details about solving the OSE by special method like CMM, here we present some typical results with the physical parameters given.

First, the Blasius boundary layer is obtained by solving Eq. (3.3) for the range $0 \leq \bar{\eta} \leq 12$ using 2400 uniformly distributed points. Subsequently, the OSE is solved when the exciter is located where the Reynolds number based on displacement thickness (δ^*) is 1500 and the physical frequency is about 3.6 Hz. The solution is shown in Fig. 3.4 at a non-dimensional time of 801.11 for the streamwise disturbance velocity at a height of $\tilde{y} = 0.278$. In calculating the physical frequency, we have assumed the boundary layer has a displacement thickness of 5mm at the exciter location. Considering the kinematic viscosity of air as 1.5×10^{-5} m^2/s., the boundary layer edge velocity (U_e) works out to be equal to 4.5 m/s. For this low

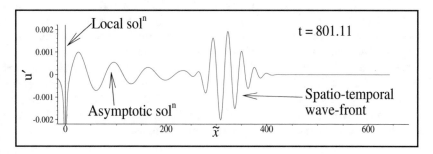

Fig. 3.4 Streamwise disturbance velocity plotted as function of $\tilde{x}$ at $\tilde{y} = 0.278$, for the case of harmonic excitation of a parallel boundary layer. The location of the exciter corresponds to $\tilde{Re} = 1500$ and non-dimensional frequency of excitation is $\bar{\omega}_0 = 0.025$, with the result shown for the viscous time scale of $t = 801.11$

frequency excitation, the solution of the OSE provides the streamwise disturbance velocity component at $t = 801.11$ in Fig. 3.4, which is referred with respect to the viscous time scale given by $\nu/U_e^2 = 7.4074 \times 10^{-7}$ s. If one computes NSE in non-dimensional form with length scale given by δ^* and velocity scale by U_e, then the corresponding convection time scale is $\delta^*/U_e = 1.11 \times 10^{-3}$ s. Thus the solution time of 801.11 shown in Fig. 3.4, is equivalent to the nondimensional time of DNS of NSE for a non-dimensional time of 0.5342 only! Thus, the solution obtained by the OSE is extremely finely-resolved time scale, while the DNS of NSE resolves the time scale quite coarsely. This conversion of different formulations and scales used for non-dimensionalization is extremely important and should be kept in mind for future reference.

The solution shown in Fig. 3.4 at $t = 801.11$ for u' at $\tilde{y} = 0.278$, is a typical response for harmonic wall excitation (for 3.6 Hz) at a location where $\tilde{Re} = 1500$, has three distinct components, with the origin for the abscissa is at the localized delta function exciter. The first component of the solution is called the *local solution*, which is seen in the immediate neighbourhood of the exciter. Even though the exciter is located at a super-critical position (as $\tilde{Re}_{cr} = 519$ for Blasius boundary layer), but the frequency is quite low, for which the linear spatial theory result [53] indicates stability. This is also supported by the result obtained by linearized receptivity analysis by BCIM. According to Abel's theorem for any excitation problem [53], the eigenvalues near the origin of the complex α-plane constitutes the *asymptotic solution*, as marked in the figure. This component of the solution corresponds mostly to the least damped mode of the eigen-spectrum obtained by spatial stability analysis for $\tilde{Re} = 1500$ and $\bar{\omega}_0 = 0.025$. We note the correspondence between linear stability theory and corresponding linearized receptivity analysis. However, Fig. 3.4 shows the third component, namely the STWF, which can only be obtained by spatio-temporal BCIM. This element has been shown in [55, 61, 62], for Blasius boundary excited at frequencies for which spatial theory indicates stability.

Decaying *asymptotic solution* and growing STWF separate from each other for spatially stable cases, while for spatially unstable cases at early times, one cannot distinguish between *asymptotic solution* and the STWF, as noted first in [57] for $\tilde{Re} = 1000$ and $\bar{\omega}_0 = 0.1$. This distinction between spatial unstable and stable cases are shown in frames (a) and (b) of Fig. 3.5, respectively, for viscous non-dimensional time of $t = 801.11$. This is for the case of $\tilde{Re} = 750$, for $\bar{\omega}_0 = 0.10$ (unstable) and $\bar{\omega}_0 = 0.15$ (stable). For the case shown in frame (a) of Fig. 3.5, the *asymptotic solution* and the STWF is fused together, while in frame (b) the *asymptotic solution* is damped. For this case, the STWF is noted at the leading edge of the disturbance packet. In Chap. 5, we will see from DNS results that the STWF keeps growing along with its higher propagation speed and for the spatially stable case, these two components of the response field separate from each other.

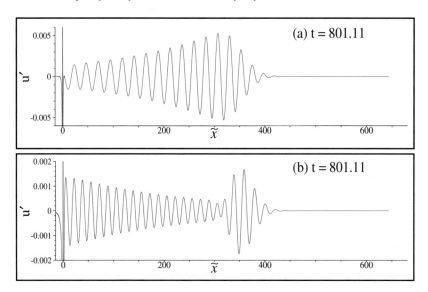

Fig. 3.5 Streamwise disturbance velocity plotted as function of $\tilde{x}$ at $\tilde{y} = 0.278$, for the case of harmonic excitation of a parallel boundary layer. The location of the exciter corresponds to $\tilde{Re} = 750$ and non-dimensional frequencies of excitation are **a** $\bar{\omega}_0 = 0.10$ and **b** $\bar{\omega}_0 = 0.15$, with the result shown for the viscous time scale of $t = 801.11$

3.3.2 Near-Field of the Localized Excitation: The Local Solution

Here, we explain the near-field of the response caused by localized delta function excitation with the help of a qualitative explanation. The details of the mathematical derivation can be found in [53], and more mathematically inclined readers can find details there. In this reference, the near-field of the response caused by localized wall excitation is shown to be due to the essential singularity of the bilateral Laplace transform of v', i.e., ϕ. In Fig. 3.6, the straight-line Bromwich contour in the α-plane is shown along with some representative discrete poles (eigenvalues). Let a circle of radius R and center at the origin cut this Bromwich contour at the points A and B, as shown Fig. 3.6. The perimeter of this circular arc above the Bromwich contour is denoted by Γ and the area bounded by Γ and the Bromwich contour AB encompasses N number of singularities at P_j. One can form a closed contour of integration which contains AB, Γ and small indented contours around the singularities connected to Γ by straight lines, as shown in Fig. 3.6. The purpose of these indented contours is to keep the singularities outside the closed domain. Thus, inside the closed contour, the integrand is analytic and Cauchy's integral theorem can be applied for integration of $\Phi(\alpha) = \phi e^{i(\alpha\tilde{x} - \omega_0 t)}$ following the closed contour excluding all singularities. Noting that $e^{i(\alpha\tilde{x} - \omega_0 t)}$ does not contribute to additional poles or singularities, we can investigate various contribution arising from ϕ. As $\phi(\alpha)$ is analytic for all points inside

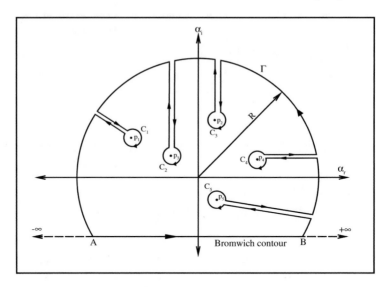

Fig. 3.6 Bromwich contour in α-plane

the closed contour in the α-plane, and if the singularities shown in Fig. 3.6 denote simple poles, then invoking Cauchy's integral theorem [34] one finds that,

$$\int_{AB} \phi(\alpha)d\alpha = -\int_{\Gamma} \phi(\alpha)d\alpha + \sum_{j=1}^{N} 2\pi i \times \text{Residue at } (P_j) \qquad (3.31)$$

In the limit of $R \to \infty$, the limits of integration on the left hand side become: $B, A \to \pm\infty -i\alpha_i$ (with α_i indicating the distance between the real α-axis from the Bromwich contour used), and the integral on the left hand side of Eq. (3.31) is Bromwich contour integral in the α-plane (Br). Hence, the Bromwich contour integral of $\phi(\alpha)$ can be written as

$$\int_{Br} \phi(\alpha)d\alpha = -\lim_{R \to \infty} \int_{\Gamma} \phi(\alpha)d\alpha + \sum_{j=1}^{N} 2\pi i \times \text{Residue at } (P_j) \qquad (3.32)$$

where N is the total number of singularities in the α-plane. The first integral on the right hand side disappears, if one can use the Jordan's Lemma [34]. However, Jordan's lemma is valid only when the degree of the denominator of the integrand $\phi(\alpha)$ is at least two orders higher than the degree of its numerator, i.e., $|\phi(\alpha)| \le k_1/|\alpha^2|$ for $|\alpha| \to \infty$ [3]. This condition has been specifically investigated in [53], for the limit of $|\alpha| \to \infty$, by obtaining a closed form analytical solution of the OSE to show that ϕ does not satisfy the condition to apply Jordan's Lemma for $|\alpha| \to \infty$.

Using singular perturbation theory [6], the closed form analytic solution for the OSE is shown as the inner solution, which is governed by the following reduced equation as,

$$\phi_i^{iv} - 2\beta^2\phi_i'' + \beta^4\phi_i = 0 \tag{3.33}$$

where, $\alpha = Re^{i\theta} = R\beta$ and $\phi = \phi_i(Y)$. Here, Y is a new independent rescaled variable given by $Y = \tilde{y}/\hat{\delta}$, with $\hat{\delta}$ is the thickness of the inner solution of OSE governed by Eq. (3.33). This equation is obtained for the distinguished limit of $\hat{\delta} = 1/R = 1/|\alpha|$. It is to be noted that Eq. (3.33) is independent of $\tilde{Re}$ and is equivalent to the Stokes problem (in the limit of $\tilde{Re} \to 0$) given by [53],

$$\nabla^4\psi = 0 \tag{3.34}$$

One can solve Eq. (3.33) for the localized delta function excitation problem [53] and the part solutions are given as,

$$\phi = (1 + \alpha\tilde{y})e^{-\alpha\tilde{y}} \quad \text{for } \alpha_r > 0 \tag{3.35}$$

$$\phi = (1 - \alpha\tilde{y})e^{\alpha\tilde{y}} \quad \text{for } \alpha_r < 0 \tag{3.36}$$

Clearly, these solutions do not satisfy the necessary condition for Jordan's lemma to be satisfied. Hence the first integral on the right hand side of Eq. (3.32) is not identically equal to zero. It has also been highlighted in [53] that this limit of the integral, i.e., is the essential singularity of ϕ ($\alpha \to \infty$) provides non-trivial contribution. This is very compatible with Tauber's theorem [53, 74], which states that the near-field solution (at $\tilde{x} \to 0$) is contributed by the point at infinity in the complex wavenumber plane, i.e., by the circular arc with $R \to \infty$. The above analytical solution of the OSE helps us obtain the near-field response field.

Thus in the near field of the exciter, one observes a highly viscous flow and it is for the same reason the near-field does not penetrate very far upstream and downstream from the location of the exciter. Most books or monographs, other than [53], does not even sketch the local solution or talk about its existence.

Performing the Bromwich contour integral, using Eqs. (3.35) and (3.36) over the semi-circle Γ and collating terms, it can be shown that the solution at the near-field of the exciter is given by [53],

$$v'(\tilde{x}, \tilde{y}, t) = \frac{e^{-i\bar{\omega}_0 t}}{2\pi} \lim_{R \to \infty} \left[e^{-R\tilde{x}} \cos R\tilde{y} + \frac{ie^{iRz}}{z}\left(1 + R\tilde{y} + \frac{i\tilde{y}}{z}\right) \right.$$
$$\left. - \frac{ie^{-iR\tilde{z}}}{\bar{z}}\left(1 + R\tilde{y} - \frac{i\tilde{y}}{\bar{z}}\right) - \frac{ie^{Rz+iz}}{z}\left(1 + \frac{i\tilde{y}}{z}\right) \right.$$

$$+iR\tilde{y}+\tilde{y}\Big)+\frac{ie^{-R\bar{z}-i\bar{z}}}{\bar{z}}\left(1-\frac{i\tilde{y}}{\bar{z}}-iR\tilde{y}+\tilde{y}\right)\Bigg]\qquad(3.37)$$

where, $z = \tilde{x} + i\tilde{y}$ and $\bar{z} = \tilde{x} - i\tilde{y}$. At the wall, i.e., for $\tilde{y} = 0$, Eq. (3.37) simplifies to

$$v'(\tilde{x},0,t)=\frac{e^{-i\bar{\omega}_0 t}}{2\pi}\left[e^{-R\tilde{x}}-\frac{2\sin R\tilde{x}}{\tilde{x}}+2e^{-R\tilde{x}}\left(\frac{\sin\tilde{x}}{\tilde{x}}\right)\right]\qquad(3.38)$$

In the limit, $R \to \infty$, the first and third terms of Eq. (3.38) do not contribute. But the second term turns out to be the Dirichlet function, which is an approximation of the Dirac-delta function, $\delta(\tilde{x})$ [74]. This clearly shows that one recovers the applied delta function at $\tilde{y} = 0$, which is the imposed boundary condition. Therefore, the delta function is totally supported by the point at infinity in the wavenumber space (which is nothing but the circular arc of Fig. 3.6, i.e., the essential singularity of the kernel of the contour integral).

The analysis in the present subsection clearly indicates that the localized delta function excitation in the physical space is supported by the essential singularity ($\alpha \to \infty$) in the image plane. This is made possible because $\phi(\tilde{y}, \alpha)$ does not satisfy the condition required for the satisfaction of Jordan's lemma. This is also consistent with the Tauber's theorem [53], as noted above. Thus, the solution very close to the exciter is due to the essential singularity of the OSE and called the *local solution*. The relevance of the eigenvalues of the OSE, located very close to the origin in the α-plane, affects the response field far away from the exciter and manifest itself as traveling waves with appropriate group velocities [53]. This part of the solution is also given by the second term in the right hand side of Eq. (3.32), which is the total contribution due to the residues calculated at all the poles.

3.3.3   The Spatio-Temporal Wave-Front (STWF)

To obtain the full receptivity solution to a linearized parallel shear-layer as depicted in Fig. 3.2, excited by a time-harmonic localized wall excitation started at $t = 0$ and located at $\tilde{x} = 0$, one solves the OSE given by Eq. (3.14) subject to the boundary conditions given by Eqs. (3.24) and (3.25). To overcome the stiffness problem one can use the CMM to solve the receptivity problem by BCIM, whose details are in [53, 57]. From the solution depicted in Figs. 3.4 and 3.5, one can see the genesis of STWF and we will provide further details on this element of the response field, while discussing DNS results in Chaps. 5 and 6. We briefly mention in passing here that a renewed quest for finding STWF for any system by delta function excitation in time, to look for tsunami-like disturbances caused by localized excitation has yielded definitive results in [10]. The present receptivity studies are with respect to time-periodic excitation, which shows the creation of all the three elements. What happens, when one excites the system instead by a localized impulse, i.e., by a wall-normal excitation velocity at the origin at $t = 0$: $v' = \delta(x)\delta(t)$? This question has been addressed in [7, 10], and here one can see the presence of STWF only.

3.4 Stability and Transition of Mixed Convection Flows

So far we have discussed various aspects of instability and receptivity of pure hydro-dynamic flows, whose linear dynamics can be followed by solving the fourth order OSE. If one includes any one of the effects due to compressibility, heat transfer or surface compliance, then the linear dynamics become more involved, as one needs to solve instead the sixth order OSE [56]. This added complexity also enhances flow instability, whether one is interested in linear analysis for small excitation amplitudes or for large amplitude excitations for which one would be required to solve the full NSE. Effects of heat transfer are studied from both linear and nonlinear instability perspectives here. Focus is on mixed convection flow at low speed, for which heat transfer can be modeled by Boussinesq approximation. One of the major difficulties of studying mixed convection flows is a lack of available canonical equilibrium flows, even for flow past a flat plate with heat transfer. For a constant external flow past a horizontal flat plate, a similarity solution is available in [49]. This has been studied in [56, 64] for instability and receptivity of this flow over horizontal flat plate, while reference is made to flow past a vertical plate also, whose equilibrium flow is obtained by non-similar approaches.

The spatial stability properties of mixed convection boundary layer, developing due to a constant external flow over a horizontal plate with heat transfer are studied here using linear and quasi-parallel flow assumptions, with the aim to find out if there is a critical buoyancy parameter that highlights the importance of heat transfer in destabilizing mixed convection flows, with buoyancy effects given by Boussinesq approximation. The undisturbed flow is the one given by the similarity solution of [49], which requires the wall temperature to vary as inverse square root of the distance from the leading edge of the plate for the similarity to hold, when the boundary layer edge velocity is held constant. Linear stability is investigated first by using the CMM, which allows finding all modes in a chosen range of the complex wavenumber plane by spatial stability analysis using grid search technique [53].

Mixed convection flows are important, as these are ubiquitous in nature and in engineering devices. At the global scale of the atmosphere, geophysical fluid dynamics depend critically on mixed convection flow instability properties. Similarly, heating and cooling in electronic devices at micro scale depend on mixed convection flow transition. Instability and receptivity studies of such flows to different types of disturbance environments is of importance, but attention is focused here only on vortical disturbance introduced at the wall for the receptivity study. The present approach of using vortical excitation shows that weak vortical excitation create thermal fluctuations (entropic disturbances).

Heat transfer effects at low speed are often modeled by buoyancy effects in one of the momentum equations, due to which instabilities in mixed convection flows differ considerably from instabilities of flows without heat transfer. A consequence of heat transfer is to induce additional pressure gradient altering the equilibrium flow, as compared to flows without heat transfer. In [13], it is noted that for natural and mixed convection flows, instability arises due to the growth of small disturbances. In this context, the similarity profile in [49] for flow past a horizontal plate becomes important, as this can be used for mixed convection flows, for constant edge velocity. For edge velocity varying as x^m, generalization is possible for the equilibrium flow, as given in [42].

Flow and heat transfer properties are more complex for mixed convection flows past inclined or horizontal plates, due to the buoyancy forces inducing streamwise pressure gradient altering equilibrium flow. Eckert and Soehngen [20] have shown flow visualization pictures indicating transition to turbulence for the natural convection problem, originating as an instability of small disturbances. For natural convection flow past inclined flat plates in [68], the authors have reported generation of an array of longitudinal vortices. In [37, 78], the authors have experimentally investigated instability of flow past an inclined plate and reported the presence of two modes, depending on the inclination angle of the plate. For inclination angles less than 14° with respect to the vertical, the authors reported wave-like instability. Such wave-like instability has also been studied in [15]. For inclination angles greater than 17° for the natural convection problem, it is experimentally observed that the disturbance field is dominated by vortices, and is often termed vortex instability. This vortex mode of instability is present for horizontal, as well as inclined plates. This has been variously studied in [25, 26, 40, 67, 75, 76], among many other studies. Alternate viewpoint in [27], classifies instability of forced-convection boundary layers over horizontal heated plates in terms of the two prototypical instabilities: Rayleigh–Benard type, usually described for a closed convection system heated from below, and the Tollmien–Schlichting type that is typical of isothermal open flows, as in wall-bounded shear layers triggered by viscous actions.

Linear analysis has been traditionally performed using temporal theory, as in [41] for mixed convection flow for isothermal vertical flat plate. Primary mean flow was obtained by the local non-similarity method and the authors reported that for assisting flows, the effect of buoyancy is to stabilize the flow. The critique of various non-similar flow descriptions used in instability studies should be kept in view [13]. For an inclined plate, the instability study has been performed in [15] and for the horizontal isothermal plate in [16]. Results of these indicated that the flow along vertical and inclined plates is more stable, when the buoyancy force aids external convection, and the stability decreases, as the inclination angle approaches the horizontal. Also noted for horizontal plates is the tendency of flow to become more unstable, when buoyancy force is directed away from the surface.

Despite the distinction between temporal and spatial methods, the neutral curve is identical and is of vital interest for instability and transition for low frequency excitation. In [29], results were reported using linear spatial theory with a parallel flow approximation for free-convection flow past heated, inclined plates. The spatial stability results of natural convection flow over inclined plates were reported in [72], providing the eigenvalue spectrum.

When the Boussinesq approximation is adopted in NSE, effects of buoyancy appears in terms of Gr/Re^2, where Gr is the Grashof number and Re is the Reynolds number defined in terms of appropriate length, velocity and temperature scales. However in [35, 69], authors have shown for boundary layers, that the equivalent buoyancy parameter changes to $K = Gr/Re^{5/2}$. Experimental investigations in [25, 75] have also demonstrated that the onset of instability always occurs at the same value of K, showing its importance as the relevant buoyancy parameter. Similarity profiles derived in [42, 49] are also given in terms of K alone. One of the main aims here is to identify a K_{cr}, beyond which the transport property changes qualitatively for a mixed convection flow past horizontal plate. In [17], the authors have noted significant buoyancy effects for $K_x \geq 0.05$ and $K_x \leq -0.03$, for aiding and opposing flows past a horizontal plate, where $K_x = Gr_x/Re_x^{5/2}$.

3.5 Governing Equations

We consider laminar, 2D flow past a hot semi-infinite plate, with the free stream velocity and temperature denoted by U_∞ and T_∞, respectively. We focus our attention on the top of the plate with a temperature distribution $T_w(x)$, which is greater than T_∞, while assuming the leading edge of the plate as the stagnation point. The governing equations are written in dimensional form (indicated by the quantities with asterisks), along with the Boussinesq approximation, for the velocity and temperature fields [24]

$$\nabla^* \cdot \overrightarrow{V}^* = 0 \tag{3.39}$$

$$\frac{D\overrightarrow{V}^*}{Dt^*} = \overrightarrow{g}_j \, \beta_t \, (T^* - T_\infty) - \frac{1}{\rho}\nabla^* p^* + \nu \, \nabla^{*2}\overrightarrow{V}^* \tag{3.40}$$

$$\frac{DT^*}{Dt^*} = \alpha \, \nabla^{*2}T^* + \frac{\nu}{C_p} \, \Phi_v + \frac{\bar{q}}{\rho C_v} \tag{3.41}$$

which give the evolution of velocity and temperature fields with space and time. In the Boussinesq assumption valid for small excursion of temperature, one retains the density as constant, except that is present in the y-momentum equation, where density variation is kept to produce buoyancy effects. Here $\frac{D}{Dt^*}$ represents the substantial derivative, $g_j = [0, \ g, \ 0]^T$ represents the gravity vector, β_t is the volumetric thermal expansion coefficient and α is the thermal diffusivity. In this analysis, both the viscous dissipation (Φ_v) and heat source terms ($\bar{q}$) will not be considered in the energy equation. To non-dimensionalize above equations, a length scale (L), a velocity scale (U_∞), a temperature scale ($\Delta T_L = T_w(L) - T_\infty$) and a pressure scale ($\rho U_\infty^2$) are adopted. Instead of prescribing L, a Reynolds number based on L is taken as 10^5 for the ensuing analysis. The non-dimensional form of these equations are,

$$\nabla \cdot \overrightarrow{V} = 0 \tag{3.42}$$

$$\frac{D\overrightarrow{V}}{Dt} = \frac{Gr}{Re^2}\,\theta\hat{j} - \nabla p + \frac{1}{Re}\nabla^2\overrightarrow{V} \tag{3.43}$$

$$\frac{D\theta}{Dt} = \frac{1}{RePr}\nabla^2\theta \tag{3.44}$$

where $\overrightarrow{V} = \frac{\overrightarrow{V}^*}{U_\infty}$, $\theta = \frac{T^*-T_\infty}{\Delta T_L}$, $Gr = \frac{g\beta_t \Delta T_L L^3}{\nu^2}$, $Re = \frac{U_\infty L}{\nu}$, $Pr = \frac{\nu}{\alpha}$ and $\hat{j}$ is the unit vector in the wall-normal direction. The Grashof number (Gr) gives the ratio of buoyancy to viscous force in the flow and the Richardson number (Ri) or Archimedes number given by, $\frac{Gr}{Re^2}$, shows the relative dominance of natural to forced convection. Positive and negative signs of the Richardson number refer to assisting and opposing flows. In mixed convection regime, Ri is of order one. The Prandtl number (Pr) used in the present study is 0.71, a value for air as the working medium. The instability of boundary layer is also dependent upon buoyancy parameter defined as

$$K = \frac{Gr}{Re^{\frac{5}{2}}} = g\,\beta_t\,[T_w(L) - T_\infty]\,(L\nu)^{\frac{1}{2}}\,U_\infty^{-\frac{5}{2}} \tag{3.45}$$

The buoyancy parameter can be defined alternatively in terms of Ri as $K = \frac{Ri}{\sqrt{Re}}$. The value of $K = 0$ indicates flow over a flat plate without any heat transfer, the Blasius flow over a flat plate without heat transfer. $K > 0$ corresponds to assisting flows in which the flow occurs above a heated flat plate, and $K < 0$ corresponds to opposing flows over a cooled flat plate. K switches sign for flow below the corresponding plates.

3.6 Equilibrium Boundary Layer Flows

For high Reynolds numbers, momentum and energy transport through gradients is limited to a narrow region inside the boundary layer. The boundary layer approximation applied to the NSE yields the following equations for 2D incompressible flow in Cartesian coordinates as,

$$\frac{\partial U}{\partial X} + \frac{\partial V}{\partial Y} = 0 \tag{3.46}$$

$$U\frac{\partial U}{\partial X} + V\frac{\partial U}{\partial Y} = -\frac{\partial P}{\partial X} + \frac{\partial^2 U}{\partial Y^2} \qquad (3.47)$$

$$0 = K\theta - \frac{\partial P}{\partial Y} \qquad (3.48)$$

$$U\frac{\partial \theta}{\partial X} + V\frac{\partial \theta}{\partial Y} = \frac{1}{Pr}\frac{\partial^2 \theta}{\partial Y^2} \qquad (3.49)$$

where $X = x^*/L$, $Y = y^*\sqrt{Re}/L$, $U = u^*/U_\infty$, $V = v^*\sqrt{Re}/U_\infty$, $\theta = \frac{T^*-T_\infty}{T_w(L)-T_\infty}$ and $P = p^*/(\rho_\infty U_\infty^2)$. The above equations are solved subject to following boundary conditions,

(i) at the wall, ($Y = 0$ and $X > 0$): $U = V = 0$ and
(ii) at the free stream, ($Y \to \infty$): $U = 1$, $p^* = p_\infty$ and $\theta = 0$.

The buoyancy parameter K appears in the Y-momentum equation as a wall-normal pressure gradient. Integrating Y-momentum equation with respect to Y and using the boundary condition at $Y \to \infty$: $P = 0$, one gets the streamwise pressure gradient as $-K\int_Y^\infty \theta_X dY$, where the subscript in θ indicates a partial derivative. Hence depending upon the sign of K, one can create either an adverse or a favorable streamwise pressure gradient in the boundary layer. It is inappropriate to call these adverse or favorable pressure gradients – as is customary for flows without heat transfer based on stability property. Here, this will be only evident from the computed stability solutions. For the present investigation of flow over a heated flat plate, K is positive, with buoyancy resulting in a pressure gradient accelerating the flow in the streamwise direction. This term, therefore, usually induces non-similarity in the governing equations.

3.6.1 Schneider's Similarity Solution

A similarity solution of the boundary layer equation derived above has been given by Schneider [49]. Introducing the stream function (Ψ) automatically satisfies the continuity equation. Substituting the expression for streamwise pressure gradient in the X-momentum equation and the energy equation, we get

$$\Psi_Y\Psi_{XY} - \Psi_X\Psi_{YY} - K\int_Y^\infty \theta_X dY = \Psi_{YYY} \qquad (3.50)$$

$$\Psi_Y\theta_X - \Psi_X\theta_Y = \frac{1}{Pr}\theta_{YY} \qquad (3.51)$$

where subscripts X and Y denote partial derivatives with respect to X and Y, respectively.

For the above system of equations to admit a similarity solution, the wall temperature distribution must be given by $\theta_w \propto X^{-1/2}$, where $\theta_w = \frac{T_w(x^*)-T_\infty}{T_w(L)-T_\infty}$. This similarity transform converts the above PDEs in X and Y, into an ODE for the similarity variable defined by, $\eta = YX^{-1/2}$. Transforming dependent variables as, $\Psi = X^{1/2}f(\eta)$ and $\theta = \theta_w \Theta(\eta)$, yield the following governing equations

$$2f''' + ff'' + K\eta\Theta = 0 \tag{3.52}$$

$$\frac{2}{Pr}\Theta'' + f\Theta' + f'\Theta = 0 \tag{3.53}$$

In the above, a prime indicates a derivative with respect to η, and these equations have to be solved subject to the following boundary conditions at $\eta = 0$: $f = f' = 0$ and $\Theta = 1$ and as $\eta \to \infty$: $f' = 1$ and $\Theta \to 0$. Equation (3.53) is integrated analytically once to obtain the following equation

$$\frac{2}{Pr}\Theta' + f\Theta = 0 \tag{3.54}$$

Thus, the similarity profile depends only on K, with dependence on Re implicit through the definition of the Y-coordinate and η. It is seen from Eq. (3.54) and the boundary condition at $\eta = 0$: $f = 0$ that irrespective of K, adiabatic condition exists all over the plate, with heat transfer occurring singularly at the leading edge only.

The similarity solution for the problem under consideration is given for f', f''', Θ and Θ' as functions of η in frames (a) to (d) of Figs. 3.7, 3.8 and 3.9, for different values of K. Here f' represents the streamwise velocity, f''' its second derivative, Θ represents the temperature, and Θ' its wall-normal derivative. Figure 3.7 shows the result for a low value of $K = 1 \times 10^{-6}$. In Fig. 3.7a, the velocity profile monotonically grows following the boundary condition at $\eta = 0$ to $\eta = \eta_{max}$, where it tends to unity. Similarly from Fig. 3.7c, the temperature profile is seen to reach the free stream condition, where it is zero according to the non-dimensionalization adopted. In Fig. 3.7b, the inset provides a detailed view of the portion enclosed, helping detect any existing inflection points. For this K, the result indicates no inflection point to be present. Figure 3.8 shows the mean flow quantities plotted for $K = 3 \times 10^{-3}$. The general trend for the variation of mean flow quantities remains the same. But Fig. 3.8b indicates an inflection point for $\eta \simeq 9$. The same trend is observed in Fig. 3.9, for $K = 9 \times 10^{-2}$, which has a greater heat transfer due to buoyancy effects. In addition, we see an overshoot in the velocity profile in Fig. 3.9a. It has been noted in Fig. 3.8 of [49] that high values of K result in a velocity overshoot within the boundary layer. These observations tell us that for low values of K, although the second derivative of streamwise velocity remains zero at the wall and at the free stream, enhanced heating of the plate surface causes an inflection point at an intermediate height. From Figs. 3.7, 3.8 and 3.9, it is seen that as K is increased, variation of mean parameters progressively occurs within a lower range of η, which means that the mean flow quantities and their gradients achieve free stream conditions, more rapidly with higher

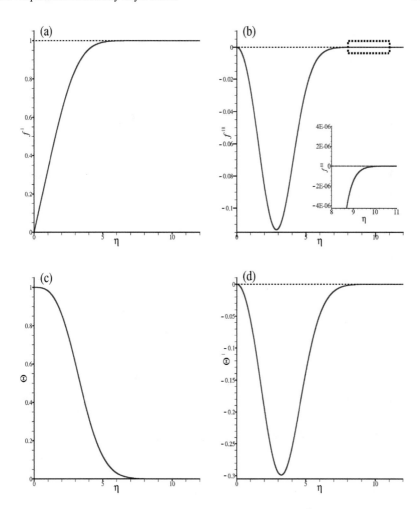

Fig. 3.7 Variation of mean flow quantities obtained for $K = 1 \times 10^{-6}$

gradients. Increasing K, i.e., increasing the temperature difference between the plate surface and the free stream, the boundary layer shrinks and the point of inflection moves closer to the wall. A similar overshoot of velocity within the boundary layer has been reported in [42] for wedge flow.

From Eq. (3.48), it is understood that the effect of heat transfer is to introduce a wall-normal pressure variation within the boundary layer. This is contrary to flows without heat transfer, where pressure is impressed upon the boundary layer by the outer inviscid flow, which remains invariant with height within the boundary layer. This aspect sets apart mixed convection boundary layers from isothermal flows. The wall-normal pressure variation within the boundary layer modifies the streamwise pressure gradient in the X-momentum equation given by

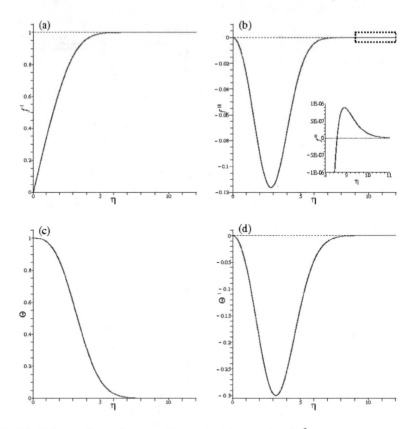

Fig. 3.8 Variation of mean flow quantities obtained for $K = 3 \times 10^{-3}$

$$-K \int_Y^\infty \theta_X \, dY$$

We also have

$$\theta = \theta_w \Theta = X^{-1/2} \Theta, \quad \eta = Y X^{-1/2}$$

Here the temperature gradient is with respect to X (keeping Y fixed), i.e., given by $\frac{\partial \theta}{\partial X}|_Y$. With wall temperature varying as $X^{-1/2}$, this would be

$$\frac{\partial}{\partial X}\left[X^{-1/2} \Theta(\eta) \right] = -\frac{1}{2} X^{-3/2} \Theta + X^{-1/2} \frac{\partial \Theta}{\partial \eta} \frac{\partial \eta}{\partial X}$$

As

$$\frac{\partial \eta}{\partial X} = -\frac{\eta}{2X}$$

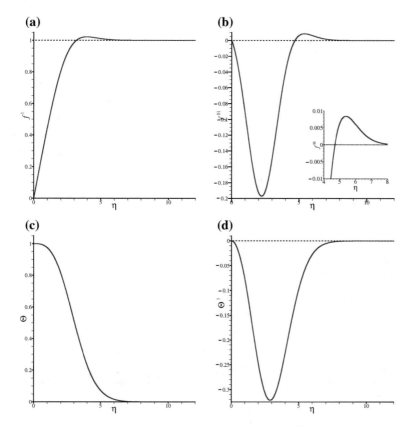

Fig. 3.9 Variation of mean flow quantities obtained for $K = 9 \times 10^{-2}$

Thus

$$\frac{\partial \theta}{\partial X}\Big|_Y = -\frac{1}{2X^{3/2}}[\Theta + \eta\,\Theta']$$

Therefore, the induced streamwise pressure gradient is given by

$$-K\int_Y^\infty \theta_X dY = -\frac{K}{2X^{3/2}}\left[\int_\eta^\infty (\Theta + \eta\,\Theta')X^{1/2}d\eta\right]$$

$$= -\frac{K}{2X}\left[\int \Theta d\eta + \eta\Theta - \int \Theta d\eta\right]$$

Using above relations and integrating the pressure gradient term, we get the following form for the streamwise pressure gradient

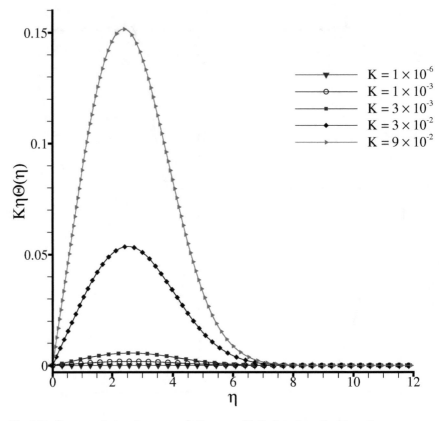

Fig. 3.10 Buoyancy-induced pressure gradient term $K\eta\Theta$ plotted as a function of η

$$- K \int_Y^\infty \theta_X \, dY = -\frac{K}{2X}\left[\eta\Theta\right]_\eta^\infty = -\frac{K}{2X}\eta\Theta \qquad (3.55)$$

The boundary condition $\Theta \to 0$ for $\eta \to \infty$ and Eq. (3.55) show that buoyancy effects do not impose a pressure gradient at the wall and at the far stream. But at all intermediate heights within the boundary layer, there is a height-dependent streamwise pressure gradient. For heated plates, K has a positive value and hence the pressure gradient within the mixed convection boundary layer is negative, which accelerates the flow. The value $K\eta\Theta$ determines the magnitude of this pressure gradient at any given streamwise location, which is plotted as a function of η for various buoyancy parameter (K) values in Fig. 3.10, certain aspects of which are noteworthy. The location of the maximum pressure gradient within the boundary layer moves towards the plate, as K is increased. Also, its magnitude is of the same order, as that of the corresponding K. The maximum value of favorable pressure gradient along with the location at which it occurs (η_{fpg}) are given in Table 3.1.

Table 3.1 Streamwise pressure gradient induced by buoyancy effects for a heated flat plate for different buoyancy parameters

Case	K	η_{fpg}	Maximum pressure gradient Parameter $(K\eta\Theta)$
1	1×10^{-6}	2.62250	1.86443×10^{-6}
2	1×10^{-3}	2.61816	1.86144×10^{-3}
3	3×10^{-3}	2.61000	5.55667×10^{-3}
4	3×10^{-2}	2.52144	5.36419×10^{-2}
5	9×10^{-2}	2.38344	1.51597×10^{-2}

In flows without heat transfer, it is well known that a favorable pressure gradient stabilizes the flow, which is impressed upon the boundary layer by the outer inviscid flow. Pressure remains invariant inside the shear layer in the wall-normal direction. But for flows with heat transfer, we note the presence of a differential streamwise pressure gradient within the boundary layer. The effect of this favorable pressure gradient is to accelerate the flow differentially within the shear layer, which results in a velocity overshoot, responsible for the presence of an inflection point. This is clearly dependent on the value of K. The higher the value of K, the higher will be the magnitude of the velocity overshoot. Also, Fig. 3.10 and Table 3.1 show that at higher values of K, the maximum of the pressure gradient moves closer to the plate surface. Figure 3.11 shows the contours of pressure gradient expression as given in Eq. (3.55) for values of K under discussion in the (x, y)-plane ($x = x^*/L$ and $y = y^*/L$). The η level at which the maximum pressure gradient occurs is also marked by a dotted line in Fig. 3.11.

The existence of an inflection point for the mean flow provides the necessary condition for inviscid instability, according to the Rayleigh–Fjørtoft theorem. Therefore, we are dealing with an equilibrium flow, which has an inherent tendency towards inviscid temporal instability. This is in stark contrast to flows without heat transfer, where phenomenologically one observed spatial growth of disturbances. Thus, one notes the propensity of the flows with heat transfer to display tendencies of both spatial and temporal linear instabilities.

Figure 3.12 shows variation of the inflection point with respect to K, indicating the inflection point to move closer to the wall, as K is increased. It helps to conclude that the mean flow becomes more and more unstable as the wall is heated, due to the presence of an inflection point nearer to the wall. However, the quantification of this should not be done using Rayleigh's stability equation derived for flows without heat transfer.

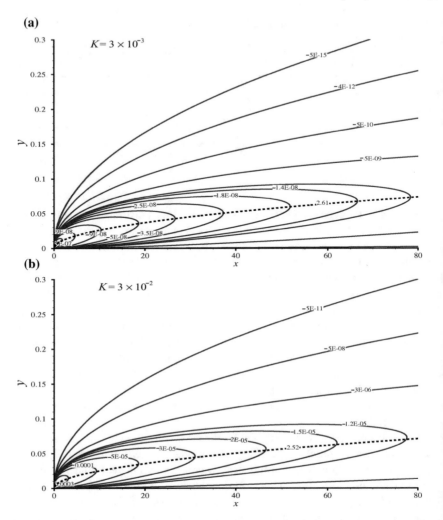

Fig. 3.11 $(-K\eta\Theta/2X)-$ contours plotted in the (x, y)-plane for the indicated values of K. η_{fpg} is marked by a dotted line

3.6.2   Ambiguities of Spatial and Temporal Linear Theories: Example of Mixed Convection Problem

Receptivity studies require computations of equilibrium flow, and whose response to deterministic excitation is sought. Wall-excitation for a ZPG boundary layers by DNS have been studied in [9, 55, 65]. The key features of this is that the same methodology is used to compute equilibrium and disturbance field, for different boundary conditions. When same methodology is used for mixed convection flows past horizontal plate, with Boussinesq approximation to model heat transfer effects

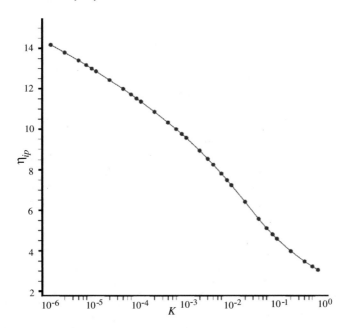

Fig. 3.12 Location of inflection point (η_{ip}) plotted as a function of K

in [66], even the equilibrium flows could not be computed. For horizontal hot flat plate with adiabatic wall conditions, the equilibrium flow is computed and its receptivity correlates with linear spatial theory for lower buoyancy parameter. However, receptivity of mixed convection flows to wall excitation for the following cases, did not allow computing equilibrium flows by DNS: (i) Adiabatic horizontal flat plate cooled significantly at the leading edge and (ii) strongly heated isothermal wedge flow for an angle of 60°. The cold plate case is interesting, as we note in next section that the linear spatial theory indicates enhanced stabilization with cooling for higher magnitude of K_i, instead for the cold plate, one notices disturbance growth outside the shear layer. The authors in [66] re-investigated various mechanisms of instability present for mixed convection flows, explaining relative roles of viscous and inviscid mechanisms for strong heat transfer effects, within the Boussinesq approximation. The main issue highlighted is: What happens to a flow, if both linear temporal and spatial theory indicate instability, with growth/decay rates completely different? This ambiguity can only be resolved either by adopting linear spatio-temporal receptivity analysis, or nonlinear, nonparallel analysis by DNS. We emphasize that DNS with high accuracy scheme is preferred in such cases. New theorems of inviscid instability are proposed [66], which are more generic than Rayleigh's and Fjørtoft's theorems for hydrodynamic instability without heat transfer. Thus, inviscid mechanism has been stated for mixed convection flows by replacing Rayleigh's equation [66].

Local instability studies have often been supplemented by linear and nonlinear global studies, as in [18, 39], with flow instabilities classified as resonator and amplifier type. It is noted that though linear local theories can account for amplifier behavior, yet it is preferable *to develop a fully nonlinear formulation involving the presence of a front separating the base state region from the bifurcated state region* [18].

The resonator dynamics imply the ability of flow to self-sustain oscillations, which is due to flow being globally unstable showing 3D stationary mode(s) [39]. For separated flows, this exhibits absolute instability, i.e., the disturbances do not convect and grow with time. In contrast, amplifier dynamics show the flow displaying large transient amplification of initial perturbation, which is often followed by convective instability. For flow over a backward facing step, Blackburn et al. [11] noted that their earlier computations reported absolute instability [4], while the *flow is actually unstable at much lower Reynolds numbers* to convective instability shown by investigating *directly the linear convective instability by means of transient-growth computations.* This highlights the central role played by accuracy of numerical methods used in computing NSE, to demarcate convective and absolute instabilities correctly. For a choice of time step, some wavenumbers can show spurious dispersion, while some specific wavenumber may indicate absolute instability (with numerical group velocity being zero). While spectral element method is known for its accuracy, in [31] a self-supported oscillation for a flow over 2D backward facing step was reported for sub-critical Re, which was not found using a better resolved DNS [32]. This problem shows the need to calibrate any method with model equation before actual simulation. In [66], different local instability mechanisms in the mixed convection flow over a flat plate and wedges are explained.

Receptivity is studied for those equilibrium flows, which are usually steady in the mean, and often defined by similarity profile(s). For flows without heat transfer, Blasius profile is for the canonical ZPG flow, whose instability has been studied extensively [2, 12, 22, 47]. We also note that use of Blasius profile is misleading for instability studies, as one excludes region near the leading edge of the flat plate, where this is not an accurate representation of the equilibrium flow. It has been conclusively shown in [60] that the leading edge is a site of disturbance energy, which creates convected disturbances that remain outside the boundary-layer. This is known as shear sheltering, which affects secondary instabilities and bypass transition over the flat plate by free stream modes and these outer disturbances interfere with growing disturbances inside the boundary layer during secondary stages. Apart from this, one also needs to incorporate rapid growth of the boundary layer, near the leading edge of the plate/ wedge. Such nonparallel and nonlinear effects change stability properties of flow without heat transfer [58], and for flow with heat transfer having adiabatic wall condition [53]. For these reasons, in [55, 58, 65], equilibrium flow has been calculated by solving unsteady NSE, including the leading edge of the flat plate. Identical approach is used in computing equilibrium and disturbance flows. We also show that for study of mixed convection flows, there is truly no similarity profiles that can be used as equilibrium flows.

3.7 Equilibrium Solution for Mixed Convection Flows: Isothermal Wall Case

In [42] the formulation used in [49] has been generalized for flow past isothermal wedge. This flow does not exhibit singular heat transfer at the leading edge. The governing equation transforms to ordinary differential equation for the external flow given by $U_e \simeq U_\infty X^n$, with the new independent variable, $\eta = Y X^{(n-1)/2}$. The wall temperature distribution is given by $\theta_w = X^{(5n-1)/2}$. The choice of $n = 1/5$ provides a wall-temperature distribution which is independent of X, which corresponds to a wedge angle of $60°$. Despite similarity, flow and heat transfer at the wall are completely different in [42], as compared to that in [49]. However, as the boundary layer edge velocity in [42] is a function of X, none of these velocity profiles directly represents similar solution. Both these flows are considered to study spatial and temporal viscous instabilities by solving NSE with different heat transfer at the wall, and help identify the active instability mechanisms for mixed convection flows.

Heat transfer modeled by Boussinesq approximation for mixed convection flows induces pressure gradient of the equilibrium flows. Such mean flows display flow instabilities, including inviscid instability, similar to that given by Rayleigh's equation for flows without heat transfer [19, 53]. This is shown here and new theorems stated with necessary conditions for temporal instability by linear inviscid mechanism. Also DNS of flows with heat transfer are provided, which show viscous and inviscid mechanisms simultaneously. In such a scenario, it is essential that we show predominance of one mechanism over the other.

The wedge flow given in [42] is a general equilibrium flow. In defining the governing equations for mixed convection flows, a buoyancy parameter is introduced as G_0 in [42] and K in [49], these symbols will be used here as K_i and K_a, respectively, with the subscript indicating isothermal and adiabatic conditions for flow over the general wedge flow defined as,

$$K_i \text{ or } K_a = \frac{Gr}{Re^{\frac{5}{2}}} = g \, \beta_t \, [T_w(x^*) - T_\infty] \, (x^* v)^{\frac{1}{2}} \, U_e^{-\frac{5}{2}}$$

where, Gr is the Grashof number. $K_a = 0$ corresponds to flow without heat transfer, which represents Falkner–Skan flow for the case of [42] and Blasius flow for the case of [49]. The cases of $K_a > 0$ correspond to assisting flows, occurring above a heated plate and the cases with $K_a < 0$ correspond to opposing flows, occurring over the top surface of a cold plate.

3.7.1 Governing Equation for Flow over Isothermal Wall

The 2D steady Navier Stokes and energy equations for incompressible mixed convection flow with Boussinesq approximation are written for a boundary layer, as in

Eqs. (3.46)–(3.49) for Schneider's similarity profile with $\theta = \frac{T^* - T_\infty}{T_w(L) - T_\infty}$. The first term on the right hand side of Eq. (3.48) is due to the free stream pressure gradient acting on the boundary layer by the edge velocity $U_e(X)$. The second term on right hand side of Eq. (3.47) represents the buoyancy-induced height dependent pressure gradient inside the shear layer. Depending upon the sign of K_a, this term can create either an adverse or favorable pressure gradient upon the boundary layer. Equation (3.48) shows these flows to display buoyancy effect by imposing a wall-normal variation of streamwise pressure gradient [53] by $K_a \int_Y^\infty \theta_X \, dY$. For flows with negative K_a, the cases in [49] have induced favorable pressure gradient. This discussion is also valid for the isothermal wedge cases, with K_a replaced by K_i.

Equations (3.46)–(3.49) are solved for wedge flow subject to the boundary conditions:

$$\text{At the wall } (Y = 0 \text{ and } X > 0) : U = V = 0,$$

$$\text{And at the free stream } (Y \to \infty) : U = U_e(X) \text{ and } \theta = 0.$$

These are further transformed to ODEs using a independent variable, $\eta = Y X^{(n-1)/2}$. We obtain equilibrium solution, following the procedures in [42, 49], with the wall temperature given by $\theta_w = X^{(5n-1)/2}$.

We introduce new variables for velocity, temperature and pressure as,

$$U = U_e f'(\eta), \quad \theta = X^{(5n-1)/2} \Theta(\eta), \quad P = X^{2n} q(\eta) \tag{3.56}$$

Introduction of non-dimensional stream function, $f(\eta)$, automatically satisfies Eq. (3.46). Equations (3.47)–(3.49) are transformed in terms of the new variables as,

$$f''' = n(f'^2 - 1) - \frac{1}{2}(n + 1)ff'' + 2n \, K_i \, q + \frac{1}{2}(n - 1)K_i \, \eta \, q' \tag{3.57}$$

$$q' = \Theta \tag{3.58}$$

$$\Theta'' = \frac{1}{2}Pr[(5n - 1)\Theta f' - (n + 1)f\Theta'] \tag{3.59}$$

which are solved subject to the boundary conditions:

$$\text{At } \eta = 0 : \ f = f' = 0 \text{ and } \Theta = 1$$

$$\text{As } \eta \to \infty : \ f' = 1 \text{ and } q = \Theta = 0$$

These equations are for mixed convection flow over a wedge, with the wedge angle: $2\gamma = 2n\pi/(n + 1)$. Derivation of these are given in Appendix A. Schneider's case [49] is obtained by putting $n = 0$, which is for mixed convection flow over a horizontal plate, with singular heat transfer at the leading edge. The wall temperature here is obtained from Eq. (3.56) by $\theta_w \sim X^{-1/2}$.

Also, using wall boundary condition in the relation of Eq. (3.56), we note that for $n = 0.2$, θ_w becomes constant with respect to X – a mixed convection flow over an isothermal wedge with the half-wedge angle (γ) as $30°$ in Fig. 3.13. For $n = 0$, we have mixed convection flow over a flat plate, and by integrating Eq. (3.59), we obtain, $\Theta' = -\frac{Pr}{2} f \Theta$. With the wall boundary condition, one notes that $\Theta'_w = 0$ at all X, for any K_a, as $f = 0$ for $\eta = 0$ for the case of [49]. Thus, all heat transfer occurs singularly from the leading edge of the horizontal flat plate.

3.7.2 Mixed Convection Governing Equations and Boundary Conditions for DNS

Study of nonlinear receptivity for mixed convection flows, starts with an equilibrium flow obtained by solving the conservation equations in (ψ, ω)-formulation, along with Boussinesq approximation. The physical plane is defined by non-dimensional Cartesian co-ordinate, for which the same reference length (L) is used to obtain X in Eqs. (3.46)–(3.49). However, Y in the same set of equations and y^* are related as $Y = y^* \sqrt{Re}$ for the boundary layer equation. Here, the equilibrium and disturbance flows are obtained from DNS of the NSE and energy equation. This is performed in the transformed (ξ, ζ)-plane. A schematic diagram of the problem is shown in Fig. 3.13. In the transformed plane, vorticity transport, stream-function and energy equations are obtained as,

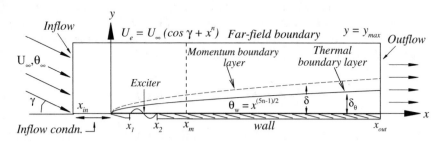

Fig. 3.13 Schematic diagram for mixed convection flow over a wedge. Identical computational domains are used for computing equilibrium flow and its receptivity to wall–excitation. [Reproduced from *Direct numerical simulation of transitional mixed convection flows: Viscous and inviscid instability mechanism*. Sengupta et al., *Physics of Fluids*, **25**, 094102 (2013), with the permission of AIP Publishing.]

$$h_1 h_2 \frac{\partial \omega}{\partial t} + h_2 u \frac{\partial \omega}{\partial \xi} + h_1 v \frac{\partial \omega}{\partial \zeta} = \frac{1}{Re_L} \left[\frac{\partial}{\partial \xi} \left(\frac{h_2}{h_1} \frac{\partial \omega}{\partial \xi} \right) + \frac{\partial}{\partial \zeta} \left(\frac{h_1}{h_2} \frac{\partial \omega}{\partial \zeta} \right) \right]$$

$$+ K_a \sqrt{Re_L} \frac{\partial}{\partial \xi} (h_2 \theta) \quad (3.60)$$

$$\frac{\partial}{\partial \xi} \left(\frac{h_2}{h_1} \frac{\partial \psi}{\partial \xi} \right) + \frac{\partial}{\partial \zeta} \left(\frac{h_1}{h_2} \frac{\partial \psi}{\partial \zeta} \right) = -h_1 h_2 \omega \quad (3.61)$$

$$h_1 h_2 \frac{\partial \theta}{\partial t} + h_2 u \frac{\partial \theta}{\partial \xi} + h_1 v \frac{\partial \theta}{\partial \zeta} = \frac{1}{Re_L Pr} \left[\frac{\partial}{\partial \xi} \left(\frac{h_2}{h_1} \frac{\partial \theta}{\partial \xi} \right) + \frac{\partial}{\partial \zeta} \left(\frac{h_1}{h_2} \frac{\partial \theta}{\partial \zeta} \right) \right] \quad (3.62)$$

In the transformed plane, the contra-variant components of the velocity vector are given by,

$$u = \frac{1}{h_2} \frac{\partial \psi}{\partial \zeta} \quad \text{and} \quad v = -\frac{1}{h_1} \frac{\partial \psi}{\partial \xi}$$

The scale factors of transformation h_1 and h_2, are given by, $h_1 = (x_\xi^2 + y_\xi^2)^{\frac{1}{2}}$ and $h_2 = (x_\zeta^2 + y_\zeta^2)^{\frac{1}{2}}$. Here, ξ is in the direction along the wall and ζ is in the wall-normal direction. Thus, if the transformations are given by, $x = x(\xi)$ and $y = y(\zeta)$, then the scale factors are, $h_1 = x_\xi$ and $h_2 = y_\zeta$. For the presented results, the parameters Re and Pr are selected as 10^5 and 0.71, respectively.

3.7.2.1 Auxiliary Conditions

For the reported receptivity to wall excitation, Eqs. (3.60)–(3.62) are solved for the equilibrium and disturbance flows. Dependent variables are split into equilibrium (denoted by overbar) and disturbance component (indicated by subscript d). For example, the vorticity is represented as $\omega = \bar{\omega} + \omega_d$. For the equilibrium flow, free stream conditions are given at the top of the domain for all ξ given by,

$$\frac{\partial \bar{\psi}}{\partial \zeta} = U_e h_2, \quad \bar{\omega} = 0 \quad \text{and} \quad \bar{\theta} = 0 \quad (3.63)$$

where $U_e = U_\infty \cos \gamma$ for $x \le 0$ and $U_e = U_\infty (\cos \gamma + x^n)$ for $x > 0$, with γ, as the semi-wedge angle. The no-slip and impervious conditions are used at the wall as the boundary conditions, which provides ψ and ω for the equilibrium flow. Thus, for $x \ge 0$, the boundary conditions at $y = 0$ are

$$\bar{\psi}_w = \bar{\psi}_o = constant \quad (3.64)$$

$$\bar{\omega}_w = -\frac{1}{h_2^2} \frac{\partial^2 \bar{\psi}}{\partial \zeta^2} \quad (3.65)$$

$$\bar{\theta}_w = x^{(5n-1)/2} \tag{3.66}$$

Ahead of the leading edge of the wedge ($x < 0$), Eqs. (3.65) and (3.66) remain same and Eq. (3.64) changes as,

$$\bar{\psi} = \bar{\psi}_o + U_\infty \, x \sin \gamma \tag{3.67}$$

The wall vorticity is obtained by noting ψ as a function of ζ and using the no-slip condition to obtain the derivative in Eq. (3.65). The wall vorticity has been obtained similarly in [53–55, 58, 65].

For equilibrium and disturbance field at the outflow, boundary conditions are used as,

$$\frac{\partial v}{\partial x} = 0 \tag{3.68}$$

$$\frac{\partial \omega}{\partial t} + U_c \frac{\partial \omega}{\partial x} = 0 \tag{3.69}$$

$$\frac{\partial \theta}{\partial t} + U_c \frac{\partial \theta}{\partial x} = 0 \tag{3.70}$$

The last two equations are the Sommerfeld outflow conditions for vorticity and temperature, respectively. This type of boundary condition is quite common and used in [12, 22, 55, 65]. This helps enforce the disturbance to convect smoothly out of the domain with the velocity, U_c. The convection speed is taken as U_e, as also used in [55, 60, 65]. One can also use filters to avoid numerical instability near the outflow, as described in Chap. 2.

To start the simulation for equilibrium flow, properties are initialized in the interior of the domain with potential flow solution, using the following initial conditions:

$$\frac{\partial \bar{\psi}}{\partial \zeta} = U_e h_2 \tag{3.71}$$

$$\bar{\omega} = 0 \tag{3.72}$$

$$\bar{\theta} = 0 \tag{3.73}$$

The computational domain is taken as, $-0.05 \leq x \leq 80; 0 \leq y \leq 2$. In the wall-normal direction, a stretched tangent hyperbolic grid is used, given by the following for the jth point as

$$y_j = y_{max} \left[1 - \frac{\tanh[b_y(1 - \zeta_j)]}{\tanh b_y} \right]$$

Such stretching helps in controlling aliasing error [54]. Here, $y_{max} = 2$ and $b_y = 2$ have been used to cluster grid and a total of 501 points are taken in the wall-normal direction. In the streamwise direction, we have two segments: the first segment from

$x_{in} = -0.05$ to $x_m = 10$, contains stretched grid distributed by tangent hyperbolic function, with the ith point obtained as,

$$x_i = x_{in} + (x_m - x_{in})\left[1 - \frac{\tanh[b_x(1 - \xi_i)]}{\tanh b_x}\right]$$

From $x_m = 10$ to $x_{out} = 80$, uniform distribution of points are used, so that 4501 points are used in the streamwise direction. The wall resolution used is, $\Delta y_w = 4.36 \times 10^{-4}$ and the grid spacing in the streamwise direction at the exciter location is taken as $\Delta x_{exc} = 0.002$, which is more than adequate with the high accuracy compact schemes used.

The diffusion operator (∇^2) in stream function equation is discretized using second order central differencing and is solved by Bi-CGSTAB algorithm [77]. First derivatives in vorticity transport and energy equations are obtained using the OUCS3

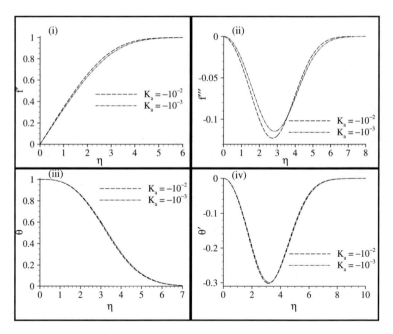

Fig. 3.14 a Variation of mean flow quantities obtained from boundary layer solution for flow past an adiabatically cooled horizontal flat plate buoyancy parameter $K_a = -0.001$ and -0.01. **b** Streamwise velocity component and temperature obtained from DNS for flow past an adiabatically cooled plate $K_a = -0.001$ and -0.01 plotted as a function of the similarity variable η at indicated x-locations along with the corresponding *similarity solution*. [Reproduced from *Direct numerical simulation of transitional mixed convection flows: Viscous and inviscid instability mechanism*. Sengupta et al., *Physics of Fluids*, **25**, 094102 (2013), with the permission of AIP Publishing.]

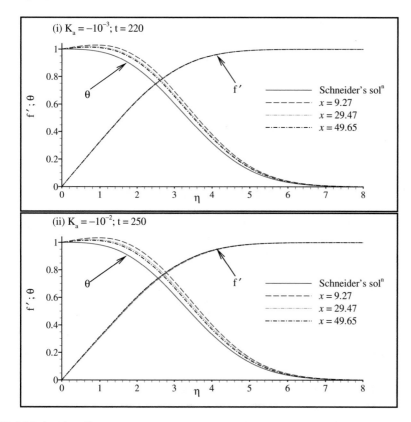

Fig. 3.14 (continued)

scheme [63]. Optimized 3-stage Runge-Kutta method (ORK3) [45] has been used for time integration. A very narrow buffer domain has been used at the outflow ($79.5 \leq x \leq 80$), where second order 1D filter is used in both x- and y-directions, that prevents reflections from the outflow boundary. Such buffer domain has been used in [12, 33, 36, 59, 70] for the solution of NSE.

Solutions of Eqs. (3.57)–(3.59) are obtained for $K_a = -0.01$ and -0.001, as shown in Fig. 3.14a for the velocity profile (f'), curvature of the velocity profile (f'''), temperature (Θ) and heat flux (Θ'). These cold plate cases have also been obtained by solving NSE given in Eqs. (3.60)–(3.62), with the help of boundary and initial conditions defined. Results from NSE are compared in Fig. 3.14a with the solution obtained using the formulation of [49], for different streamwise stations. Results of NSE match very well with Schneider's profile. Temperature profiles show mismatch due to the singularity at $x = 0$ for the wall temperature used in the formulation of [49]. Similar results for hot plate cases have been shown in [53], which showed better match between DNS and the solution of Eqs. (3.57)–(3.59).

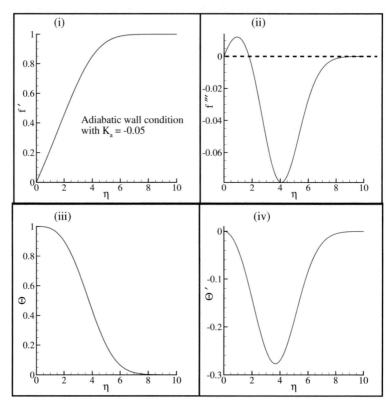

Fig. 3.15 **a** Variation of mean flow quantities obtained from *similarity solution* for the case of flow over adiabatic flat plate with buoyancy parameter $K_a = -0.05$. **b** Variation of mean flow quantities obtained from *similarity solution* for the case of flow over isothermal wedge with buoyancy parameter $K_i = 0.04$. [Reproduced from *Direct numerical simulation of transitional mixed convection flows: Viscous and inviscid instability mechanism.* Sengupta et al., *Physics of Fluids,* **25,** 094102 (2013), with the permission of AIP Publishing.]

The DNS reproduces the boundary layer solution [49] to fair degree of accuracy, in Fig. 3.14b. However for higher degree of cooling cases, the DNS does not even produce steady solution. One such cases of adiabatic cold plate is shown in Fig. 3.15a for $K_a = -0.05$ for the solution of Eqs. (3.57)–(3.59). In frame (ii) of Fig. 3.15b, one notices existence of a inflection point inside the boundary layer at $\eta \approx 1.5$. Drawing analogy with flow without heat transfer in [53], existence of inflection point was identified as an indicator for inviscid instability following Rayleigh's theorem [19, 53]. Another case of heated isothermal plate results are shown in Fig. 3.15b for $K_i = 0.04$, obtained by solving Eqs. (3.57)–(3.59). One notices an inflection point distinctly at the outer inviscid part of the flow ($\eta \approx 6.3$).

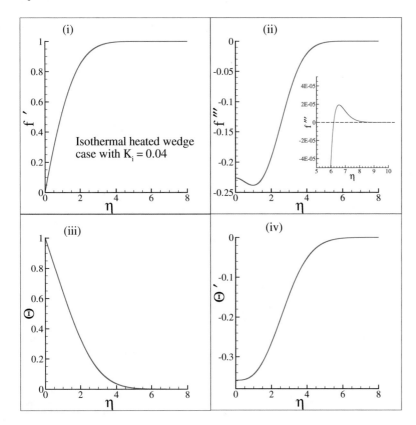

Fig. 3.15 (continued)

Computed solution of Eqs. (3.60)–(3.62) are shown in Fig. 3.16, for the indicated time frames corresponding to the case shown in Fig. 3.15a of adiabatic cold plate with $K_a = -0.05$. Vorticity contours shown up to $t = 90$, display accumulating fluctuations near the leading edge, which is convected downstream along the edge of the boundary layer. Fluctuations also build up progressively inside the boundary layer simultaneously. Despite the inflection point being deep inside the boundary layer, the fluctuations appear initially outside the shear layer. Such behavior is reminiscent of disturbance energy growth mechanism explained in [60] for bypass transition via vortex-induced instability. There the disturbances originated from the leading edge for a ZPG flow without heat transfer. This justifies solving the full NSE including the leading edge of the plate to investigate mechanisms of instability.

Solution of NSE, as the vorticity contours are shown for the isothermal hot wedge case of Fig. 3.15b ($K_i = 0.04$) in Fig. 3.17. Here also, one notices unsteady fluctuations near the outer edges of the momentum boundary layer, with no such fluctuations

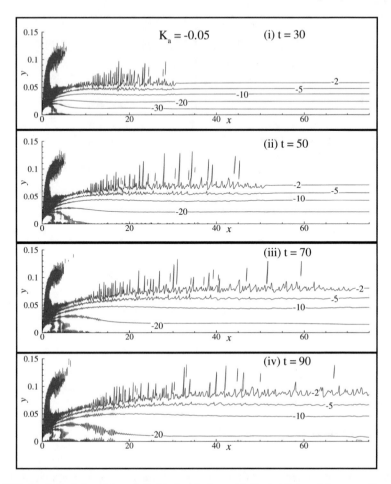

Fig. 3.16 Vorticity (ω) contours plotted in (x, y)-plane at indicated times while computing equilibrium flow cold adiabatic plate with buoyancy parameter $K_a = -0.05$. [Reproduced from *Direct numerical simulation of transitional mixed convection flows: Viscous and inviscid instability mechanism*. Sengupta et al., *Physics of Fluids*, **25**, 094102 (2013), with the permission of AIP Publishing.]

inside the shear layer, as seen in Fig. 3.16. It is noted in [66] that *there are other qualitative differences between the cases in Figs. 3.17 and 3.16. First, the disturbances do not originate from the leading edge. Secondly, fluctuations reveal significantly taller vertical structures with higher intensity. Such differences are also perplexing, because adiabatic plate with higher cooling from the leading edge induces favorable pressure gradient. Despite this, the observed intense fluctuations, which do not allow to compute even the equilibrium flows, prompted to study flow instability and their propensity to very low amplitude small scale disturbances, which are due to numerical error, as noted also in Kaiktsis et al.* [32]. *The same numerical methods have been*

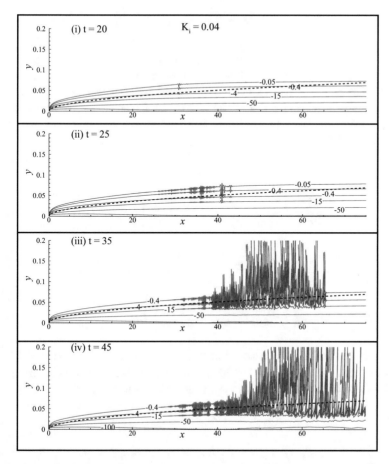

Fig. 3.17 Vorticity (ω) contours plotted in (x, y)-plane at indicated times while computing equilibrium flow of heated isothermal wedge with buoyancy parameter $K_i = 0.04$. The dotted line in each frame indicate the location in y direction where Fjørtoft integrand Γ changes sign from positive to negative. [Reproduced from *Direct numerical simulation of transitional mixed convection flows: Viscous and inviscid instability mechanism.* Sengupta et al., *Physics of Fluids,* **25**, 094102 (2013), with the permission of AIP Publishing.]

used very successfully for receptivity study of ZPG boundary layer in [55, 65]. *Same methodologies have been used to study instability of adiabatic plate with heated leading edge, without any problem* [53, 73]. *In these successful cases, equilibrium flows have been obtained by DNS and whose receptivity to vortical wall excitation studied.* Such counter-intuitive behavior of the solution of NSE prompted the authors in [66] to relook at the linear viscous flow instability mechanisms for both spatial and temporal growths.

3.7.3 Linear Viscous Instability: Spatial and Temporal Routes

Linear viscous instabilities via spatial and temporal routes for the equilibrium flow
obtained as *similarity solutions* of mixed convection flows are studied next. Detailed
method of linear theory is described in [53, 73] for spatial analysis, and a brief account
of the same is provided here. As before. to obtain the equations for linear stability
analysis, consider the nondimensional NSE and energy equation with Boussinesq
approximation given by,

$$\frac{\partial \tilde{u}}{\partial \tilde{x}} + \frac{\partial \tilde{v}}{\partial \tilde{y}} = 0 \qquad (3.74)$$

$$\frac{\partial \tilde{u}}{\partial \tilde{t}} + \tilde{u}\frac{\partial \tilde{u}}{\partial \tilde{x}} + \tilde{v}\frac{\partial \tilde{u}}{\partial \tilde{y}} = -\frac{\partial \tilde{p}}{\partial \tilde{x}} + \frac{1}{\tilde{Re}}\left[\frac{\partial^2 \tilde{u}}{\partial \tilde{x}^2} + \frac{\partial^2 \tilde{u}}{\partial \tilde{y}^2}\right] \qquad (3.75)$$

$$\frac{\partial \tilde{v}}{\partial \tilde{t}} + \tilde{u}\frac{\partial \tilde{v}}{\partial \tilde{x}} + \tilde{v}\frac{\partial \tilde{v}}{\partial \tilde{y}} = Ri\ \tilde{\theta} - \frac{\partial \tilde{p}}{\partial \tilde{y}} + \frac{1}{\tilde{Re}}\left[\frac{\partial^2 \tilde{v}}{\partial \tilde{x}^2} + \frac{\partial^2 \tilde{v}}{\partial \tilde{y}^2}\right] \qquad (3.76)$$

$$\frac{\partial \tilde{\theta}}{\partial \tilde{t}} + \tilde{u}\frac{\partial \tilde{\theta}}{\partial \tilde{x}} + \tilde{v}\frac{\partial \tilde{\theta}}{\partial \tilde{y}} = \frac{1}{\tilde{Re}\ Pr}\left[\frac{\partial^2 \tilde{\theta}}{\partial \tilde{x}^2} + \frac{\partial^2 \tilde{\theta}}{\partial \tilde{y}^2}\right] \qquad (3.77)$$

where, Ri and $\tilde{Re}$ are Richardson and Reynolds numbers, respectively. For linear
instability analysis, flow properties including velocity, pressure and temperature are
split into a mean and a fluctuating part, with variables represented as

$$\tilde{\mathbf{z}}(\tilde{x}, \tilde{y}, \tilde{t}) = \mathbf{Z}(\tilde{x}, \tilde{y}) + \varepsilon\bar{\mathbf{z}}(\tilde{x}, \tilde{y}, \tilde{t}) \qquad (3.78)$$

where, $\tilde{\mathbf{z}} = [\tilde{u},\ \tilde{v},\ \tilde{p},\ \tilde{\theta}]^T$ represents the total quantities; $\mathbf{Z} = [U,\ V,\ P,\ T]^T$ represents the mean components and $\bar{\mathbf{z}} = [\bar{u},\ \bar{v},\ \bar{p},\ \bar{\theta}]^T$ represents fluctuating components. Mean components are obtained after proper transformations of the mean flow
equations described before.

For 2D local linear instability studies, parallel flow assumption is invoked by
requiring $U = U(\tilde{y}), V = 0, P = P(\tilde{y})$ and $T = T(\tilde{y})$. For non-dimensionalization,
suitable length, velocity and temperature scales chosen are δ^* (local displacement
thickness), shear layer edge velocity U_e and $\Delta T(x^*)$ (difference between local plate
and free stream temperatures, i.e., $\Delta T(x^*) = T_w(x^*) - T_\infty$). The non-dimensional
numbers in Eqs. (3.74)–(3.77) are defined as, $\tilde{Re} = U_e\delta^*/\nu$ and $Ri = \tilde{Re}/\tilde{Gr}^2$ with
Grashof number as $\tilde{Gr} = g\beta_t\Delta T(x^*)\delta^{*3}/\nu^2$. Using Fourier-Laplace transform, linear analysis is carried out with the perturbation quantities given by,

$$[\bar{u},\ \bar{v},\ \bar{p}, \bar{\theta}] = \int [\tilde{g}(\tilde{y}),\ \phi(\tilde{y}),\ \pi(\tilde{y}),\ h(\tilde{y})]\ e^{i(k\tilde{x}-\beta\tilde{t})}\ dk d\beta \qquad (3.79)$$

Here, $k = k_{real} + ik_{img}$ and $\beta = \beta_0 + i\beta_i$, are complex wavenumber and circular
frequency, respectively. For spatial instability studies one considers wavenumber k

to be complex, whereas β is treated as real. Spatial growth of disturbances is noted for k_{img} being negative. For temporal instability, wavenumber k is real ($k = k_{real}$), while β is complex. Positive values of β_i signify temporal growth of disturbances. By substituting Eq. (3.78) into Eqs. (3.74)–(3.77) and retaining $O(\varepsilon)$ terms, linearized equations are obtained. Further, substitution of Eq. (3.79) into linearized equations, results in the ODEs, governing the disturbance amplitude functions (with $\tilde{y}$ as the independent variable) which can be simplified to,

$$i(kU - \beta)(k^2\phi - \phi'') + ikU''\phi = Ri\ k^2h - \frac{1}{\tilde{Re}}(\phi^{iv} - 2k^2\phi'' + k^4\phi) \quad (3.80)$$

$$i(kU - \beta)h + T'\phi = \frac{1}{\tilde{Re}Pr}(h'' - k^2h) \quad (3.81)$$

Where prime indicate derivative with respect to $\tilde{y}$. These are the sixth order OSE for mixed convection flows. The system given by Eqs. (3.80) and (3.81) are solved subject to six boundary conditions,

$$\text{at} \quad \tilde{y} = 0: \quad \phi, \phi' = 0; \quad h = 0 \quad (3.82)$$
$$\text{and as} \quad \tilde{y} \to \infty: \quad \phi, \phi', h \to 0 \quad (3.83)$$

General solution to the ODEs are given as

$$\phi = a_1\phi_1 + a_2\phi_2 + a_3\phi_3 + a_4\phi_4 + a_5\phi_5 + a_6\phi_6 \quad (3.84)$$
$$h = a_1h_1 + a_2h_2 + a_3h_3 + a_4h_4 + a_5h_5 + a_6h_6 \quad (3.85)$$

Adopted boundary conditions imply disturbances which decay in the far stream ($\tilde{y} \to \infty$). However in the free stream, h decouples from ϕ, as $U \approx 1$ and U'', $T' \approx 0$. Thus in the free stream, energy equation Eq. (3.81), reduces to

$$h'' - [i\tilde{Re}Pr(k - \beta) + k^2]h = 0 \quad (3.86)$$

whose solution at the free stream is given by the characteristic modes as

$$h_\infty = a_5e^{-S\tilde{y}} + a_6e^{S\tilde{y}} \quad (3.87)$$

where $S = \sqrt{k^2 + i\tilde{Re}Pr(k - \beta)}$. To satisfy free stream boundary condition, we must have $a_6 = 0$ for real$(S) > 0$. It can be shown that as $\tilde{y} \to \infty$, $h_{1\infty} = h_{3\infty} = 0$. At the free stream, $U = 1$, and all mean flow derivatives are zero, and linearized momentum equation reduces to

$$\phi^{iv} - [2k^2 + i\tilde{Re}(k - \beta)]\phi'' + [k^4 + i\tilde{Re}(k - \beta)k^2]\phi = Ri\ \tilde{Re}\ k^2h \quad (3.88)$$

Even in the free stream, momentum equation is not decoupled from the thermal field, with the latter providing forcing on the former. Solution of Eq. (3.88) is a sum of homogeneous solution and a particular integral, which after removing exponentially growing terms (by enforcing $a_2 = a_4 = 0$) gives

$$\phi_\infty = a_1 e^{-k\tilde{y}} + a_3 e^{-Q\tilde{y}} + a_5 \tilde{\Gamma} e^{-S\tilde{y}} \tag{3.89}$$

where $Q = \sqrt{k^2 + i \tilde{Re}(k - \beta)}$ and $\tilde{\Gamma} = Ri \, \tilde{Re} \, k^2 / [S^4 - (k^2 + Q^2)S^2 + k^2 Q^2]$, with real parts of k, Q and S as positive. Hence, the general solution of the coupled sixth order OSEs, subject to the prescribed homogeneous boundary conditions at free stream given by Eq. (3.83) are

$$\phi = a_1 \phi_1 + a_3 \phi_3 + a_5 \phi_5 \tag{3.90}$$

$$h = a_1 h_1 + a_3 h_3 + a_5 h_5 \tag{3.91}$$

where $h_{1\infty} = h_{3\infty} = 0$, $h_{5\infty} = e^{-S\tilde{y}}$, and $\phi_{1\infty} = e^{-k\tilde{y}}$, $\phi_{3\infty} = e^{-Q\tilde{y}}$ and $\phi_{5\infty} = \tilde{\Gamma} e^{-S\tilde{y}}$. As the three fundamental solutions decay exponentially at three different widely separate rates, Eqs. (3.80) and (3.81) represent a set of stiff ODEs. One adopts stiff solvers like the CMM [1, 44, 53] to solve the equations. As noted [53] in CMM, one solves a set of auxiliary equations derived from the original equation, in terms of second-compound variables, well-defined combinations of ϕ_j, h_j and their higher derivatives. The second compound variables grow/ decay exponentially at comparable rates ([53] provides details of CMM). The sixth order OSE is converted into an initial value problem in CMM with twenty (20), second-compounds, with initial conditions defined in free stream. The eigenvalues of Eqs. (3.80) and (3.81) are found by integrating these twenty equations from the free stream to the wall, subject to initial conditions described in [53, 73]. Satisfying the dispersion relation obtained from the wall-boundary condition given by Eq. (3.82), one obtains the eigenvalues, which represent complex wavenumber (k), for a particular combination of $\tilde{Re}$, β_0 and K_i (or K_a) for spatial stability analysis and complex frequency β in temporal analysis for fixed values of $\tilde{Re}$, k_{real} and K_i (or K_a). For disturbance components of temperature and velocity, wall-boundary condition in Eq. (3.82), can be written in terms of fundamental solution components as given below (which in turn provides the dispersion relation),

$$a_1 \phi_1 + a_3 \phi_3 + a_5 \phi_5 = 0 \tag{3.92}$$
$$a_1 \phi_1' + a_3 \phi_3' + a_5 \phi_5' = 0 \tag{3.93}$$
$$a_1 h_1 + a_3 h_3 + a_5 h_5 = 0 \tag{3.94}$$

Characteristic determinant of the above linear system combined with definition of second-compounds in [53] provides the dispersion relation

$$D_r + i D_i = y_3 = 0 \quad \text{at} \quad \tilde{y} = 0 \tag{3.95}$$

where [73],

$$y_3 = \begin{vmatrix} \phi_1 & \phi_3 & \phi_5 \\ \phi_1' & \phi_3' & \phi_5' \\ h_1 & h_3 & h_5 \end{vmatrix}. \tag{3.96}$$

3.7.4 Linear Spatial Viscous Theory for Mixed Convection Flows

Linear spatial viscous instability of mixed convection flows over adiabatic cold plate and isothermal heated wedge are studied first, where wavenumber k is complex and frequency β is real. Obtained growth rate contours (k_{img}) are plotted with the corresponding neutral curves ($k_{img} = 0$) in Fig. 3.18a, b.

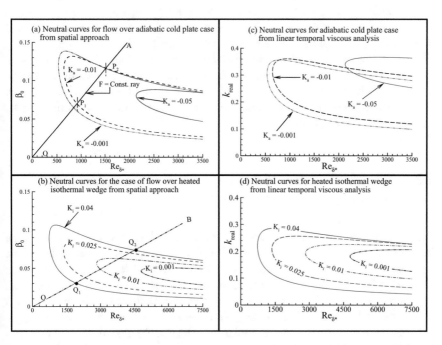

Fig. 3.18 **a, b** Neutral curve obtained from linear spatial analysis shown in ($Re_{\delta*}$, β_0)-plane buoyancy parameter for: **a** flow over cold adiabatic plate and **b** flow over heated isothermal wedge. Frames **c, d** show neutral curve obtained from linear temporal instability analysis shown in the ($Re_{\delta*}$, k_{real})-plane for indicated values of buoyancy parameter for the cases shown in frames (**a**) and (**b**). [Reproduced from *Direct numerical simulation of transitional mixed convection flows: Viscous and inviscid instability mechanism.* Sengupta et al., *Physics of Fluids*, **25**, 094102 (2013), with the permission of AIP Publishing.]

In Fig. 3.18a, b, neutral curves ($k_{img} = 0$ contour) are plotted in ($Re_{\delta*}$, β_0)-plane, for flow over cold adiabatic plate and over a heated isothermal wedge for different values of buoyancy parameter, respectively. The Reynolds number is $Re_{\delta*} = \frac{U_e \delta^*}{\nu}$ and β_0 is the real part of β. These neutral curves are obtained for the *similarity solution* in [49]. The flow is spatially unstable inside the neutral curve. Figure 3.18a shows the case of flow over adiabatic cold plate, with the critical Reynolds number $(Re_{\delta*})_{cr}$ increasing with increased cooling. The highest β_0, above which all disturbances are stable, is termed here as the critical circular frequency $(\beta)_{cr}$. One notes that $(\beta)_{cr}$ decreases with increase in cooling. The region inside the neutral curve becomes wider for a given $Re_{\delta*}$ when cooling is reduced.

The ray OA in Fig. 3.18a correspond to a constant non-dimensional physical frequency defined as follows:

$$F = \beta/Re_{\delta*} = 2\pi \nu \tilde{f}/U_e^2$$

where $\tilde{f}$ is the excitation frequency in Hertz. When the flow is excited by a source at constant frequency, one tracks the disturbance following the ray corresponding to $F = constant$ in this theory. The ray OA enters the neutral curve for $K_a = -0.01$ at P_1 in Fig. 3.18a, and P_2 represents the point where it exits the neutral curve. In Fig. 3.18b, neutral curves for flows over hot isothermal wedge have been plotted in ($Re_{\delta*}$, k_{real})-plane, for different K_i. In this case, $(Re_{\delta*})_{cr}$ decreases and $(\beta)_{cr}$ increases, with increasing value of K_i. Thus, heating the plate destabilizes the flow more by the linear local viscous spatial theory. Also, the neutral curve becomes wider with increasing value of K_i.

3.7.5 Linear Temporal Viscous Theory for Mixed Convection Flows

For linear temporal viscous instability, wavenumber k is real ($k = k_{real}$), and the frequency β is complex. Disturbance will grow with time, if β_i is positive. Here, one finds eigenvalues $\beta = \beta_0 + i\beta_i$, for chosen combinations of $Re_{\delta*}$, k_{real} and buoyancy parameter K_a or K_i. In Fig. 3.18c, d, the neutral curves ($\beta_i = 0$) obtained from temporal analysis are shown in ($Re_{\delta*}$, k_{real})-plane for adiabatic cold plate (frame (c)) and isothermal heated wedge (frame (d)). From Fig. 3.18c one notes that $(Re_{\delta*})_{cr}$ decreases with increased cooling, similar to corresponding results of spatial viscous instability analysis shown in Fig. 3.18a. For isothermal heated wedge cases shown in Fig. 3.18d, temporal analysis predicts $(Re_{\delta*})_{cr}$ to increase with increased heating, as is concluded from corresponding plots shown in Fig. 3.18b. One also notes from different frames of Fig. 3.18 that for a chosen buoyancy parameter K_a or K_i, $(Re_{\delta*})_{cr}$ obtained from both spatial and temporal instability analysis are identical.

Temporal viscous growth rate is provided in Fig. 3.19, with $c_i (= \beta_i / k_{real})$-contours plotted in ($Re_{\delta*}$, k_{real})-plane for the indicated cases. This parameter is used

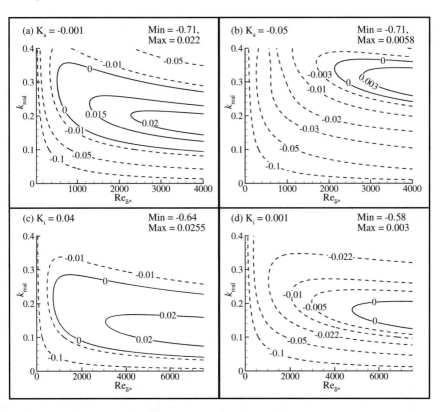

Fig. 3.19 Temporal growth rate contours $c_i = \beta_i/k_{real}$ plotted in ($Re_{\delta*}$, k_{real})-plane for indicated mixed convection cases. [Reproduced from *Direct numerical simulation of transitional mixed convection flows: Viscous and inviscid instability mechanism.* Sengupta et al., *Physics of Fluids*, **25**, 094102 (2013), with the permission of AIP Publishing.]

for inviscid temporal instability studies. The maximum and minimum values of c_i are indicated in each frame. One notes that the adiabatic cold plate case with $K_a = -0.05$ displays lesser temporal growth rate than $K_a = -0.001$ case, i.e., cooling decreases instability. For the isothermal heated wedge, one notes that heating enhances temporal growth rate.

Thus the viscous temporal analysis for mixed convection flows over both adiabatic cold plate and isothermal heated wedge show identical trend, as one notes for spatial analysis. Both the viscous linear theories do not explain the instabilities noted in Figs. 3.16 and 3.17 for DNS, with enhanced instability noted for heating and cooling increased.

The mixed convection flow instabilities are not explained well by local analysis. The local linear viscous theory has several drawbacks. The spatial theory posed as signal problem creates output at the frequency of excitation, which has been shown to be incorrect in [55, 65]. Linear theory based on parallel flow approximation is not accurate near the leading edge of the plate. Results in Fig. 3.16, clearly show onset of disturbance near the proximity of the leading edge, as noted earlier for vortex-induced instability [60]. These linear analyses viewing instability either as a spatial or a temporal growth problem has been shown as inadequate for Blasius boundary layer [55, 65]. To incorporate the streamwise variation of the flow, one can perform either global linear studies [5, 11] or use fully nonlinear formulation [18, 39, 55, 65]. The full simulations provide results including nonlinear effects, and help in relating the results with the local analysis for some cases. Results in [66] has shown that there are many unexplained facets of local analysis for mixed convection flows. Ambiguities of linear theories have been clearly seen with respect to results in Figs. 3.16 and 3.17, even in not being able to compute the equilibrium flow. Another aspect which has been completely overlooked is the inviscid instability mechanisms in mixed convection flows, as compared to such well developed analysis for flows without heat transfer. These suggest that a spatio-temporal receptivity study of a mixed convection boundary layer to a deterministic excitation is essential to fully understand various instability mechanisms based on solving NSE and few representative cases are explored next.

3.8   Receptivity of Mixed Convection Flows

From Figs. 3.16 and 3.17, it is noted that in some mixed convection cases it is not even possible to obtain equilibrium flows by solving the NSE, for those special cases for which one obtains a steady *similarity solutions* [42, 49]. DNS produces steady equilibrium flows, only when heat transfer is lower. In this section, we consider few cases for which mean flow is obtained by solving NSE and their receptivity to wall excitation is followed. These correspond to the cases of (i) adiabatic cold plate with $K_a = -0.01$ and -0.001 and (ii) the isothermal hot wedge with $K_i = 0.025$. Receptivity of such cases are studied for wall excitation by simultaneous blowing and suction (SBS) strip. Such vortical excitation has been applied for flows without heat transfer in [9, 22, 55]. The SBS strip provides monochromatic harmonic excitation, and a brief description of the SBS strip is provided. The surface temperature distribution is kept unaltered at the exciter, while the expressions for wall stream function and wall vorticity are changed as,

$$\psi_w = \bar{\psi}_o + \psi_{wp} \tag{3.97}$$

$$\omega_w = -\frac{1}{h_1 h_2} \frac{\partial}{\partial \xi} \left(\frac{h_2}{h_1} \frac{\partial \psi}{\partial \xi} \right) - \frac{1}{h_2^2} \frac{\partial^2 \psi}{\partial \eta^2} \tag{3.98}$$

Expression for wall vorticity contains additional ξ derivatives, because $\frac{\partial v}{\partial x} \neq 0$ at the wall, when exciter is switched on. Wall perturbation stream function ψ_{wp} in Eq. (3.97) is derived from the imposed wall velocity condition given by,

$$u_d = 0, \qquad v_d = A(x) \sin(\beta_L t) \tag{3.99}$$

where β_L is the non-dimensional exciter frequency based on the reference length L. The amplitude of exciter is obtain as $A(x) = \alpha_1 A_m(x)$, where α_1 is an amplitude control parameter. For an exciter located between x_1 and x_2, the amplitude distribution for the exciter is given by $A_m(x)$,

For $x_1 \leq x \leq x_{st}$

$$A_m = 15.1875 \left(\frac{x - x_1}{x_{st} - x_1} \right)^5 - 35.4375 \left(\frac{x - x_1}{x_{st} - x_1} \right)^4 + 20.25 \left(\frac{x - x_1}{x_{st} - x1} \right)^3 \tag{3.100}$$

and for $x_{st} \leq x \leq x_2$

$$A_m = -15.1875 \left(\frac{x_2 - x}{x_2 - x_{st}} \right)^5 + 35.4375 \left(\frac{x_2 - x}{x_2 - x_{st}} \right)^4 - 20.25 \left(\frac{x_2 - x}{x_2 - x_{st}} \right)^3 \tag{3.101}$$

where $x_{st} = \frac{x_1 + x_2}{2}$.

3.8.1 Receptivity of Cold Adiabatic Flat Plate Cases

A specific excitation applied to a cold adiabatic flat plate has been studied in [66], with the frequency of excitation considered is $F = 9.76 \times 10^{-5}$, and with an amplitude control parameter of $\alpha_1 = 0.001$. Thus, the maximum excitation amplitude is 0.1% of the free stream speed. The cases studied are for $K_a = -0.001$ and -0.01 with the exciter at $x = 1.5$ (where $Re_{\delta*} = 670$ for $K_a = -0.01$), and we noted in Fig. 3.18a, P_1 corresponds to $Re_{\delta*} = 744$, while P_2 is at $Re_{\delta*} = 1256$. It is evident from Fig. 3.18a that the onset of growth for the lower cooling case occurs slightly early, while the point at which the ray exits the neutral curve is almost the same. In Fig. 3.20a, results for $K_a = -0.01$ and -0.001 cases are shown side by side for the disturbance streamwise velocity (u_d), at $y = 0.0061$, which demonstrate same elements of the disturbance field as reported in [55, 65], for flows without heat transfer. The equilibrium flow is obtained before the excitation is switched on at $t = 0$ and this is subtracted from the instantaneous solution for the excited flow to obtain the disturbance field.

Results are shown for $K_a = -0.01$ on the left column of Fig. 3.20a, showing a nascent STWF visible at $t = 85$, ahead of the TS wave-packet visible at the left of the panel. Features of the disturbance field are: (i) Fixed streamwise extent of the TS wave-packet at all times and (ii) the growing STWF that convects downstream. The STWF has been noted in [62] from the solution of OSE, implying its origin by linear mechanism, is also noted here in the time evolution of u_d up to $t = 190$.

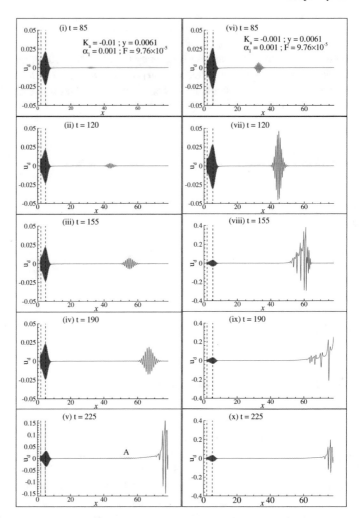

Fig. 3.20 **a** Disturbance streamwise velocity u_d at a height of $y = 0.0061$ plotted at indicated times for flow past a horizontal adiabatic cold plate with $K_a = -0.001$ (frames $(i - v)$) and -0.01 (frames $(vi - x)$) is excited by SBS strip placed at $x = 1.5$. Here, non-dimensional frequency and amplitude of excitation are $F = 9.76 \times 10^{-5}$ and $\alpha_1 = 0.001$, respectively. The vertical dashed lines in the frames indicate onset and decay of disturbance locations (as P_1 and P_2 in Fig. 3.18a) as per local linear spatial instability theories. **b** u_d is plotted at a height of $y = 0.0061$ for the indicated times for flow past an isothermal heated wedge with $K_i = 0.025$ excited by SBS strip placed at $x = 21.0$. Here, $F = 1.65 \times 10^{-5}$ and $\alpha_1 = 0.001$, are used and the vertical dashed lines in the frames indicate onset and decay of disturbance locations (as Q_1 and Q_2 in Fig. 3.18b) obtained by linear spatial instability theory. Reproduced from [Direct numerical simulation of transitional mixed convection flows: Viscous and inviscid instability mechanism. Sengupta et al., Physics of Fluids, **25**, 094102 (2013)], with the permission of AIP Publishing

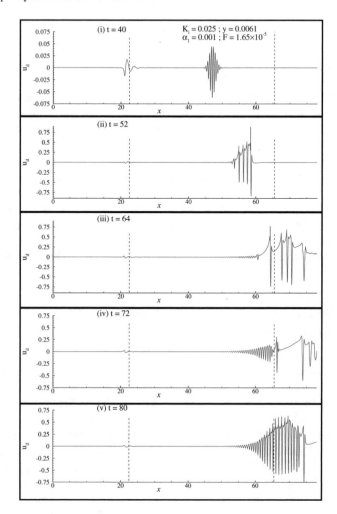

Fig. 3.20 (continued)

STWF experiences nonlinear growth, as noted in the frame at $t = 225$. This also shows the creation of an upstream disturbance packet, marked as A. In [55, 65], dynamics and self-regeneration mechanism of STWF have been described for flows without heat transfer. The lower cooling case of $K_a = -0.001$ shown in the right frames of Fig. 3.20a show that the instability is stronger and occurs earlier, implying that the cooling stabilizes viscous modes, as noted in the last section, as the results for $K_a = -0.001$ show earlier onset of the STWF and larger amplitude in the top frames. With higher growth, nonlinearity also sets in earlier, as noted at $t = 155$. The y-axis is magnified for the lower K_a case, to indicate its higher growth rate.

3.8.2 Receptivity of Hot Isothermal Wedge Case

Another case of receptivity is studied for hot isothermal wedge case, with the SBS excitation frequency of $F = 1.65 \times 10^{-5}$ and $\alpha_1 = 0.001$. In Fig. 3.18b, this case has been marked by the ray OB, with the exciter located at a position where $Re_{\delta^*} = 2190$, while point of entry inside the neutral curve at Q_1 corresponds to $Re_{\delta^*} = 2282$ and point of exit at Q_2 correspond to $Re_{\delta^*} = 4333$. This is evidently at lower frequency for the isothermal heated wedge, and the flow remains stable up to a longer streamwise stretch. In Fig. 3.20b, one notes near-absence of TS wave-packet, while the STWF is visible at an earlier time. Also, the growth rate for the STWF is higher and therefore the onset of nonlinear distortion of the STWF occurs much earlier. Additionally, one notices the presence of very strong secondary upstream STWF created by the primary STWF leaving the computational domain. The low frequency disturbances have stronger instability, and such response fields are qualitatively different from moderate to higher frequency excitation cases for flows without heat transfer [7, 65].

3.8.3 DNS of Instability of Mixed Convection Flows: New Theorems of Instability

So far we have noted contradictory observations for mixed convection flow instabilities, both for heating and cooling at the wall. For example, adiabatic flow past a horizontal plate indicate enhanced stability with cooling in Fig. 3.18a, c. However, when the cooling is more intense, we are even unable to compute the equilibrium flow, e.g., for $K_a = -0.05$. But the lower cooling cases allow one to calculate equilibrium flow and study its receptivity, and the observations from linear viscous instability studies are compatible with DNS of the NSE. Same observation holds good for computing mean flow and study receptivity of lesser heated isothermal wedges. For the increased heating of $K_i = 0.04$ case, the mean flow is not obtained by DNS. All these evidences suggest that there can be additional instability mechanisms for mixed convection flows, apart from that is obtained by solving the OSE by spatial or temporal theories. From DNS shown in Figs. 3.16 and 3.17, we note in the course of failed attempt in computing an equilibrium flow, disturbances appear *naturally* outside the boundary layer, indicating temporal growth without significant convection. With passage of time, disturbances spread over larger streamwise stretch, without penetrating inside the boundary layer. Such growths are reminiscent of inviscid instability mechanism arising from Rayleigh's stability equation and the theorems stating necessary conditions provided by Rayleigh and Fjørtoft [19, 53] for flows without heat transfer. The corresponding theorems for inviscid temporal instabilities for mixed convection flow have been reported for the first time in [66], and is explained in the following.

   The inviscid governing equations for disturbances in mixed convection flows modeled by Boussinesq approximation are given in non-dimensional form as

$$\frac{\partial \tilde{u}}{\partial \tilde{x}} + \frac{\partial \tilde{v}}{\partial \tilde{y}} = 0, \tag{3.102}$$

$$\frac{\partial \tilde{u}}{\partial \tilde{t}} + \tilde{u}\frac{\partial \tilde{u}}{\partial \tilde{x}} + \tilde{v}\frac{\partial \tilde{u}}{\partial \tilde{y}} = -\frac{\partial \tilde{p}}{\partial \tilde{x}}, \tag{3.103}$$

$$\frac{\partial \tilde{v}}{\partial \tilde{t}} + \tilde{u}\frac{\partial \tilde{v}}{\partial \tilde{x}} + \tilde{v}\frac{\partial \tilde{v}}{\partial \tilde{y}} = Ri\ \tilde{\theta} - \frac{\partial \tilde{p}}{\partial \tilde{y}}, \tag{3.104}$$

$$\frac{\partial \tilde{\theta}}{\partial \tilde{t}} + \tilde{u}\frac{\partial \tilde{\theta}}{\partial \tilde{x}} + \tilde{v}\frac{\partial \tilde{\theta}}{\partial \tilde{y}} = 0. \tag{3.105}$$

Scales for length, velocity, pressure and temperature used to non–dimensionalize these equations are the local displacement thickness (δ^*), shear layer edge velocity U_e, ρU_e^2 and $\Delta T(x^*) = (T_w(x^*) - T_\infty)$, respectively. The Richardson number $Ri = \frac{\tilde{Gr}}{\tilde{Re}^2}$ is retained in this linearized analysis for the inviscid mechanism, as the buoyancy effect is not a lower order perturbation. This is due to the fact that the boundary layer grows as $O(Re^{-1/2})$ and to have equal effects of free and forced convection in mixed convection flows, we must have $K_i = Ri/Re^{1/2} \approx O(1)$, which necessitates retaining the buoyancy effects noted in wall-normal direction.

Once again, parallel flow approximation for the equilibrium flow is used, and the perturbation quantities are expressed by Fourier-Laplace transform given in Eq. (3.79). With these, Eqs. (3.102)–(3.105) are simplified to get a single equation in terms of ϕ (Fourier-Laplace transform of $\tilde{v}$) as

$$\frac{d^2\phi}{d\tilde{y}^2}(U-c) + \phi\left[k^2(c-U) + \frac{Ri\frac{dT}{d\tilde{y}}}{(U-c)} - \frac{d^2U}{d\tilde{y}^2}\right] = 0 \tag{3.106}$$

where, $c = \beta/k$ is the complex phase speed for temporal instability. This is the equivalent equation to study inviscid temporal instability of mixed convection flows. For such studies, one takes k as real, ($k = k_{real}$) and β as a complex quantity. Hence, $c = c_r + i\ c_i$, with c_r and c_i denoting the phase speed and temporal growth rate (if $c_i > 0$) of the disturbance field.

Rayleigh's and Fjørtoft's theorems for flows without heat transfer are stated without solving this equation explicitly, to state necessary conditions. We will obtain robust conditions for temporal instability, by multiplying Eq. (3.106) with ϕ^* (complex conjugate of ϕ) and integrating over all possible limit $(0, \infty)$ to get

$$\int_0^\infty \left[\phi^*\frac{d^2\phi}{d\tilde{y}^2} + |\phi|^2\left(-k^2 + \frac{Ri\frac{dT}{d\tilde{y}}}{(U-c)^2} - \frac{\frac{d^2U}{d\tilde{y}^2}}{(U-c)}\right)\right]d\tilde{y} = 0 \tag{3.107}$$

This equation is simplified by using integration by parts and using boundary conditions: $\phi(0) = 0$ and $\phi \to 0$, as $\tilde{y} \to \infty$ to get

$$\int \left[\left| \frac{d\phi}{dy} \right|^2 + k^2 |\phi|^2 \right] d\tilde{y} + \int \frac{|\phi|^2}{|U - c|^2} \left[\frac{d^2 U}{d\tilde{y}^2} (U - c^*) - \frac{Ri \frac{dT}{d\tilde{y}}}{|U - c|^2} (U - c^*)^2 \right] d\tilde{y} = 0$$

(3.108)

where c^* is the complex conjugate of c, i.e., $c^* = c_r - ic_i$. The imaginary part of Eq. (3.108) is given as

$$\int \frac{c_i |\phi|^2}{|U - c|^2} \left[\frac{-2Ri \frac{dT}{d\tilde{y}} (U - c_r)}{|U - c|^2} + \frac{d^2 U}{d\tilde{y}^2} \right] dy = 0$$

(3.109)

For temporal instability, one must have $c_i \neq 0$, and hence to have the integral to be zero in Eq. (3.109), one must require the term within the third brackets to change sign, somewhere in the flow domain. We define this as

$$\Phi = \left[\frac{-2Ri \frac{dT}{d\tilde{y}} (U - c_r)}{|U - c|^2} + \frac{d^2 U}{d\tilde{y}^2} \right]$$

(3.110)

For flows without heat transfer ($Ri = 0$), one gets back the well known Rayleigh's inflection point theorem, with $\Phi \equiv \frac{d^2 U}{d\tilde{y}^2}$. Thus, we state an extension of the Rayleigh's theorem for inviscid temporal instability by [66]:

Mixed Convection Flow Theorem I - The necessary condition for inviscid instability of mixed convection flows described by the velocity and temperature profiles, $U(\tilde{y})$ and $T(\tilde{y})$, the integrand Φ as defined in Eq. (3.110) should be zero somewhere in the domain.

The consequence of Mixed Convection Flow Theorem I follows as: (i) The velocity and temperature fields have all viscous information of the parallel flow and can be used without the need for any similarity profile, and hence DNS data can also be used with a local parallel flow approximation to study the resultant inviscid instability; (ii) Unlike flow cases without heat transfer, existence of an inflection point of velocity profile alone does not define this temporal instability.

Existence of a point, where Φ changes sign depends on the complex phase speed c, making the analysis more involved and can be performed parametrically. Noting that Φ remains bounded for non-neutral cases, there is a critical layer for mixed convection problem also for neutral disturbances where $U(\tilde{y}_{cr}) = c_r$.

Mixed Convection Flow Theorem I is obtained here from the imaginary part of Eq. (3.108) and is identical to the procedure used in Rayleigh's theorem. It was subsequently modified by Fjørtoft, who considered the real part of Eq. (3.108) also in deriving another necessary condition.

Similarly, we obtain another necessary condition for mixed convection flows, along the procedure used by Fjørtoft for flows without heat transfer. This necessary condition uses the real part of Eq. (3.108) for temporal instability of the mixed convection flows. The real part of Eq. (3.108) is given by,

$$\int \frac{|\phi|^2}{|U-c|^2} \left[\frac{-Ri \frac{dT}{d\tilde{y}} \{(U-c_r)^2 - c_i^2\}}{|U-c|^2} + \frac{d^2U}{d\tilde{y}^2}(U-c_r) \right] d\tilde{y}$$

$$+ \int \left[\left| \frac{d\phi}{d\tilde{y}} \right|^2 + k^2|\phi|^2 \right] d\tilde{y} = 0 \quad (3.111)$$

In this equation, the second integral is always positive, so the integrand of the first integral must be negative in some interval within the flow to have the net integral to vanish. Let us indicate a height $\tilde{y} = \tilde{y}_s$, where $\Phi(\tilde{y}_s) = 0$ and denote $U(\tilde{y}_s) = U_s$. Multiplying Eq. (3.109) by $(c_r - U_s)$ and adding it with the first integral of Eq. (3.111) and simplifying, we get

$$\int |\phi|^2 \left[\frac{Ri \frac{dT}{d\tilde{y}}}{|U-c|^2} \{(U_s - c_r)^2 + c_i^2 - (U - U_s)^2\} + \frac{d^2U}{d\tilde{y}^2}(U - U_s) \right] d\tilde{y} < 0$$

$$(3.112)$$

We define the integrand within the third bracket as

$$\Gamma_F = \left[\frac{Ri \frac{dT}{d\tilde{y}}}{|U-c|^2} \{(U_s - c_r)^2 + c_i^2 - (U - U_s)^2\} + \frac{d^2U}{d\tilde{y}^2}(U - U_s) \right] \quad (3.113)$$

Defining Γ_F as the *Fjørtoft's* integrand, we proceed and state the following theorem for inviscid temporal instability of mixed convection flow as [66]:

Mixed Convection Flow Theorem II - The necessary condition for the inviscid instability for a mixed convection parallel flow described by velocity and temperature profiles $U(\tilde{y})$ and $T(\tilde{y})$ is that the integrand Γ_F (in Eq. (3.113)) should be negative somewhere in the interior of the domain.

This is corresponding to Fjørtoft's theorem, and is valid for inviscid temporal instability of mixed convection flows. As noted for Φ, Γ_F is also a function of c, which means that one can inspect parametrically the likelihood of such inviscid instabilities by tracking different temporal growth rates (c_i) and associated time/length scales (c_r).

The utilities of Mixed Convection Flow Theorems I and II, is to help explain why in some cases of heat transfer, the equilibrium flow given by [42, 49] cannot be realized from the time-accurate solution of NSE. We are specifically interested for the cases in (i) Figs. 3.15a and 3.16 for the cold adiabatic plate case of $K_a = -0.05$ and (ii) Figs. 3.15b and 3.17 for the hot isothermal wedge flow case of $K_i = 0.04$. The inviscid instability will be predicted by Mixed Convection Flow Theorems I and II, when $\Phi = 0$ and $\Gamma_F < 0$ conditions are satisfied inside the flow field.

We have plotted Φ as a function of $\tilde{y}$ in Fig. 3.21a, for flow over an adiabatic cold plate, with $K_a = -0.05$. The similarity profiles obtained as functions of the similarity variable η have been converted to functions of $\tilde{y}$ as $\tilde{y} = \eta/C_\delta$, where $C_\delta = \int_0^\infty (1 - f') d\eta$. The variation of Φ with η are shown in frame (i) for the fixed value of $c_i = 0.01$. The three curves correspond to three different values of

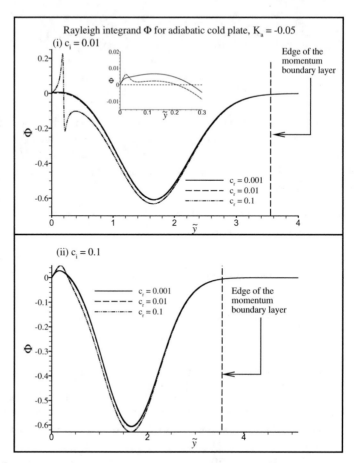

Fig. 3.21 a Rayleigh integrand Φ for flow over adiabatic plate parameter $K_a = -0.05$ for
i $c_i = 0.01$ and ii $c_i = 0.1$. **b** Fjørtoft integrand Γ_F for flow over adiabatic plate with buoyancy
parameter $K_a = -0.05$ for i $c_i = 0.01$ and ii $c_i = 0.1$. Reproduced from [Direct numerical simulation of transitional mixed convection flows: Viscous and inviscid instability mechanism. Sengupta
et al., Physics of Fluids, **25**, 094102 (2013)], with the permission of AIP Publishing

$c_r = 0.001, 0.01$ and 0.1. One notes for the lower values of c_r, Φ crosses zero near
the wall, as shown in the inset of this frame. For the highest frequency of $c_r = 0.1$,
Φ flips sign at one value of η discontinuously. In the next frame (ii) of Fig. 3.21a,
similar curves are plotted for the higher growth rate of $c_i = 0.1$. The zero-crossing
is noted for all the three values of c_r shown in this frame. These imply that for all
the three time scales considered (varying by a factor of 100), are unstable for both
$c_i = 0.01$ and 0.1. Such temporal growth is seen in the outer part of the shear layer,
near the leading edge of the plate in Fig. 3.16. This is a plausible explanation for

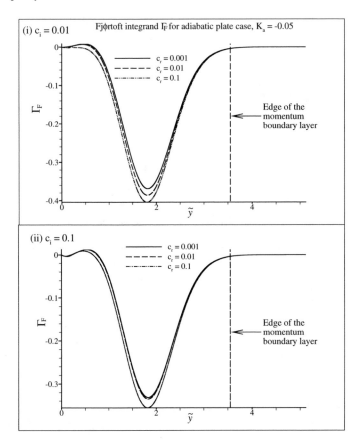

Fig. 3.21 (continued)

failing to obtain a mean flow for $K_a = -0.05$, as due to inviscid temporal instability for very high cooling rate, which however was shown to be stable for any cooling by linear viscous instability theories.

Similarly, Γ_F has been plotted as a function of $\tilde{y}$ in Fig. 3.21b, for $K_a = -0.05$. The necessary condition for inviscid instability according to Mixed Convection Flow Theorem II is for Γ_F becoming negative in the flow domain. The frames in Fig. 3.21b are plotted again for $c_i = 0.01$ and 0.1; with both the cases showing inviscid temporal instability criterion to be satisfied. This satisfaction of both the necessary conditions of Theorems I and II, clearly explain why an equilibrium flow is not computed for this case. It is also equally important to note that when the flow is both temporally and spatially unstable, the stronger one will dominate, as in the present case the spatial instability is weaker while the temporal instability is stronger and is readily evident while performing DNS of this case, with results shown in Fig. 3.16.

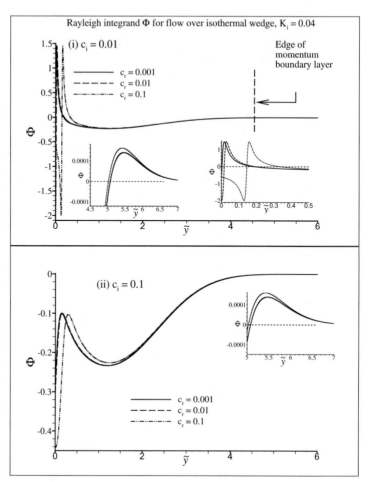

Fig. 3.22 **a** Rayleigh integrand Φ for flow over isothermal wedge with buoyancy parameter $K_i = 0.04$ for i $c_i = 0.01$ and ii $c_i = 0.1$. **b** Fjørtoft integrand Γ_F for flow over isothermal wedge with buoyancy parameter $K_i = 0.04$ for i $c_i = 0.01$ and ii $c_i = 0.1$. [Reproduced from *Direct numerical simulation of transitional mixed convection flows: Viscous and inviscid instability mechanism.* Sengupta et al., *Physics of Fluids*, **25**, 094102 (2013), with the permission of AIP Publishing.]

Rayleigh integrand, Φ has been plotted for flow over an isothermal heated wedge with $K_i = 0.04$ in Fig. 3.22a for $c_i = 0.01$ and 0.1 corresponding to $c_r = 0.001$; 0.01; 0.1. Existence of multiple zero-crossing points, where $\Phi = 0$, can be seen in frame (i) of Fig. 3.22a. One notes from the figure that as c_r decreases, the first zero-crossing point moves towards the wall. But for all the values of c_r, there also exists a zero-crossing point around $\tilde{y} \simeq 5.2$, which is closer to the edge of the momentum boundary layer (approximately at $\tilde{y} = 4.54$). Thus, the inner zero-crossing point indicates temporal instability inside the shear layer, and affects the boundary layer

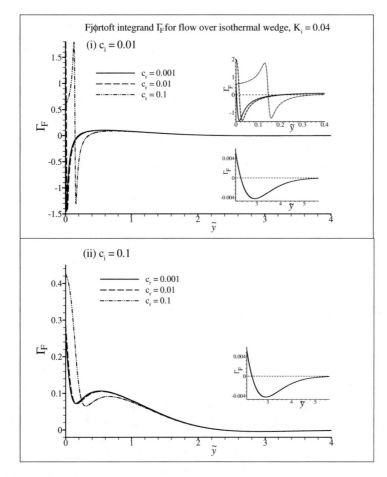

Fig. 3.22 (continued)

temporally. The outer zero-crossing point is in the inviscid part of the flow and indicates an inviscid mechanism of temporal instability. This indication of inviscid instability is clearly seen in the inset of frame (i) of Fig. 3.22a. For the higher value of $c_i = 0.1$ in frame (ii) of Fig. 3.22a, only an outer zero-crossing point exists around $\tilde{y} \simeq 5.5$. This indicates stronger temporal instability of mixed convection flows outside the shear layer for all frequencies corresponding to $c_r < 1$.

   Fjϕrtoft's integrand, Γ_F of Eq. (3.113) is plotted in Fig. 3.22b, as a function of $\tilde{y}$ for $K_i = 0.04$ (heated isothermal wedge). In calculating Γ_F, the mean speed U_s corresponding to the outer zero-crossing point of Fig. 3.22b is used. For $c_r = 0.01$, we see $\Gamma_F < 0$ in two ranges, one closer to the wall, while the other is near the edge of the boundary layer. For higher value of $c_r = 0.1$, $\Gamma_F < 0$ occurs slightly away from the wall, as compared to the lower values of c_r. Also, $\Gamma_F < 0$ at similar wide range of $\tilde{y}$, for all values of c_r shown.

In Fig. 3.21a, only one zero-crossing point exists for the three c_r's shown in the figures for both the growth rates, unlike the case of isothermal heated wedge. So for the adiabatic cold plate case, the temporal instabilities occur within the shear layer by the inviscid mechanism and is dominated by viscous action by the equilibrium velocity and temperature profiles obtained by the NSE. One notes that $\Gamma_F < 0$ for an extended range of $\tilde{y}$ within the shear layer, for all combinations of c_r and c_i in Fig. 3.21b. The necessary condition of $\Gamma_F < 0$ predicts temporally unstable disturbances to occur inside the boundary layer. The condition of $\Gamma_F < 0$ is expected to be stronger, than the condition given by $\Phi = 0$, as the condition with Γ_F provides a necessary condition by using both the real and imaginary parts of Eq. (3.108).

For a combination of K_i / K_a and c_i, there exists a critical value $(c_r)_{crit}$, such that for $c_r > (c_r)_{crit}$, no zero-crossing point of Φ exists in the flow. For a fixed frequency excitation, c_r is directly proportional to length scale. So critical c_r represents a critical length scale of disturbances $l_{crit} = \frac{2\pi (c_r)_{crit}}{\beta}$, for a given β. Scales larger than l_{crit}, do not suffer temporal instability. There exists positive c_i for disturbances with $c_r < (c_r)_{crit}$, which satisfy the necessary condition for temporal instability.

The two inviscid instability theorems help explain the instability in Figs. 3.16 and 3.17, when full NSE is solved. The instability in Fig. 3.16 for the adiabatic cold plate, displays high wavenumber/ frequency instability at the edge of the shear layer originating from the leading edge. Additionally, relatively lower wavenumber/ frequency disturbances inside the boundary layer originates at the leading edge. These are apparent from Fig. 3.21a, b, which show the occurrence of $\Phi = 0$ (according to Theorem I) closer to the wall for higher c_r. As $c_r = \beta/k$, this implies lower k component disturbances near the wall, as seen in Fig. 3.16. Similarly, from Fig. 3.21b, one notes $\Gamma_F < 0$ occurring for a wide range of heights, with maximum negative value occurring for $\tilde{y} \simeq 2$, implying maximum inviscid instability at the edge of the shear layer for all wavenumber components, as evident in Fig. 3.16. Comparing these, one justifies the noted instability, and the length/ time scales for the case of isothermal heated wedge, which occurs predominantly in the outer edge of the shear layer. In Fig. 3.17, we have also drawn a dotted line above which Γ_F is negative for high wavenumbers, as noted in Fig. 3.22b.

In Fig. 3.23, $(c_r)_{crit}$ is plotted in (K_i, c_i)- and (K_a, c_i)-planes in frames (a) and (b), respectively. One distinguishes instabilities indicated in frames of Fig. 3.23. For the flow over isothermal wedge, the line AB represents the $(c_r)_{crit}$ values above which all c_r, including the critical value are temporally unstable. Thus the region below AB indicates instability for all c_r less than equal to 1, implying that for higher values of K_i, one gets higher growth rates for larger ranges of length scales. Similarly, for the region above CD, $(c_r)_{crit} = 0.0001$ everywhere, which experiences instability only for very small length scales, but with larger growth rates. Similar conclusions are obtained for adiabatic cold plate shown in Fig. 3.23b, indicates enhanced temporal instability with cooling rate increased.

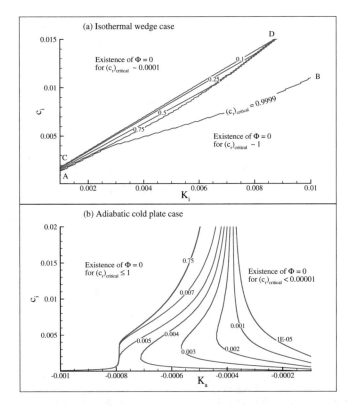

Fig. 3.23 Critical value $(c_r)_{crit}$ plotted for cases of **a** flow over isothermal wedge in (K_i, c_i)-plane and **b** for flow over adiabatic cold plate in (K_a, c_i)-plane. Here, $(c_r)_{crit}$ represents the critical phase speed such that for all $c_r > (c_r)_{crit}$, no zero-crossing point of Φ exists in the flow. [Reproduced from *Direct numerical simulation of transitional mixed convection flows: Viscous and inviscid instability mechanism.* Sengupta et al., *Physics of Fluids*, **25**, 094102 (2013), with the permission of AIP Publishing.]

The necessary condition given by $\int |\phi|^2 \Gamma_F d\tilde{y} < 0$ is a consequence of Theorem II, and is investigated for the cold adiabatic plate. As $|\phi|^2$ is strictly non-negative and less than one, this condition is reduced to a more conservative inequality given by $I_\Gamma = \int \Gamma_F d\tilde{y} \le 0$. In Fig. 3.24, the contours of I_Γ in (c_r, c_i)-plane for the indicated values of K_a are shown. For $K_a = -0.05$, temporal instability is indicated by this criterion for $c_r \le 0.892$. Also for a fixed value of c_r, all c_i's are indicated as possible in the plotted range. Strength of the instability is indicated by the minimum value in the domain of interest. It is noted in all the frames, that the maximum instability is obtained for $c_r \to 0$, i.e., for vanishing length scales. Such instabilities have been noted in Fig. 3.16, for very small length scales near the outer edge, closer to the leading edge. This criterion depends on the value of U_s corresponding to Theorem I,

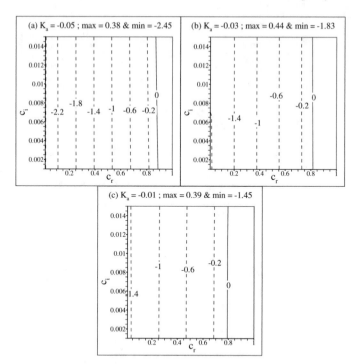

Fig. 3.24 Contours of $I_\Gamma = \int \Gamma_F d\tilde{y}$ plotted in (c_r, c_i)-plane for flow past a adiabatic cold plate with indicated values of buoyancy parameter K_a. [Reproduced from *Direct numerical simulation of transitional mixed convection flows: Viscous and inviscid instability mechanism*. Sengupta et al., *Physics of Fluids*, **25**, 094102 (2013), with the permission of AIP Publishing.]

and is important to use the actual velocity profiles obtained from DNS. The snapshots in Fig. 3.16 indicate disturbances in the outer layer, which spread in the streamwise direction. Nonetheless, Theorems I and II are vital aids in explaining instabilities that prevent one from even obtaining an equilibrium flow.

The role of inviscid instability mechanism in destabilizing the flow is emphasized, by solving Eq. (3.106) for flows past adiabatic cold plate with ($K_a = -0.001, -0.05$) and isothermal heated wedge cases with ($K_i = 0.001, 0.05$). Equation (3.106) is integrated using fourth-order Runge-Kutta method from $\tilde{y}_{max} = 12$ to wall, for a particular choice of k_{real}. Integration of Eq. (3.106) is performed by specifying $\phi_\infty = e^{-k\tilde{y}}$ and $\phi'_\infty = -ke^{-k\tilde{y}}$ at $\tilde{y} = \tilde{y}_{max}$. Direct integration of Eq. (3.106) is possible, as it does not have any stiffness problem that is inherent with the OSE. While integrating Eq. (3.106), $c_i = 0$ is avoided, i.e., neutral disturbances, as it would give rise to solution blow-up near the critical layer. Here, a legitimate eigenvalue would correspond to satisfaction of homogeneous boundary condition at the wall, i.e., when $\phi_{real} + i\phi_{img} = 0$ at $\tilde{y} = 0$. Thus in Fig. 3.25, the contours of $\phi_{real} = 0$ (solid lines) and $\phi_{img} = 0$ (broken lines) are plotted at the wall, in (c_r, c_i)-plane, for adiabatic cold plate cases with indicated buoyancy parameters. Only two values are chosen:

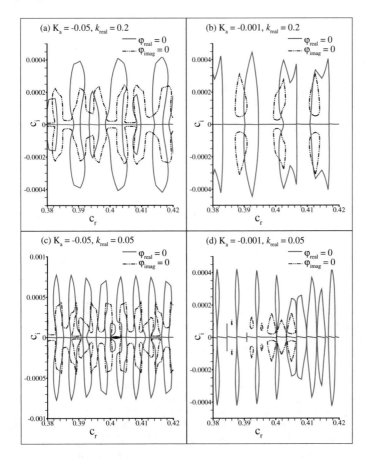

Fig. 3.25 Contours of $\phi_{real} = 0$ (solid lines) and $\phi_{img} = 0$ (broken lines) plotted at the wall in (c_r, c_i)-plane for flows past adiabatic cold plate with indicated buoyancy parameter K_a and disturbance wavenumber k_{real}. [Reproduced from *Direct numerical simulation of transitional mixed convection flows: Viscous and inviscid instability mechanism.* Sengupta et al., *Physics of Fluids*, **25**, 094102 (2013), with the permission of AIP Publishing.]

$k_{real} = 0.2$ (frames (a) and (b)) and $k_{real} = 0.05$ (frames (c) and (d)). In this figure, plotted range are only for $0.38 \le c_r \le 0.42$, for ease of understanding. Intersection of these two curves indicate an eigenvalue corresponding to inviscid instability mechanism. One notes from Fig. 3.25 that as one reduces the cooling rate, the number of inviscid eigenvalues also reduces. The symmetry of the contours in Fig. 3.25, is due to the fact that, if $c = c_r + ic_i$ is an eigenvalue of Eq. (3.106) with eigenfunction ϕ, $c^* = c_r - ic_i$ is also an eigenvalue of Eq. (3.106) with eigenfunction ϕ^*, where ϕ^* is complex conjugate of ϕ. First ten most unstable inviscid eigenvalues are listed in Table 3.2. One observes from Table 3.2 that for $K_a = -0.05$, lower wavenumber disturbances are more unstable than higher wavenumber disturbances. However for lower cooling rate of $K_a = -0.001$, the higher wavenumber disturbances exhibit

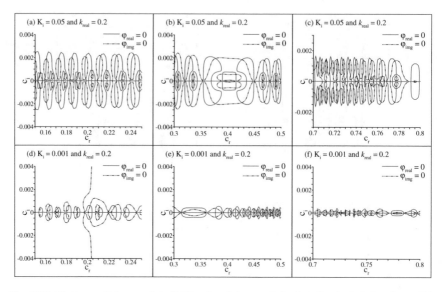

Fig. 3.26 Contours of $\phi_{real} = 0$ (solid lines) and $\phi_{img} = 0$ (broken lines) plotted at the wall in (c_r, c_i)-plane for flows past isothermal hot plate with indicated buoyancy parameter K_i. We have chosen disturbance wavenumber $k_{real} = 0.2$. [Reproduced from *Direct numerical simulation of transitional mixed convection flows: Viscous and inviscid instability mechanism*. Sengupta et al., *Physics of Fluids*, **25**, 094102 (2013), with the permission of AIP Publishing.]

Table 3.2 First ten most dominant unstable modes from inviscid instability analysis tabulated for flows past adiabatic cold plate with $K_a = -0.05$ and $K_a = -0.001$. We have tabulated only for disturbance wavenumber of $k_{real} = 0.2$ and 0.05. [Reproduced from *Direct numerical simulation of transitional mixed convection flows: Viscous and inviscid instability mechanism*. Sengupta et al., *Physics of Fluids*, **25**, 094102 (2013), with the permission of AIP Publishing.]

| | $K_a = -0.05$ | | $K_a = -0.001$ | |
Mode No.	$k_{real} = 0.2$	$k_{real} = 0.05$	$k_{real} = 0.2$	$k_{real} = 0.05$
1	(0.192, 4.19e-4)	(0.332, 4.24e-4)	(0.263, 3.41e-4)	(0.595 1.51e-4) Other modes are not clearly distinguishable
2	(0.223, 3.84e-4)	(0.337, 4.23e-4)	(0.289, 2.96e-4)	
3	(0.246, 3.57e-4)	(0.342, 4.18e-4)	(0.411, 2.56e-4)	
4	(0.265, 3.29e-4)	(0.321, 4.18e-4)	(0.421, 2.52e-4)	
5	(0.284, 2.94e-4)	(0.306, 4.18e-4)	(0.312, 2.49e-4)	
6	(0.459, 2.58e-4)	(0.316, 4.16e-4)	(0.362, 2.31e-4)	
7	(0.489, 2.58e-4)	(0.348, 4.16e-4)	(0.331, 2.23e-4)	
8	(0.505, 2.55e-4)	(0.327, 4.12e-4)	(0.348, 2.23e-4)	
9	(0.526, 2.48e-4)	(0.295, 4.12e-4)	(0.348, 2.23e-4)	
10	(0.445, 2.45e-4)	(0.311, 4.06e-4)	(0.377, 2.16e-4)	

pronounced instability. Figure 3.25 together with Table 3.2, establish unambiguously that cooling destabilizes the flow via the inviscid instability mechanism, which cannot be explained by the OSE, but by Eq. (3.106). In Fig. 3.26, $\phi_{real} = 0$ and $\phi_{img} = 0$ contours are plotted at the wall for flows past isothermal heated wedge with $K_i = 0.05$ and $K_i = 0.001$. As noted in DNS, inviscid instability mechanism also shows that heating destabilizes the flow shown in Fig. 3.26 and Table 3.2. Therefore, for mixed convection flow past a hot plate, destabilization of flow is indicated simultaneously through viscous and inviscid mechanisms, with the latter dominating over the former.

Having established the centrality of DNS for mixed convection flow, in the following we return to hydrodynamic instability and use DNS to highlight those aspects of fluid flow, which depend upon the solution of the full NSE.

3.9 Nonlinear and Nonparallel Effects: Receptivity by DNS

The experimental verification of TS waves in Schubauer and Skramstad [50] heralded renewed flow instability studies. This experiment identified the region where the ZPG boundary layer is unstable for high to moderate frequencies. In Fig. 3.27, the experimental neutral curve is identified by discrete symbols. The right half of theoretically predicted thumb-shaped region (the neutral curve shown by continuous line) in the $(Re_{\delta^*}, \omega_0)$-plane matches well with experiment [48]. As the theoretical analysis is based on parallel flow assumption, it's results require interpretation for disturbance propagation in an actual growing boundary layer. It can be shown that a

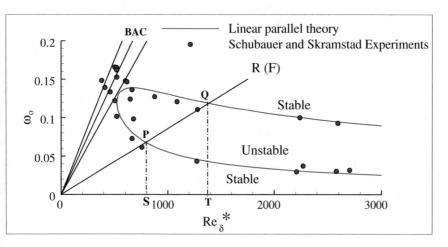

Fig. 3.27 Theoretical neutral curve in $(Re_{\delta^*}, \omega_0)$-plane. According to linear theory with parallel flow assumption, three frequencies shown corresponding to OA, OB and OC are all stable. The discrete symbols are the experimental data from Schubauer and Skramstad [50]

constant physical frequency excitation follows a ray starting from the origin, as shown in Fig. 3.27. In this figure, three rays A, B and C correspond to the nondimensional frequencies of excitation $F = 3.0 \times 10^{-4}$, 3.5×10^{-4} and 2.5×10^{-4}, respectively. It is noted in Fig. 3.27, that the experimental results of [50] reveal discrepancy at (lower Re_{δ^*} - high ω_0) combinations. Experimental critical Re_{δ^*} is significantly lower from the linear instability theory value. In all the experimental cases, the boundary layer was presumably excited at same physical location, although it is not recorded in the paper. According to linear instability theory, the three frequencies corresponding to A, B and C in Fig. 3.27, create disturbances those are stable, while the experimental results indicate a finite length where the TS wave is unstable.

While the experiment qualitatively verified central features of instability theory, there are features in the experiment that instability theory is incapable of explaining. One realizes that the stability theory for a parallel flow model predicts the system response with respect to small perturbations that is only valid at a distance far away from the leading edge of the plate ($x \to \infty$). The created waves (either spatially stable or unstable) are associated with wavenumbers, very close to the origin of the complex wavenumber plane. This was clearly explained in [51], with the help of Abel's and Tauber's theorems given in [74]. These relate the poles (eigenvalues) near the origin of the complex wavenumber plane with the asymptotic behaviour in the physical plane. Similarly, the essential singularity of the Fourier-Laplace transform in the wavenumber plane determines the local solution, in the vicinity of the exciter in the physical plane as explained in Sect. 3.3.2.

In this section, nonparallel and nonlinear features of instability of Blasius boundary layer is shown in detail, obtained with the help of high accuracy DNS. To calculate the receptivity of a ZPG boundary layer, NSE has been solved in [22]. The exciter is placed at a location close to the leading edge of the plate, where nonparallelism of the flow is dominant. SBS exciter used in [22, 58] has amplitudes of the excitation varying from a very small to large values (0.2–10% of U_∞). The SBS exciter causes vortical excitation at the wall. The frequency of excitation is varied from a very stable (according to linear spatial theory) to moderate value, for the purpose of quantifying the nonlinear and nonparallel effects to resolve conflicting claims in the literature about the critical Reynolds number of the Blasius profile [22, 23, 46] with the help of DNS, avoiding restrictive assumptions of signal problem [57] and recording effects of nonparallelism and nonlinearity directly.

In Fig. 3.28, the input spectrum of the SBS strip exciter is shown for $\alpha_1 = 0.05$ and 0.10 for the exciter located between $x_1 = 0.22$ ($Re_{\delta^*} = 255$) and $x_2 = 0.264$ ($Re_{\delta^*} = 284$). The spectrum shows the wide range of wavenumbers excited by the exciter with wide band. The response of the fluid dynamical system depends upon its inherent transfer function.

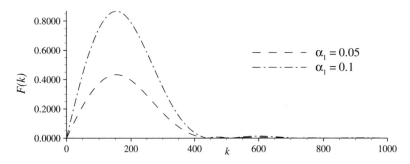

Fig. 3.28 Fourier transform of the input disturbance by the SBS strip given by Eqs. (3.100) and (3.101) for the amplitudes of excitation $\alpha_1 = 0.05$ and 0.1. The exciter is located from $x_1 = 0.22$ to $x_2 = 0.264$

3.9.1 Computational Domain and Grid Resolution

Particular focus here is on non-dimensional frequency of $F = 3.0 \times 10^{-4}$ on the asymptotic solution. Effects of higher frequency excitations (for $F \geq 3.0 \times 10^{-4}$) and various amplitudes of excitation cases are also investigated. Focus of present section is solely on TS wave-packet and not on the STWF, for which one requires a much longer domain. Results here employs a stretched grid in x- and y-directions having 1600 and 600 points, respectively. The schematic diagram of the problem is shown in Fig. 3.13, with $\gamma = 0$. All simulations here include the leading edge of the plate as explained before. The reference velocity and length scales are chosen to be U_∞ and L, respectively. The Reynolds number for all simulations are for $Re_L = \frac{U_\infty L}{\nu} = 10^5$ and the computational domain extends from $-0.05L$ to $10L$ in the streamwise, and 0 to L in the wall-normal directions. For this computational domain, L comes out to be approximately as $60\delta_D^*$, where δ_D^* is the displacement thickness at the outflow of the computational domain (i.e., at $x = 10L$).

Tangent hyperbolic grid stretching is adopted in both streamwise and wall-normal direction, as given by equations following Eq. (3.73), to cluster points near the leading–edge and near the wall of the plate, respectively [21]. The grid mapping is selected in such a way that the wall-resolution is given by: $\Delta y_0 = 2.455 \times 10^{-4}$ and the smallest grid size in the streamwise direction is given by: $\Delta x_0 = 9.218 \times 10^{-4}$.

3.9.2 The Equilibrium Flow

In Fig. 3.29, the similarity functions $f'(\bar{\eta})$, $f''(\bar{\eta})$ and $f'''(\bar{\eta})$ corresponding to the equilibrium flow as obtained by DNS is plotted in frames (a), (b) and (c), respectively, as a function of the similarity variable $\bar{\eta} = y\sqrt{\frac{U_\infty}{\nu x}}$, at indicated streamwise locations. The similarity functions $f'(\bar{\eta})$, $f''(\bar{\eta})$ and $f'''(\bar{\eta})$ obtained from directly integrating

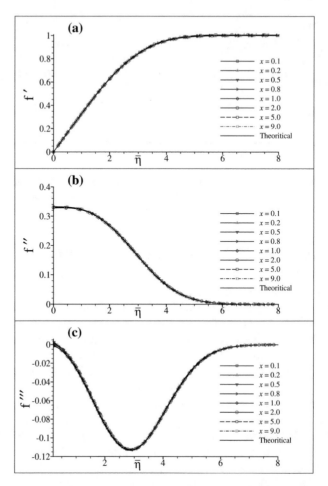

Fig. 3.29 The similarity functions **a** $f'(\bar{\eta})$, **b** $f''(\bar{\eta})$ and **c** $f'''(\bar{\eta})$ to the equilibrium flow obtained by DNS, plotted as a function of the similarity variable $\bar{\eta}$, at indicated streamwise locations. The corresponding solution of the Blasius equation is also shown for comparison

the Blasius equation $f''' + \frac{1}{2}ff'' = 0$, is also plotted in the respective frames for comparison. The functions $f'(\bar{\eta})$ and $f''(\bar{\eta})$ show excellent match for all the values of η and all the streamwise locations shown. The function f''', however, shows slight departure from that predicted by the similarity solution for very small values of η. This discrepancy is an outcome of the shortcomings of the similarity assumption. One should note that for stability calculations, f''' is very important, as it represents the term $\frac{d^2u}{dy^2}$, which appears in OSE given in Eq. (3.14). However, it is noted that the results here are obtained by solving NSE and is not restricted by any limiting assumptions.

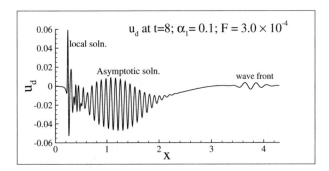

Fig. 3.30 Streamwise disturbance velocity u_d obtained as a receptivity solution at $t = 8$, for $y = 0.008$, identifying the local, asymptotic and wave-front components due to a strip excitation near the leading edge. The frequency of excitation of the SBS strip on the wall is $F = 3.0 \times 10^{-4}$

3.9.3 Typical Morphology of the Disturbance Field

A typical receptivity solution to localized wall excitation at a fixed frequency is shown in Fig. 3.30 at $t = 8$, with the solution obtained by solving NSE for the frequency corresponding to OA in Fig. 3.27. Here, $t = 0$ corresponds to the onset of the excitation. According to linear theory, this solution should have been completely damped, but the presented solution in Fig. 3.30, at a particular time indicate a streamwise stretch where the solution first grows and subsequently decays. The height at which this solution is recorded corresponds to $y = 0.008$. This is a point very close to the wall and all distances are non-dimensionalized.

The structure shown in Fig. 3.30 is generic and has distinct features, some of which are also shared by the solution of the problem obtained using the OSE in [62]. Solutions in [52, 57] correspond to unstable cases and the leading STWF of Fig. 3.30 was not distinctly noted. However in [62], response field for spatially stable cases displayed a distinct STWF, different from the following TS waves, as seen in Fig. 3.30. To understand the various components of the response shown in Fig. 3.30, the Fourier transform of the response is shown plotted in Fig. 3.31. In this figure, the main peak corresponds to the wave-packet representing the TS waves. As shown in [57, 62], apart from the TS mode, there are only a few additional modes with very high spatial damping rates and these are also noted among the different peaks of Fig. 3.31. The exciter is located very near to the leading edge of the flat plate, and the computed solution in Fig. 3.30 shows large peaks in the physical plane for the local solution and the asymptotic solution emerges from it in a continuous manner. This local solution corresponds to the marked distant peak in the legend of Fig. 3.31. The wave front corresponds to the local maximum close to the origin in Fig. 3.31. We also note that the solution of NSE is for the excitation, for which the linearized solution corresponds to damped TS waves. When such an excitation is applied, the local and the asymptotic solution remain stationary in space, at the location depicted in Fig. 3.30. However, the STWF continuously propagate downstream.

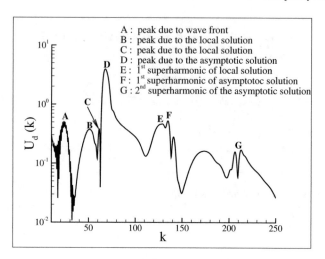

Fig. 3.31 Fourier transform of the receptivity solution shown in Fig. 3.30, identifying the wavenumbers of the local, asymptotic and wave front components in the legend

3.9.4 Receptivity to SBS Excitation: Nonlinear and Nonparallel Effects

In this section, nonlinear and nonparallel effects on the receptivity of the Blasius boundary layer is studied, when the flow is harmonically excited by SBS exciter with $F = 3 \times 10^{-4}, \alpha_1 = 0.05$. The exciter is located between $x_1 = 0.22$ and $x_2 = 0.264$. In Fig. 3.32, the streamwise component of the disturbance velocity u_d is plotted as a function of streamwise co-ordinate x, at indicated heights for this case. In showing the signal in the left side frames of Fig. 3.32, the origin refers to the leading edge of the flat plate. From the solution of Eqs. (3.60) and (3.61) without the Boussinesq term, one obtains instantaneous values of vorticity and velocity components. The disturbance quantities are obtained (as displayed in Fig. 3.32) by subtracting the mean from the total quantity. This corresponds to a spatially stable case in linear theory.

The displayed variations of u_d with x, at different heights over the plate indicate an interesting phenomenon. This solution corresponds to the time $t = 20$ after the onset of excitation. While the linear normal mode analysis, predicts this flow field to be stable, one notices that barring the top and bottom-most left-side frames in Fig. 3.32, the disturbances actually grow spatially at intermediate heights (frames (c), (e) and (g)). All the frames clearly indicate the presence of multiple modes which are also directly evident from the multiple peaks of the Fourier transform of these signals shown in the right-side frames of Fig. 3.32. The Fourier amplitudes show the dominant normal mode near the wall, while multiple dominant modes are seen in the third and fourth frames showing the transform. Also, one notices the first super harmonic of the asymptotic solution in the right side frames of Fig. 3.32, which is clearly shown in Fig. 3.31.

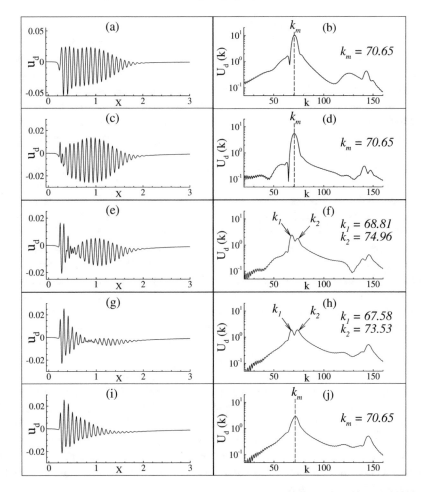

Fig. 3.32 u_d plotted as a function of x for $t = 20$ at **a** $y = 0.00521$, **c** 0.00662, **e** 0.00808, **g** 0.00958 and **i** 0.01137. The signals are for the SBS strip excitation amplitude of $\alpha_1 = 0.05$, and non-dimensional frequency, $F = 3.0 \times 10^{-4}$ with exciter location between $x = 0.22$ and $x = 0.264$. Fourier transform of the signals shown in right frames

Frames (c), (e) and (g) in Fig. 3.32, clearly show different streamwise stretches at different heights, over which the disturbance actually grows with x. The Fourier transform of the signal in frames (b), (d) and (j) display a dominant peak at $k = k_m = 70.65$. From the linear stability analysis at $x = 0.3$ ($Re_{\delta*} = 298$) corresponding to the present excitation frequency, one should have a wave with wavenumber $\alpha = 0.225$. Here, the wavenumber α is based on the local δ^*, whereas k_m is based on the reference length scale L. The dominant wavenumber of the signal based on δ^* and obtained from DNS is given as $\alpha_m = \frac{k_m\sqrt{Re_{\delta*}}}{1.72\sqrt{x}}$ and at $x = 0.3$, $\alpha_m = 0.21$. One also notices, the presence of two adjacent modes at $k = k_1$ and $k = k_2$, for the signals at

$y = 0.00808$ and $y = 0.00958$. The lower mode at $k = k_1$ corresponds to a mode that moves downstream at a faster speed than the mode at $k = k_2$. It is also noted that $k_1 + k_2 \simeq k_m$. The presence of the mode at $k = k_2$ becomes weaker as the height increases. However, there is another damped mode that dominates with increasing height over the plate. It is noted that in earlier nonparallel studies (as referred to in [22]), various investigators have tried to explain experimental data that indicated instability at particular heights for the streamwise disturbance velocity components.

It has been noted in [57, 61, 62] that there are no particular physical reasons as to why one has to adopt either spatial or temporal framework in studying boundary layer stability. This is also readily apparent from the information in Figs. 3.32 and 3.33. An interesting aspect of the computed results is noted with respect to time variation- as shown in the left frames of Fig. 3.33 for signals collected at a distance of $x = 0.3$ from the leading edge, at the indicated heights. Despite the fact that the fluid dynamic system is excited at a single constant frequency ($\bar{\omega}_0$), the Fourier transform in the right frames of Fig. 3.33 indicates presence of superharmonics at all heights and in different proportions. From Fig. 3.27 one clearly notes the corresponding normal modes to be more stable from linear stability point of view for these additional superharmonics. Of specific interest is the data for $y = 0.00662$ and 0.00808, which show the presence of strong second and third harmonics.

In Fig. 3.32 also, we noted most pronounced instability at these heights. Thus, the disturbance growth at these heights must be related to nonlinear effect and is experienced at intermediate heights only.

Apart from the fact that the growth is noted at intermediate heights, we also note similarity of flow field at heights very close to the wall, and those in the outer edge of the boundary layer. This aspect is depicted in Fig. 3.34, where spatial variation of the streamwise disturbance velocity is shown for heights those are either very near the wall or are in the outer part of the shear layer. Corresponding Fourier transform of these signals are shown in the right frames of Fig. 3.34. At these heights, signals display a range of x for which they remain neutral, while in the outer part, the signals decay with streamwise distance.

Thus, Figs. 3.32, 3.33 and 3.34 show that this particular case represents nonlinear receptivity at intermediate heights; neutral stability near the wall and in the outer part of the boundary layer.

Presence of growing solution at intermediate heights is further investigated by plotting the mean and disturbance component of streamwise velocity (obtained from the DNS) with height at $x = 0.3$ (close to the exciter), in Fig. 3.35a. In the same figure, the band of heights at which the direct simulation indicates growth is identified by drawing two horizontal lines 1a and 1b. It is already noted that there are multiple spatial modes (including the normal mode, identified as the TS mode in the present figure) and superharmonics of the excitation frequency present in the simulated cases (due to nonlinearity). Their effects are different at different heights, as shown in Figs. 3.32, 3.33 and 3.34. For example in Fig. 3.33, one can clearly note the presence of dominant superharmonics at $y = 0.00662$ and 0.00808. While these are also noted at different heights, and play secondary roles there.

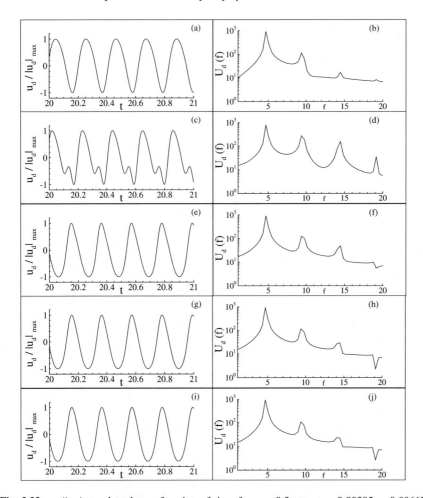

Fig. 3.33 $u_d/|u_d|_{max}$ plotted as a function of time for $x = 0.3$ at **a** $y = 0.00385$, **c** 0.00662, **e** 0.00808, **g** 0.00958 and **i** 0.0144. The figure is shown for the SBS strip excitation amplitude of $\alpha_1 = 0.05$ and non-dimensional frequency $F = 3.0 \times 10^{-4}$ with exciter location between $x_1 = 0.22$ and $x_2 = 0.264$. Fourier transforms plotted for the signals in right frames. Here, $|u_d|_{max}$ is the maximum absolute value of u_d for $20 \le t \le 21$ at the corresponding (x, y)-location

In discussing this aspect of Fig. 3.33, we reason that these are essentially nonlinear and nonparallel effects. In Fig. 3.35a, the intermediate height of maximum TS wave-packet growth is attempted to be linked with critical layers of the first and second TS mode and their superharmonics. From the inviscid stability analysis [19, 28], the height where $U(y) = \frac{\bar{\omega}_{0j}}{\alpha_j}$ is known to be the critical layer, where α_j and $\bar{\omega}_{0j}$ are the wavenumber and circular frequency corresponding to the jth TS mode. These critical layers for the TS and the second modes are identified in Fig. 3.35a by lines 2–5, respectively. It is seen that the maximum growth height is located closer to the critical

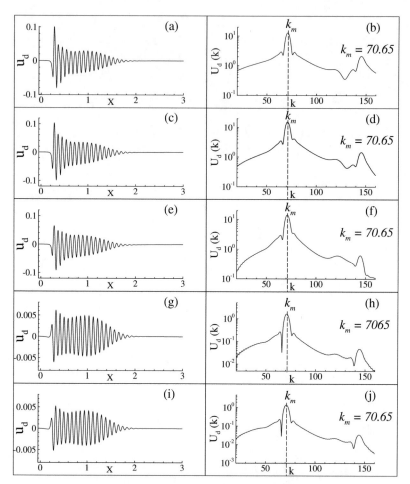

Fig. 3.34 u_d plotted as a function of x for $t = 20$ at **a** $y = 0.00124$, **c** 0.00252, **e** 0.00385, **g** 0.03 and **i** 0.032. The ZPG flow past a flat-plate is excited by a SBS strip located between $x_1 = 0.22$ and $x_2 = 0.264$ with $\alpha_1 = 0.05$ and the frequency $F = 3.0 \times 10^{-4}$. Fourier transform of the signals are shown in right frames

heights of the TS and second modes, corresponding to the first superharmonics, at which the linearized convection terms disappear in the governing OSE for the TS mode-fundamental frequency combination. However, this intermediate height is closest to the point where u_d variation with respect to y changes sign. This point is further substantiated in Fig. 3.35b, where u_d is plotted as a function of y at indicated streamwise locations at time-instants, when corresponding u_d at inner maxima attains its maximum value. The range of y, where spatial growth of the TS wave-packet is observed, is also marked in the figure. It can be clearly seen that this range of y lies in a zone, where u_d changes sign. The location where u_d changes sign is also the location where absolute value of the disturbance normal velocity v_d is maximum. This can be explained from linear stability theory with parallel flow approximation, as $u_d = \int \phi'(\tilde{y}) e^{i(\alpha\tilde{x} - \omega_0 t)} d\alpha$ and $v_d = \int \phi(\tilde{y}) e^{i(\alpha\tilde{x} - \omega_0 t)} d\alpha$. Hence, $|\phi(\tilde{y})|$ is maximum at the height $\tilde{y}$ where, $\phi'(\tilde{y}) = 0$. For actual nonparallel flow, though this analysis does not exactly hold good, but even then v_d attains its maxima near about the y-location, where u_d changes sign. This can be noted clearly from Fig. 3.35c, where v_d is plotted as a function of y at indicated x-locations at time instants, when its peak value is maximum. The conclusions that one makes from the above discussion are enumerated below.

(1) One obtains spatial growth of disturbance streamwise velocity component at the intermediate heights even for a non-dimensional frequency of excitation, F, that does not intersect the neutral curve.

(2) This range of intermediate heights is located around the point, where u_d changes sign and absolute value of v_d is maximum.

(3) The dominant presence of second superharmonic of the excitation frequency $\bar{\omega}_0$ is noted at this range of intermediate heights, where spatial growth of u_d is noted. This feature is shown in Fig. 3.33.

It should be pointed out that, such height–dependent variation of the neutral points were also observed in [22], where "an ambiguous behaviour" of u_d was noted to exists as the height approaches "the location of the $180°$ phase shift of the u_d eigenfunction". No detailed analysis of this height dependent behaviour was presented in [22].

3.9.5 Effects of Nonlinearity: Role of Amplitude of Excitation

Here, the role of the amplitude of excitation α_1 is investigated for the exciter located between 0.22 and 0.264. The non-dimensional excitation frequency is $F = 3.0 \times 10^{-4}$. The amplitude parameter α_1 is varied from a very small value of 0.002 (i.e., excitation amplitude to be 0.2% of U_∞) to a high value of 0.1 (excitation amplitude is 10% of U_∞). The variable x_{ex} is used subsequently to denote the center of the exciter. In Sect. 3.9.6, various amplitudes of excitation results are compared for $\alpha_1 = 0.002, 0.005$ and 0.01. When $\alpha_1 = 0.1$, one notices the induction of micro-separation bubbles on the wall, which is discussed in Sect. 3.9.7.

3.9.6 Comparison of Results for $\alpha_1 = 0.002, 0.005$ and 0.01

A more elaborate comparison is shown in Fig. 3.36 where $\hat{u}_d$ corresponding to the local and asymptotic solution are plotted as a function of x at ten indicated heights for (a) $\alpha_1 = 0.002$, (b) $\alpha_1 = 0.005$ and (c) $\alpha_1 = 0.01$, with exciter located between 0.22 and 0.264 and frequency of excitation is $F = 3.0 \times 10^{-4}$. In Fig. 3.36 normalization of u_d is done with respect to the $\alpha_1 = 0.01$ case as $\hat{u}_d(x, y, t; \alpha_1 =$

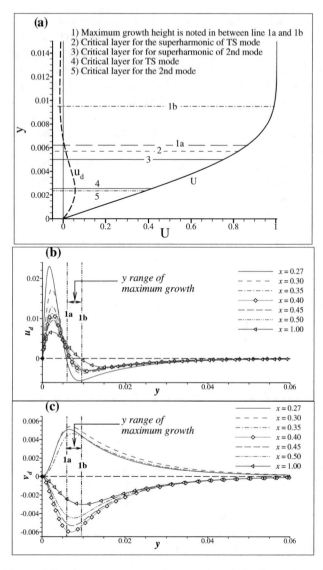

Fig. 3.35 **a** Mean and disturbance components of streamwise velocity shown along with estimates of critical layer for the second spatial mode. **b** u_d and **c** v_d plotted as function of y for indicated streamwise locations at time-instants when corresponding u_d at inner maxima attains its maximum value. In all the three frames, the lines 1a and 1b indicates the range of heights where maximum local spatial growth is noted. Presented data are for the SBS excitation with $\alpha_1 = 0.05$ and $F = 3.0 \times 10^{-4}$ with exciter location between 0.22 and 0.264

$\bar{\alpha}_1) = (0.01/\bar{\alpha}_1)u_d(x, y, t; \alpha_1 = \bar{\alpha}_1)$. The results shown are for the solution obtained at $t = 20$. The ten heights shown are the 15, 20, 25, 30, 35, 40, 65, 75, 85, and 95th point of the grid in the wall-normal (y) direction. From these plots, one can find

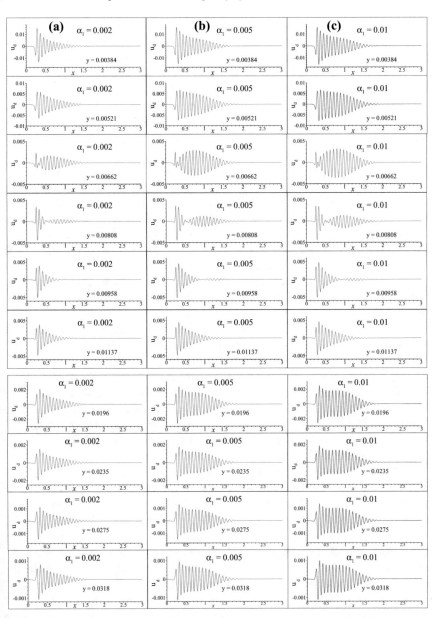

Fig. 3.36 Normalized disturbance streamwise velocity u_d at $t = 20$ and indicated heights plotted of x for **a** $\alpha_1 = 0.002$, **b** $\alpha_1 = 0.005$ and **c** $\alpha_1 = 0.01$. The normalization is done with respect to $\alpha_1 = 0.01$ case. The exciter is located between 0.22 and 0.264 which excites the flow with $F = 3.0 \times 10^{-4}$

the effects of nonlinearity and nonparallelism on the TS wave-packet, obtained from DNS of the receptivity problem. One notices from Fig. 3.36 that at $y = 0.00348$ and 0.00521, u_d for $\alpha_1 = 0.002$ displays monotonic decay with respect to x. However, the $\alpha_1 = 0.01$ case displays near-neutral behaviour at these heights for $0.6 \leq x \leq 0.75$ and $\alpha_1 = 0.005$ case shows reduced decay rate up to $x \leq 0.75$. At the same time one observes from Figs. 3.32 and 3.34, marginal instability up to $x \leq 1.0$ for $\alpha_1 = 0.05$ for $y = 0.00348$ and 0.00521. Since, this marginal instability shows up for increased amplitudes of excitation, hence one can definitely say that these effects are completely due to nonlinearity of the governing equation. One also observes that the stretch of the perturbed region also increases with the increase in amplitude of excitation.

One can also observe that the asymptotic solution displays a distinct spatial growth phase at $y = 0.00662$, 0.00808 and 0.00958 for all α_1 cases plotted here. For $\alpha_1 = 0.002$, this spatial growth of u_d at $y = 0.00958$ is not distinctly identifiable from Fig. 3.36. Spatial growth of u_d correspond to TS wave-packet, at those heights, where it is also noted for $\alpha_1 = 0.05$ in Fig. 3.32. However, the extent of the spatial growth of the asymptotic solution reduces with the reduction in amplitude of the excitation. It is the downstream end, where the growth concludes, and moves upstream with the reduction in excitation amplitude. The upstream end of the onset of growth is located almost at the same position for α_1 cases studied.

The heights $y = 0.0196, 0.0235, 0.0275$ and 0.0318 are at the near proximity of the edge of the boundary layer profile close to the exciter location. The features are similar to the behaviour of u_d, which are obtained very close to the wall ($y \leq 0.00521$). One notices a zone where disturbances evolve near-neutrally for $\alpha_1 = 0.005$ ($0.6 \leq x \leq 0.8$). Mild growth for $\alpha_1 = 0.01$ in this streamwise extent can also be noted from this figure. One also observes perceptible spatial growth of u_d for $\alpha_1 = 0.05$ at these heights from Fig. 3.34. However, any such streamwise zone of near-neutral behaviour or growth of u_d is clearly absent for $\alpha_1 = 0.002$, for which u_d strictly decays at these heights, are shown in Fig. 3.36.

In Fig. 3.37, time variation of u_d is plotted at $x = 0.35$, for $20 \leq t \leq 21$ and indicated values of α_1. In this figure, u_d is normalized with respect to its maximum absolute value $|u_d|_{max}$. Figure 3.37 shows the dominant presence of second superharmonics of the excitation frequency, i.e., $3\bar{\omega}_0$, even for $\alpha_1 = 0.002$ at $y = 0.00662$. In contrast, time variation of u_d at other heights for $\alpha_1 = 0.002$ display an almost sinusoidal variation indicating negligible presence of superharmonics. It should be noted that $y = 0.00662$ is the height, where maximum spatial growth of the asymptotic solution is noted in Fig. 3.36, even for $\alpha_1 = 0.002$. This feature of the dominant presence of second superharmonic at around the height of $y = 0.00662$ is also noted for $\alpha_1 = 0.005$ and $\alpha_1 = 0.01$ in Fig. 3.37b, c, respectively. This feature is also noted for $\alpha_1 = 0.05$ and is already shown in Fig. 3.33. This was already noted for a height very close to the point, where u_d and v_d variation with respect to y changes sign and attains its maximum value, respectively, in Fig. 3.35. Similar features are also noted here in Fig. 3.37 even for the smallest amplitude of excitation of $\alpha_1 = 0.002$.

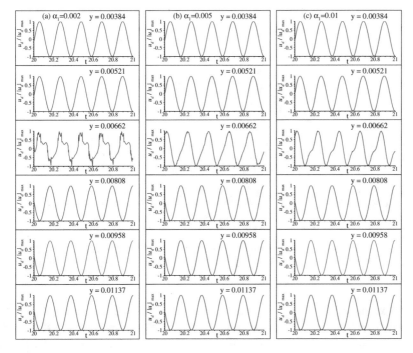

Fig. 3.37 $u_d / |u_d|_{max}$ plotted as a function of time for $20 \le t \le 21$ and $x = 0.35$ at indicated heights for **a** $\alpha_1 = 0.002$, **b** $\alpha_1 = 0.005$, **c** $\alpha_1 = 0.01$. The exciter is located between 0.22 and 0.264 and $F = 3.0 \times 10^{-4}$. $|u_d|_{max}$ is the maximum absolute value of u_d for $20 \le t \le 21$ at corresponding (x, y)-location

3.9.7 Results for $\alpha_1 = 0.1$

To understand the role of the amplitude of excitation of the SBS strip, another case is computed with identical excitation frequency and location, but with excitation amplitude of $\alpha_1 = 0.1$, i.e., the excitation amplitude is 10% of U_∞. This excitation amplitude causes induction of micro-separation near the exciter on the plate. Resultant unsteady separation bubbles are confined to small height at the wall - as shown in the stream function contours in Fig. 3.38 at $t = 20$. Only a few representative contours have been shown plotted in this figure, which clearly show the micro-bubbles near the exciter up to $x \le 1.6$ only. Effects of the present bubbles are noted in other contours shown at other heights.

To understand the dynamics of this case better, instantaneous snapshot at $t = 20$ is shown for u_d variation with x at three heights in the left frames of Fig. 3.39. In the frames on the right hand side, corresponding Fourier transform is shown plotted to help identify various components of the solution in the physical space. Presence of separation bubbles causes fluid particles outside the bubbles to traverse in the wall-normal direction more than in the streamwise direction, and this is reflected in the lower value of u_d for the higher α_1 case, as can be ascertained by comparing Figs. 3.39

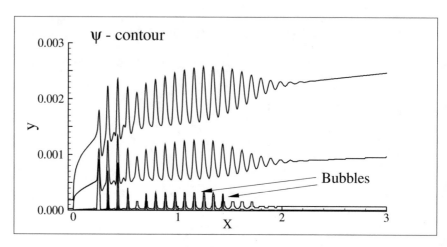

Fig. 3.38 ψ-contours shown in the physical plane for $t = 20$ for the case of $\alpha_1 = 0.10$; $F = 3.0 \times 10^{-4}$ located between $x_1 = 0.22$ and $x_2 = 0.264$. Note the micro-bubbles near the exciter on the plate

with 3.32 and 3.34. Also, these micro-bubbles interfere with the local solution that is clearly seen in the Fourier transform of Fig. 3.39b, as compared to that shown in Fig. 3.32, for the peak corresponding to the local solution near $k = 145$. For the data shown at the second height ($y = 0.000245$), the local solution peak is more pronounced as compared to the local solution peaks for $y = 0.00521$ and $y = 0.035$ in the right frames of Fig. 3.39. We conjecture that with increasing width of the exciter and increasing amplitude of excitation further, separation bubbles will be more pronounced, which could interfere destructively with the TS waves - as noted in this case with low intensity. However, presence of TS waves is unmistakable for $\alpha_1 = 0.1$. One also notices significant spatial growth of u_d at $y = 0.035$ in Fig. 3.39, than for u_d at $y = 0.032$ for $\alpha_1 = 0.05$ in Fig. 3.34. This increase in the extent of the spatial growth of u_d at the edge of the shear layer with increase in the amplitude of excitation was also noted for lower α_1 cases.

The disturbance streamwise velocity u_d at $t = 15$, normalized with respect to $\alpha_1 = 0.1$ case, for $y = 0.00252$ (the first grid line above the plate) is plotted in Fig. 3.40 to analyze effects of amplitude of excitation. In this figure for $\alpha_1 = 0.01$, u_d is multiplied by ten-times and the same is multiplied by two times for $\alpha_1 = 0.05$, to compare with the case of $\alpha_1 = 0.1$. There are distinct differences in all the cases. The lowest α_1 case shows near-neutral variation of u_d in $0.6 \le x \le 0.8$. While the higher α_1 cases show distinct spatial growth of u_d. The onset of spatial growth of u_d is located almost at the same position for $\alpha_1 = 0.05$ and $\alpha_1 = 0.1$ cases, while the extent of spatial growth increases for $\alpha_1 = 0.1$. Also, u_d for the higher α_1 cases show more dominant presence of superharmonics near the exciter, due to the induction of unsteady separation micro-bubbles on the wall. The results in Fig. 3.41 are for the same excitation parameters and at the same location of the exciter, but are recorded at higher height for $y = 0.00808$. At this height, significant spatial growth is noted for all the α_1 cases considered till now. As noted for smaller α_1 cases, here also one

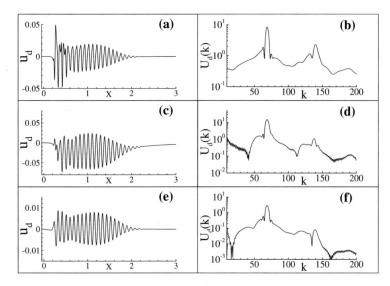

Fig. 3.39 u_d plotted for $y = 0.00245$; 0.00521 and 0.035, respectively, from top to bottom at $t = 20$. Results presented are for the SBS strip excitation with $\alpha_1 = 0.10$; $F_f = 3.0 \times 10^{-4}$ and the exciter located between $x_1 = 0.22$ and $x_2 = 0.264$. **b** Fourier transform of the signals shown in **a**

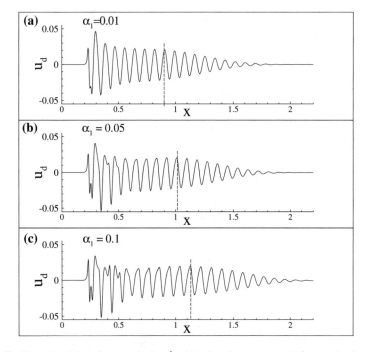

Fig. 3.40 Normalized disturbance velocity $\hat{u}_d$ plotted against x at $t = 15$ for amplitudes of excitation at $y = 0.00252$, which is the first grid line above the plate

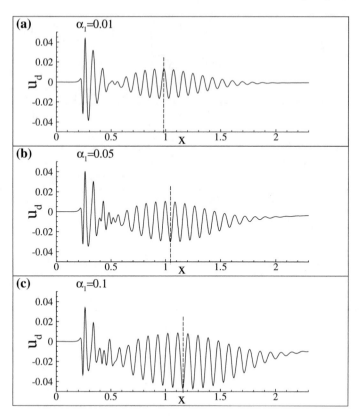

Fig. 3.41 Normalized disturbance velocity $\hat{u}_d$ (normalized with respect to $\alpha_1 = 0.01$ case) plotted against x at $t = 15$ for amplitudes of excitation at $y = 0.0081$

Table 3.3 Onset and end of x-locations for the spatial growth of u_d at $y = 0.00662$ and $y = 0.00808$ for different values of excitation amplitude α_1

α_1	Onset of spatial growth at $y = 0.00662$	End of spatial growth at $y = 0.00662$	Onset of spatial growth at $y = 0.00808$	End of spatial growth at $y = 0.00808$
0.002	0.312	0.548	0.586	0.851
0.005	0.311	0.766	0.591	0.938
0.01	0.318	0.843	0.577	0.971
0.05	0.323	0.941	0.556	1.035
0.1	0.325	1.148	0.572	1.148

observes: (1) The onset of spatial growth is located almost at the same position, even for $\alpha_1 = 0.1$ and (2) extent of spatial growth increases with increase in α_1. The onset and end-location of the spatial growth of u_d at $y = 0.00662$ and $y = 0.00808$ are tabulated in Table 3.3 for different values of α_1.

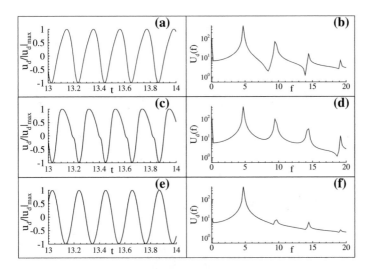

Fig. 3.42 **a** Time variation of $u_d/|u_d|_{max}$ plotted at $x = 0.3$ and $y = 0.00124; 0.00662$ and 0.034 for $\alpha_1 = 0.1$, $F_f = 3.0 \times 10^{-4}$. The exciter located between 0.22 and 0.264. In frame **b**, Fourier transform of the signal shown in **a**. Here, $|u_d|_{max}$ is the maximum absolute value of u_d for $13 \leq t \leq 14$ at the corresponding (x, y)-location

In the left frames of Fig. 3.42, the time series for $u_d/|u_d|_{max}$ is shown at $x = 0.3$ for the case of $\alpha_1 = 0.1$ at the indicated heights. Here, $|u_d|_{max}$ is the maximum absolute value of u_d, during $13 \leq t \leq 14$, at the heights, $y = 0.00124, 0.00662$ and 0.034. These time series are multi-periodic, as evident from the Fourier transform shown in the corresponding right frames. Once again, the dominant presence of second and third superharmonics are noticeable in the interior of the shear layer (at around $y = 0.00662$), similar to the lower α_1 cases. The dominant superharmonics are not seen for $y = 0.034$ and above, for this α_1 case.

On the left column of the Fig. 3.43, the time series for the disturbance vorticity (ω_d) is shown at a distance of $x = 1.34$ from the leading edge of the plate, for $\alpha_1 = 0.1$ and $F = 3.0 \times 10^{-4}$ at the indicated heights, which are very close to the wall. The time series indicate variations to be different at different heights and it is seen from the Fourier transforms shown on the right frames of Fig. 3.43. One notices the first superharmonic of F very clearly and a weak second superharmonic for heights closest to the wall. When the time series are recorded closer to the exciter $(x = 0.27)$, as shown in Fig. 3.44, then the results are seen to be significantly different for the same heights over the plate. For the lowest height, the amplitude for the fundamental is seen to be three times higher at this closer location from the exciter and the first superharmonic is also of comparable magnitude at this distance, with more harmonics seen in the spectrum. For the other two heights also, one notices clearly the presence of higher harmonics.

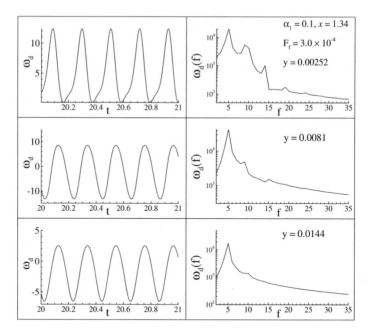

Fig. 3.43 Time variation of ω_d plotted (left) for $\alpha_1 = 0.1$ and $F_f = 3.0 \times 10^{-4}$ at $x = 1.34$ for the three indicated heights close to the plate. Corresponding Fourier transforms shown on the right

3.9.8   Wall-Normal Variation of u_d and v_d

In this sub-section, the wall-normal variation of u_d and v_d are explored for $F = 3.0 \times 10^{-4}$ and the exciter located between 0.22 and 0.264. This is investigated in Figs. 3.45 to 3.48, where normalized u_d and v_d are plotted as a function of $\bar{y} = y/\delta^*$ at $x = 0.3$ and 0.5, respectively, from $t = 20.25$ to $t = 20.45$, at a time interval of 0.02. All these time instants lie within one fundamental non-dimensional time-period of excitation $T = \frac{2\pi}{\omega_0}$, which is 0.209 for these cases. In essence, Figs. 3.45, 3.46, 3.47 and 3.48 show the variation of $u_d(y)$ and $v_d(y)$ with respect to time, within one fundamental time-period T, at two particular streamwise locations. Two x-stations chosen represent locations close to the exciter ($x = 0.3$) and in the middle part of the asymptotic solution ($x = 0.5$). The u_d plotted in Figs. 3.45, 3.46 are normalized with respect to the $\alpha_1 = 0.01$ case. Similar normalization is done for v_d also in Figs. 3.47 to 3.48.

Figure 3.45 reveal that the inner maximum not only oscillates with respect to y, but displays wider variation in its shape, as clearly noted at $t = 20.31$, 20.33, 20.41 and 20.43. At $t = 20.31$, 20.41 and 20.43 one notices the splitting of the inner maximum to create more than one peaks inside the shear layer. Such splitting of inner maximum is not predicted at all by the linear stability analysis. Existence of more than one peaks inside the shear layer are observed for $\alpha_1 = 0.002$, the lowest amplitude of excitation case considered here. Fluctuation of u_d near the inner maximum is

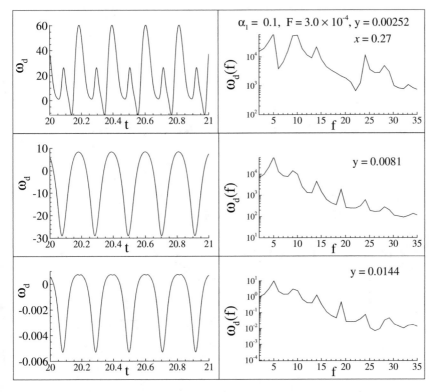

Fig. 3.44 Time variation of ω_d plotted at $x = 0.27$ for $\alpha_1 = 0.1$ and $F_f = 3.0 \times 10^{-4}$ (left) for the heights shown in Fig. 3.43. Corresponding Fourier transforms are shown on the right

more prominent as one moves downstream, noted in Fig. 3.46. One also notices the difference in u_d for various amplitudes of excitation cases displaying effects of nonlinearity. The asymptotic solution corresponding to $\alpha_1 = 0.1$ case creates multiple unsteady separation bubbles on the plate, as shown in Fig. 3.38. This is responsible for larger departure of u_d variation with respect to y from the linear theory, as well as from that obtained for other lower α_1-cases, evident in Fig. 3.46. This is also to be noted that there is a striking similarity and match between the u_d profiles between $\alpha_1 = 0.002$, 0.005 and 0.01 cases, for all the time instants and the streamwise locations shown in these figures. This implies that nonparallelism of the base flow and the spatio-temporal nature of the perturbations, rather than nonlinearity did play a dominant role in determining the u_d profiles for these α_1-cases. However, effects of nonlinearity built-up and modify the profiles for higher amplitude cases, which ultimately induced micro-separation bubbles on the wall for $\alpha_1 = 0.1$ case. However, the outer maximum, in comparison displays lesser variation with respect to time for all the three streamwise locations considered here, as it is located at the edge of the shear layer.

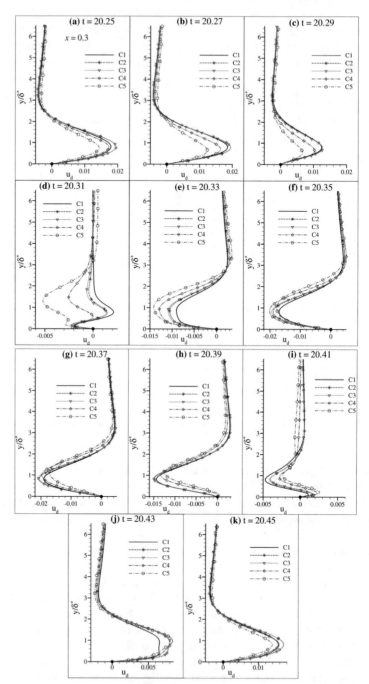

Fig. 3.45 Normalized streamwise disturbance velocity u_d plotted as a function of y/δ^* at $x = 0.3$ and indicated time instants. Here, $C1$, $C2$, $C3$, $C4$ and $C5$ represent the cases with $\alpha_1 = 0.002$, 0.005, 0.01, 0.05 and 0.1, respectively. The exciter is put between 0.22 and 0.264 with $F = 3.0 \times 10^{-4}$

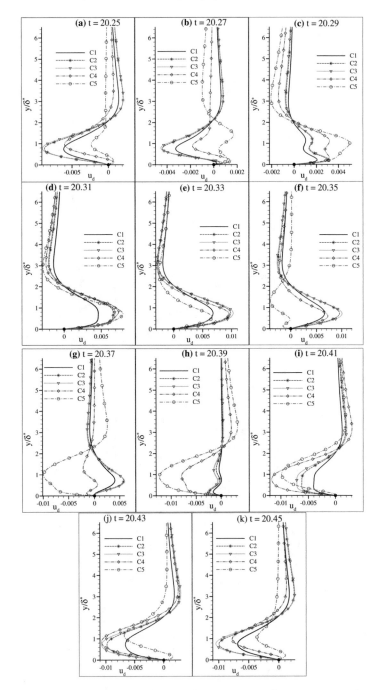

Fig. 3.46 Normalized streamwise disturbance velocity u_d plotted as a function of y/δ^* at $x = 0.5$ and indicated time instants. Here, $C1$, $C2$, $C3$, $C4$ and $C5$ represent the cases with $\alpha_1 = 0.002$, $0.005, 0.01, 0.05$ and 0.1, respectively. The exciter is between 0.22 and 0.264 with $F = 3.0 \times 10^{-4}$

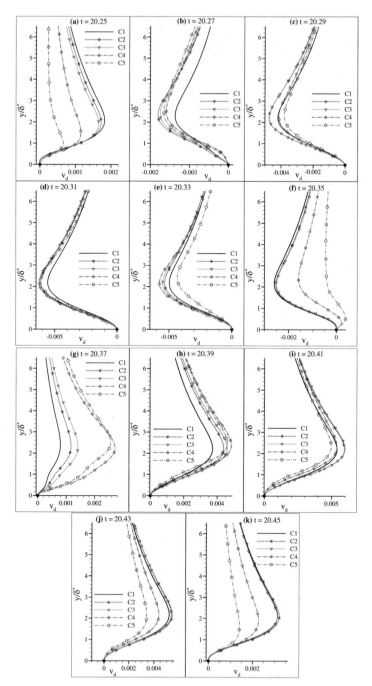

Fig. 3.47 Normalized wall-normal disturbance velocity v_d plotted as a function of y/δ^* at $x = 0.3$ and indicated time instants. Here, $C1$, $C2$, $C3$, $C4$ and $C5$ represent the cases with $\alpha_1 = 0.002$, $0.005, 0.01, 0.05$ and 0.1, respectively. The exciter is between 0.22 and 0.264 with $F = 3.0 \times 10^{-4}$

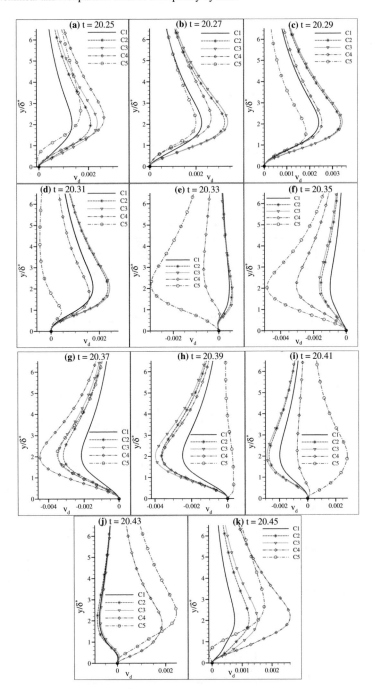

Fig. 3.48 Normalized wall-normal disturbance velocity v_d plotted as a function of y/δ^* at $x = 0.5$ and indicated time instants. Here, $C1$, $C2$, $C3$, $C4$ and $C5$ represent the cases with $\alpha_1 = 0.002$, $0.005, 0.01, 0.05$ and 0.1, respectively. The exciter is between 0.22 and 0.264 with $F = 3.0 \times 10^{-4}$

One observes also significant effects of nonlinearity for the profile of v_d shown
in Figs. 3.47 and 3.48 for $x = 0.3$ and 0.5. Presence of multiple maxima can also be
noted for these cases at certain time instants as shown in Figs. 3.47f and 3.48a, e, k.
However, for cases with $\alpha_1 = 0.002, 0.005$ and 0.01, similar variation of v_d is noted
for all time instants form these figures.

3.9.9   Further Evidences of Nonparallel, Nonlinear Effects and Bypass Transition

An interesting fact arises when one compares u_d and ω_d, which show completely different kind of spatial behavior at two different heights from the plate. Spatial growth
of u_d at intermediate heights has already been observed. This is further shown in
Fig. 3.49 by plotting u_d for three different non-dimensional frequencies of excitation, at $y = 0.00662$. The exciter is located between 0.22 and 0.264 which excites
the flow with $\alpha_1 = 0.01$. In the frames of Fig. 3.49, a vertical dotted line indicates
the location up to which the solution grows, beyond that line the solution decays.
One observes spatial growth of u_d, whose extent tend to increase with decrease in
the non-dimensional frequency of excitation F.

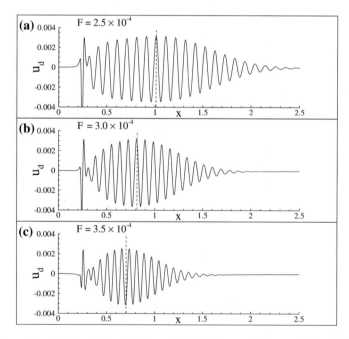

Fig. 3.49 u_d plotted against x at $t = 15$ for $y = 0.00662$, $\alpha_1 = 0.01$ and the indicated frequencies
of excitation

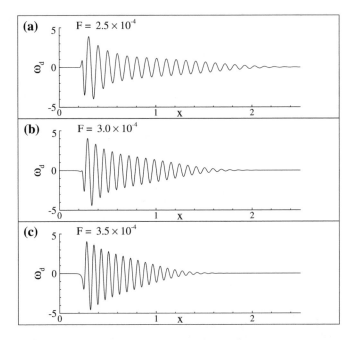

Fig. 3.50 ω_d plotted as a function of x at $t = 15$ for $y = 0.0066$ with $\alpha_1 = 0.01$ for the displayed frequencies

The interesting point emerges when ω_d is plotted against x at the same height and at same time, as shown in Fig. 3.50. For all the three frequencies, ω_d decays in contrast to the growing u_d shown in Fig. 3.49. It is worth remembering that the vorticity represents the rotational part of the flow field, while the velocity mainly contains the translational part of the flow field.

However, a completely contrasting scenario is noted at $y = 0.0112$ in Figs. 3.51 and 3.52. At this height, ω_d indicates growth, while u_d is stable, as shown in Fig. 3.52. Note the absence of local solution and the distinct growth of ω_d in Fig. 3.51. In Fig. 3.52, one can note that ω_d shows monotonic decay for the two higher frequencies, whereas there is a very small growth in that part of the signal shown in between the two vertical lines for the lowest frequency of $F = 2.5 \times 10^{-4}$.

In all the cases reported so far, ZPG boundary layer is investigated by exciting it using a SBS harmonic exciter of finite width (from $x_1 = 0.22$ to $x_2 = 0.264$). Nonparallel and nonlinear effects of receptivity specifically for high frequency disturbances is recorded. Next, another case is studies, where the exciter width is increased further, in both the directions, by considering the strip between $x_1 = 0.2$ to $x_2 = 0.29$. Having extended it further towards the leading edge of the plate, more nonparallel effects are introduced and by extending it in the downstream direction also, more energy inside the shear layer is imparted. In Fig. 3.53, the stream function and vorticity contours for these two cases of different width of the exciter are compared. In

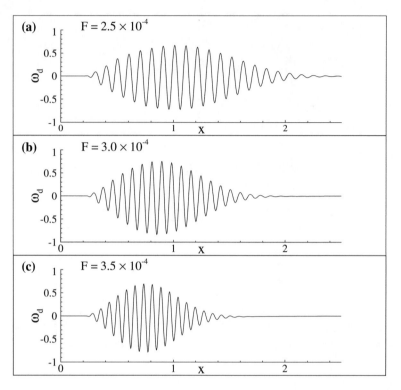

Fig. 3.51 ω_d plotted against x at $t = 15$ for the indicated frequencies of excitation at $y = 0.0112$ with $\alpha_1 = 0.01$

both the cases, the amplitude of excitation is taken as $\alpha_1 = 0.1$ and the frequency of excitation given by $F = 3.0 \times 10^{-4}$.

From the ψ-contours in (a), the presence of a few small bubbles in the immediate downstream of the exciter, spaced at a gap similar to the wavelength of TS waves is noted. These bubbles are responsible for setting up of wavy disturbance field. Note that the wall-normal direction is stretched in all the frames shown in Fig. 3.53. The vorticity contours exhibit a two-tier structure with the lower layer showing an upstream propagating tendency, while the upper tier shows vertical direction of the ejected vortices (those are slightly tilted towards downstream direction). For the case of wider excitation strip in Fig. 3.53c, the stream function contours display bubbles, over the streamwise stretch up to $x = 3.3$, as compared to the case of shorter exciter which shows the visible undulation up to $x = 1.7$ only. Also the bubbles are of significantly larger dimension in the wall-normal direction. These bubbles are not steady and the ensuing unsteady downstream motion is what is termed as bypass transition in [8, 60]. In [60], the unsteady separation on the surface of the plate was initiated by a disturbance from the free stream by creating induced adverse pressure gradient.

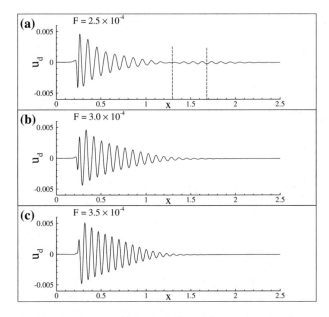

Fig. 3.52 u_d plotted against x at $t = 15$ for the indicated frequencies of excitation at $y = 0.0112$ with $\alpha_1 = 0.01$

3.9.10 Effects of the Location of Exciter

Next, the effects of the location of the exciter is described for $F = 3.0 \times 10^{-4}$ and $\alpha_1 = 0.01$. Six different exciter locations are chosen for this purpose. All the exciters are of identical width, $w_{ex} = 0.05$. In Table 3.4, the x-location of the center of the SBS exciter, corresponding Re_{δ^*}, non-dimensional wavenumber α_r, and spatial growth rate α_i at the exciter location, as obtained from the linear stability theory, are tabulated. Since, α_r and α_i are defined with respect to the local displacement thickness δ^*, k_r and k_i, (the non-dimensional wavenumber and spatial growth rate), with respect to the reference length scale L, are also listed in Table 3.4.

One sees from Table 3.4, that the TS wave-packet caused by the exciter at $x_{ex} = 0.1$, should decay faster than any other exciter locations discussed in this section. However, this does not happen for the receptivity cases computed here. In Fig. 3.54a, b, wall-normal variation of $|u_d|/|u_d|_{max}$ and $|v_d|/|v_d|_{max}$ are plotted for all the exciter locations at some representative times, after the complete establishment of the TS wave-packet. The profiles are plotted at a location of $x = 0.075$ downstream of x_{ex}. One notes from Fig. 3.54a, b that the height where $u_d = 0$ and v_d is maximum lies in the range $0.005 \leq y \leq 0.01$. Hence, one expects u_d to display growth in this range of heights as explained in Sect. 3.9.4.

For the subsequent discussions, the onset and end-location of the spatial growth or near-neutral behavior of the u_d corresponding to the TS wave-packet at a particular height is referred as TS_{onset} and TS_{end}, respectively. Hence, here TS_{onset} and TS_{end}

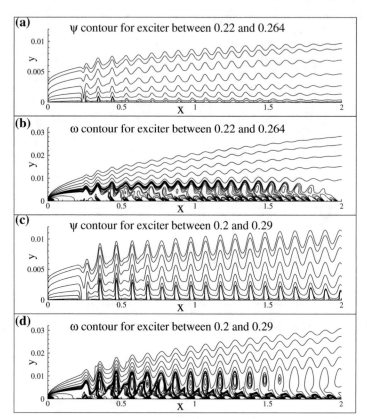

Fig. 3.53 Stream function and vorticity contours for the two different strip exciter cases having different width. **a** and **b** correspond to exciter located between 0.22 and 0.264; **c** and **d** correspond to exciter located between 0.2 and 0.29

Table 3.4 Location of the exciter x_{ex}, corresponding Re_{δ^*}, non-dimensional wavenumbers α_r; k_r and spatial growth rates α_i; k_i at the exciter location as obtained from the linear stability theory are tabulated. Here, α_r and α_i are defined with respect to local δ^*, whereas k_r and k_i is defined with respect to L. For all the cases listed here, $F = 3.0 \times 10^{-4}$ and $\alpha_1 = 0.01$

x_{ex}	Re_{δ^*}	α_r	α_i	k_r	k_i
0.1	172.00	0.1377	0.037790	80.0581	21.9709
0.25	271.96	0.2079	0.019461	76.4451	7.1558
0.4	344.00	0.2559	0.009830	74.3895	2.8576
0.625	430.00	0.3149	0.003061	73.2326	0.7119
0.8	486.49	0.3546	0.002609	72.8895	0.5363
1.0	543.91	0.3955	0.006259	72.7142	1.1507

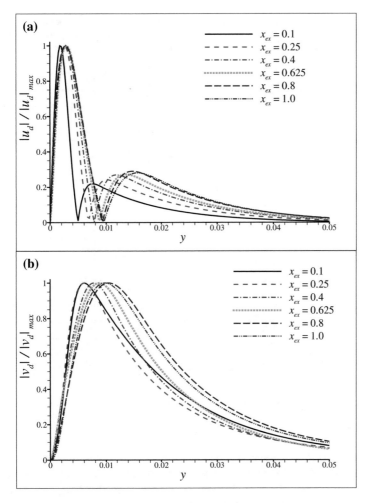

Fig. 3.54 **a** $|u_d|/|u_d|_{max}$ and **b** $|v_d|/|v_d|_{max}$ plotted as a function of y for the indicated exciter locations times after the complete establishment of the TS wave-packet. The profiles are plotted at a location of $x = 0.075$ downstream of the center of the exciter. Here, $F = 3.0 \times 10^{-4}$ and $\alpha_1 = 0.01$

are only the functions of height and x_{ex}. In Table 3.5, TS_{onset} and TS_{end} for u_d at $t = 20$ are tabulated as a function of y for the indicated exciter locations. From Table 3.5, one notes the existence of three distinct ranges of height. The first range extends up to $y \simeq 0.00956$ from the wall, whereas the second range lies in $0.01113 \leq y \leq 0.01274$. The third range lies at the top of the second range. For $x_{ex} = 0.1$ and 0.25, TS_{onset} and TS_{end} occurs at identical locations for $0.00662 \leq y \leq 0.00956$. Other exciter locations also progressively display similar values of TS_{onset} and TS_{end}, as the height approaches 0.00956. Monotonic spatial decay is observed for all the

Table 3.5 The onset and end of the spatial growth or the near-neutral behaviour of the TS wave-packet listed for all heights, for the indicated locations of exciter. Here, $F = 3.0 \times 10^{-4}$ and $\alpha_1 = 0.01$. The blank cells indicate those heights, where no spatial growth is noticed

$100 \times y$	$x_{ex} = 0.1$	$x_{ex} = 0.25$	$x_{ex} = 0.4$	$x_{ex} = 0.625$	$x_{ex} = 0.8$	$x_{ex} = 1.0$
0.521	0.168	0.326	0.49	0.75	–	–
	0.875	0.878	0.855	0.889	–	–
0.662	0.308	0.301	0.49	0.715	–	–
	0.874	0.888	0.908	0.908	–	–
0.808	0.54	0.54	0.52	0.65	0.85	–
	1.02	1.07	1.05	1.05	1.05	–
0.958	0.948	0.94	0.94	0.93	0.95	1.12
	1.29	1.28	1.28	1.28	1.29	1.26
1.113	–	–	–	–	–	–
	–	–	–	–	–	–
1.274	–	–	–	–	–	–
	–	–	–	–	–	–
2.549	0.48	0.49	0.52	0.74	0.83	–
	0.95	0.93	0.93	0.93	0.93	–
2.759	0.52	0.53	0.54	0.70	0.88	–
	0.92	0.97	0.93	0.96	0.97	–
2.966	0.47	0.48	0.52	0.76	0.88	–
	0.96	0.97	0.96	0.97	0.97	–
3.188	0.516	0.53	0.52	0.69	0.88	–
	0.96	0.97	0.96	0.96	0.97	–

exciter locations in the second range of heights ($0.1113 \leq y \leq 0.1274$). The heights corresponding to the third range are closer to the shear layer edge, where one observes that first five exciter locations display similar locations for TS_{end}, while the first three cases display similar locations of TS_{onset}. However the $x_{ex} = 1.0$ case, does not display any spatial growth of the TS wave-packet at these heights, because the exciter itself is located downstream of corresponding TS_{end} location for the other exciter location indicated in the table.

3.9.11 Effects of Frequency of Excitation

Here, the effects of frequency of excitation are considered for the exciter of width $\Delta x_{ex} = 0.05$ located at $x_{ex} = 0.25$ ($Re_{\delta^* ex} = 271.96$), with amplitude of excitation $\alpha_1 = 0.01$. Here, six different frequencies of excitation, with values of $F = 1.5 \times 10^{-4}, 2.0 \times 10^{-4}, 2.5 \times 10^{-4}, 3.5 \times 10^{-4}, 4.0 \times 10^{-4}, 4.5 \times 10^{-4}, 4.0 \times 10^{-4}$ and 5.5×10^{-4} are considered. Out of these six frequencies, only the

Table 3.6 Non-dimensional frequency of excitation F, corresponding non-dimensional wavenumbers α_r; k_r and spatial growth rates α_i; k_i at the exciter location as obtained from the linear stability theory along with the entry (x_{N1}) and exit (x_{N2}) point of the $F = Const.$ line into the neutral curve are tabulated. Here, α_r and α_i is defined with respect to local δ^*, whereas k_r and k_i is defined with respect to reference length scale L. For all the cases listed here, the exciter is located at $x_{ex} = 0.25$ with $\alpha_1 = 0.01$

$F_f \times 10^4$	α_r	α_i	k_r	k_i	x_{N1}	x_{N2}
1.5	0.12055	0.02846	44.16325	10.42750	1.102	1.644
2.0	0.15191	0.02514	55.65174	9.20941	0.985	1.2775
2.5	0.18099	0.02207	66.30495	8.08661	–	–
3.0	0.20879	0.01929	76.49143	7.06664	–	–
3.5	0.23585	0.01687	86.40424	6.17933	–	–
4.0	0.26246	0.01490	96.15182	5.45846	–	–
4.5	0.28878	0.01347	105.79572	4.93468	–	–
5.0	0.31491	0.01265	115.36928	4.63332	–	–
5.5	0.34090	0.01249	124.89009	4.57481	–	–

$F = 1.5 \times 10^{-4}$ and $F = 2.0 \times 10^{-4}$ line intersects the neutral curve, whereas the straight line corresponding to the other frequencies stay outside the neutral curve. In Table 3.6, the non-dimensional wavenumbers α_r; k_r and spatial growth rates α_i; k_i at the exciter location, as obtained from the linear stability theory are tabulated, along with the entry (x_{N1}) and exit (x_{N2}) point of the $F = Const.$ line into the neutral curve. The table includes the $F = 3.0 \times 10^{-4}$ case as well, for the purpose of comparison, whose receptivity results have already been discussed.

As noted in Table 3.6, the cases of $F = 1.5 \times 10^{-4}$ and 2.0×10^{-4} lines intersect the neutral curve at $(x_{N1}, x_{N2}) = (1.102, 1.644)$ and $(x_{N1}, x_{N2}) = (0.985, 1.2775)$, respectively. Other frequency cases do not intersect the neutral curve at all. In Fig. 3.55, the computed receptivity results are shown at $y = 0.0254961$ by plotting u_d at $t = 20$ as a function of x. The vertical lines drawn in frames (a) and (b), indicate intersection points of the neutral curve with corresponding $F = Const.$ lines. In Table 3.7, the TS_{onset} and TS_{end} of the TS wave-packet are listed for several heights for indicated frequencies of excitation cases. In this table, the blank cells indicate heights where no spatial growth is observed.

From this table one notes that the strict spatial growth of the amplitude of the TS wave-packet for $F = 1.5 \times 10^{-4}$, 2.0×10^{-4} and 2.5×10^{-4} takes place for three different ranges of height. In the first range of heights, spatial growth of u_d takes place for these frequencies. This range of heights extend from the wall to $y \simeq 0.016$, 0.01273 and 0.0113 for $F = 1.5 \times 10^{-4}$, 2.0×10^{-4} and 2.5×10^{-4}, respectively, as noted in Table 3.7. Both TS_{onset} and TS_{end} move downstream with increase in height in this range of heights for these frequencies. Also the extent of the zone of spatial growth $(TS_{end} - TS_{onset})$ is also seen to reduce with increase in height in this range. Beyond this range, there exists another range of heights where no spatial growth for u_d is noted for these frequencies. This second range of heights are close to the y-location, where respective u_d changes sign. Above this range of heights, the

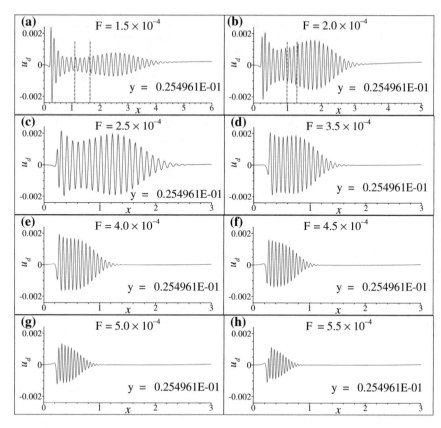

Fig. 3.55 u_d plotted as a function of x for the indicated values of F at $y = 0.02549$ and $t = 20$. Here, the exciter of width $\Delta x_{ex} = 0.05$ is located at $x_{ex} = 0.25$ with $\alpha_1 = 0.01$. The vertical straight lines in **a** and **b** indicate the entry and exit point of the corresponding $F_f = Const.$ line into the neutral curve

spatial growth of u_d is again observed for these frequencies. This range starts from $y \simeq 0.02549, 0.0161$ and 0.01439 for $F = 1.5 \times 10^{-4}, 2.0 \times 10^{-4}$ and 2.5×10^{-4}, respectively.

For $F \geq 3.0 \times 10^{-4}$, there exists a range of heights close to the wall, where no spatial growth is noted. However, spatial growth of u_d is resumed above this range, for these frequencies, as noted from Table 3.7. The terminal height of this second range of spatial growth for $3.0 \times 10^{-4} \leq F \leq 5.5 \times 10^{-4}$ is noted to decrease with increase in the frequency. For $3.0 \times 10^{-4} \leq F \leq 4.0 \times 10^{-4}$, once again a range of heights exists, where monotonic spatial decay of the amplitude of the TS wave-packet is noted. One notes from Table 3.7 that the extent of this third range of heights increases with increase in F from 3.0×10^{-4} to 4.0×10^{-4}. The spatial growth of the TS wave-packet resumes above this third range of heights for $3.0 \times 10^{-4} \leq F \leq 4.0 \times 10^{-4}$. However, no spatial growth of the TS wave-packet can be observed for all the heights above $y \simeq 0.00662$ for $F \geq 4.5 \times 10^{-4}$.

Table 3.7 The onset and end of the spatial growth or the near-neutral behaviour of the TS wave-packet for all the heights shown in Fig. 3.55 for indicated $F \times 10^4$ cases. The blank cells indicate heights, where no spatial growth is observed

$100 \times y$	1.5	2.0	2.5	3.0	3.5	4.0	4.5	5.0	5.5
0.252	1.082	0.916	0.771	–	–	–	–	–	–
	2.607	1.701	1.198	–	–	–	–	–	–
0.384	1.233	0.903	0.741	–	–	–	–	–	–
	2.606	1.694	1.181	–	–	–	–	–	–
0.521	1.341	0.854	0.758	0.326	0.408	0.289	0.307	0.319	0.308
	2.612	1.677	1.125	0.878	0.611	0.509	0.471	0.422	0.392
0.662	1.365	0.760	0.299	0.303	0.333	0.369	0.410	0.486	0.569
	2.613	1.692	1.186	0.888	0.761	0.661	0.622	0.652	0.597
0.808	1.316	0.442	0.481	0.539	0.615	0.739	–	–	–
	2.629	1.804	1.235	1.071	0.911	0.929	–	–	–
0.958	0.569	0.696	0.749	0.939	–	–	–	–	–
	2.643	1.723	1.386	1.276	–	–	–	–	–
1.113	0.854	0.996	1.253	–	–	–	–	–	–
	2.691	1.918	1.699	–	–	–	–	–	–
1.273	1.250	1.545	–	–	–	–	–	–	–
	2.818	2.163	–	–	–	–	–	–	–
1.439	1.741	–	0.621	0.563	0.487	–	–	–	–
	2.960	–	0.873	0.760	0.645	–	–	–	–
1.610	2.487	0.905	0.614	0.536	0.471	0.434	–	–	–
	3.351	1.334	1.110	0.881	0.670	0.549	–	–	–
1.786	–	0.880	0.622	0.536	0.463	0.438	–	–	–
	–	1.526	1.174	0.909	0.732	0.555	–	–	–
2.549	1.132	0.851	0.629	0.521	0.464	0.436	–	–	–
	2.551	1.769	1.277	0.931	0.733	0.550	–	–	–

References

1. Allen, L., & Bridges, T. J. (2002). Numerical exterior algebra and the compound matrix method. *Numerische Mathematik, 92*, 197–232.
2. Alizard, F., & Robinet, J. (2007). Spatially convective global modes in a boundary layer. *Physics of Fluids, 19*, 114105.
3. Arfken, G. (1985). *Mathematical methods for physicists* (3rd ed.). Orlando: Academic Press.
4. Barkley, D., Gomes, M. G. M., & Henderson, R. D. (2002). Three-dimensional instability in flow over a backward-facing step. *Journal of Fluid Mechanics, 473*, 167–190.
5. Barkley, D., Blackburn, H. M., & Sherwin, S. J. (2008). Direct optimal growth analysis for timesteppers. *International Journal for Numerical Methods in Fluids, 57*, 1435–1458.
6. Bender, C. M., & Orszag, S. A. (1987). *Advanced mathematical methods for scientists and engineers*. Singapore: McGraw Hill Book Co., International Edition.
7. Bhaumik, S. (2013). Direct numerical simulation of inhomogeneous transitional and turbulent flows. Ph. D. thesis, I. I. T. Kanpur

8. Bhumkar, Y. G. (2011). High performance computing of bypass transition. Ph.D. thesis, I. I. T. Kanpur
9. Bhaumik, S., & Sengupta, T. K. (2014). Precursor of transition to turbulence: Spatiotemporal wave front. *Physical Review E*, *89*(4), 043018.
10. Bhaumik, S., & Sengupta, T. K. (2017). Impulse response and spatio-temporal wave-packets: The common feature of rogue waves, tsunami and transition to turbulence. *Physics of Fluids*, *29*, 124103.
11. Blackburn, H. M., Barkley, D., & Sherwin, S. J. (2008). Convective instability and transient growth in flow over a backward-facing step. *Journal of Fluid Mechanics*, *603*, 271–304.
12. Brandt, L., & Henningson, D. S. (2002). Transition of streamwise streaks in zero-pressure-gradient boundary layers. *Journal of Fluid Mechanics*, *472*, 229–261.
13. Brewstar, R. A., & Gebhart, B. (1991). Instability and disturbance amplification in a mixed-convection boundary layer. *Journal of Fluid Mechanics*, *229*, 115–133.
14. Cebeci, T., & Bradshaw, P. (1977). *Momentum transfer in boundary layers*. Washington, DC: Hemisphere Publishing Corporation.
15. Chen, T. S., & Moutsoglu, A. (1979). Wave instability of mixed convection flow on inclined surfaces. *Numerical Heat Transfer*, *2*, 497–509.
16. Chen, T. S., & Mucoglu, A. (1979). Wave instability of mixed convection flow over a horizontal flat plate. *International Journal of Heat and Mass Transfer*, *22*, 185–196.
17. Chen, T. S., Sparrow, E. M., & Mucoglu, A. (1977). Mixed convection in boundary layer flow on a horizontal plate. *ASME Journal of Heat Transfer*, *99*, 66–71.
18. Chomaz, J. M. (2005). Global instabilities in spatially developing flows: Non-normality and nonlinearity. *Annual Review of Fluid Mechanics*, *37*, 357–392.
19. Drazin, P. G., & Reid, W. H. (1981). *Hydrodynamic stability*. UK: Cambridge University Press.
20. Eckert, E. R. G. & Soehngen, E. (1951). Interferometric studies on the stability and transition to turbulence in a free-convection boundary-layer. *Proceedings of the General Discussion on Heat Transfer, ASME and IME London* (Vol. 321) (1951)
21. Eiseman, P. R. (1985). Grid generation for fluid mechanics computation. *Annual Review of Fluid Mechanics*, *17*, 487–522.
22. Fasel, H., & Konzelmann, U. (1990). Non-parallel stability of a flat-plate boundary layer using the complete Navier-Stokes equations. *Journal of Fluid Mechanics*, *221*, 311–347.
23. Gaster, M. (1974). On the effect of boundary-layer growth on flow stability. *Journal of Fluid Mechanics*, *66*(3), 465–480.
24. Gebhart, B., Jaluria, Y., Mahajan, R. L., & Sammakia, B. (1988). *Buoyancy-induced flows and transport*. Washington, DC: Hemisphere Publications.
25. Gilpin, R. R., Imura, H., & Cheng, K. C. (1978). Experiments on the onset of longitudinal vortices in horizontal Blasius flow heated from below. *ASME Journal of Heat Transfer*, *100*, 71–77.
26. Haaland, S. E., & Sparrow, E. M. (1973). Vortex instability of natural convection flows on inclined surfaces. *International Journal of Heat Mass Transfer*, *16*, 2355–2367.
27. Hall, P., & Morris, H. (1992). On the instability of boundary layers on heated flat plates. *Journal of Fluid Mechanics*, *245*, 367–400.
28. Heisenberg, W. (1924). Über stabilität und turbulenz von flüssigkeitsströmen. *Annalen der Physik Leipzig*, *379*, 577–627 (Translated as 'On stability and turbulence of fluid flows'. NACA Tech. Memo. Wash. No 1291 1951)
29. Iyer, P. A., & Kelly, R. E. (1974). The instability of the laminar free convection flow induced by a heated, inclined plate. *International Journal of Hear Mass Transfer*, *17*, 517–525.
30. Jain, M. K., Iyengar, S. R. K. & Jain, R. K. (2003). *Numerical methods for scientific and engineering computation*. New Delhi: New Age International
31. Kaiktsis, L., Karniadakis, G. M., & Orszag, S. A. (1991). Onset of three-dimensionality, equibria and early transition in flow over a backward-facing step. *Journal of Fluid Mechanics*, *231*, 501–528.
32. Kaiktsis, L., Karniadakis, G. M., & Orszag, S. A. (1996). Unsteadiness and convective instabilities in two-dimensional flow over a backward-facing step. *Journal of Fluid Mechanics*, *321*, 157–187.

33. Kloker, M., Konzelmann, U., & Fasel, H. (1993). Outflow boundary conditions for spatial Navier-Stokes simulations of transitional boundary layers. *AIAA Journal, 31*, 620.
34. Kreyszig, E. (1999). *Advanced engineering mathematics*. Singapore: Wiley.
35. Leal, L. G. (1973). Steady separated flow in a linearly decelerated free stream. *Journal of Fluid Mechanics, 59*, 513–535.
36. Liu, Z., & Liu, C. (1994). Fourth order finite difference and multigrid methods for modeling instabilities in flat plate boundary layer-2D and 3D approaches. *Computers and Fluids, 23*, 955–982.
37. Lloyd, J. R., & Sparrow, E. M. (1970). On the instability of natural convection flow on inclined plates. *Journal of Fluid Mechanics, 42*, 465–470.
38. Lord, R. (1880). On the stability or instability of certain fluid motions. *Scientific Papers, 1*, 361–371.
39. Marquet, O., Sipp, D., Chomaz, J. M., & Jacquin, L. (2008). Amplifier and resonator dynamics of a low Reynolds-number recirculation bubble in a global framework. *Journal of Fluid Mechanics, 605*, 429–443.
40. Moutsoglu, A., Chen, T. S., & Cheng, K. C. (1981). Vortex instability of mixed convection flow over a horizontal flat plate. *ASME Journal of Heat Transfer, 103*, 257–261.
41. Mucoglu, A., & Chen, T. S. (1978). Wave instability of mixed convection flow along a vertical flat plate. *Numerical Heat Transfer, 1*, 267–283.
42. Mureithi, E. W., & Denier, J. P. (2010). Absolute-convective instability of mixed forced-free convection boundary layers. *Fluid Dynamics Research, 372*, 517–534.
43. Ng, B. S., & Reid, W. H. (1980). On the numerical solution of the Orr-Sommerfeld problem: Asymptotic initial conditions for shooting method. *Journal of Computational Physics, 38*, 275–293.
44. Ng, B. S., & Reid, W. H. (1985). The compound matrix method for ordinary differential systems. *Journal of Computational Physics, 58*, 209–228.
45. Rajpoot, M. K., Sengupta, T. K., & Dutt, P. K. (2010). Optimal time advancing dispersion relation preserving schemes. *Journal of Computational Physics, 229*(10), 3623–3651.
46. Saric, W. S., & Nayfeh, A. H. (1975). Nonparallel stability of boundary-layer flows. *Physics of Fluids, 18*(8), 945–950.
47. Schlatter, P., & Örlü, R. (2012). Turbulent boundary layers at moderate Reynolds numbers. *Journal of Fluid Mechanics, 710*, 5–34.
48. Schlichting, H. (1933). Zur entstehung der turbulenz bei der plattenströmung. *Nach. Gesell. d. Wiss. z. Gött., MPK,42*, 181–208
49. Schneider, W. (1979). A similarity solution for combined forced and free convection flow over a horizontal plate. *International Journal of Heat and Mass Transfer, 22*, 1401–1406.
50. Schubauer, G. B., & Skramstad, H. K. (1947). Laminar boundary layer oscillations and the stability of laminar flow. *Journal of Aerosol Science, 14*(2), 69–78.
51. Sengupta T. K. (1990). Receptivity of a growing boundary layer to surface excitation. (Unpublished manuscript).
52. Sengupta, T. K. (1991). Impulse response of laminar boundary layer and receptivity. In C. Taylor (Ed.), *Proceedings of the 7th International Conference Numerical Methods in Laminar and Turbulent Layers*. Stanford University
53. Sengupta, T. K. (2012). *Instabilities of flows and transition to turbulence*. Florida, USA: CRC Press, Taylor & Francis Group.
54. Sengupta, T. K. (2013). *High accuracy computing methods: Fluid flows and wave phenomenon*. USA: Cambridge University Press.
55. Sengupta, T. K., & Bhaumik, S. (2011). Onset of turbulence from the receptivity stage of fluid flows. *Physical Review Letters, 154501*, 1–5.
56. Sengupta, T. K., & Venkatasubbaiah, K. (2006). Spatial stability for mixed convection boundary layer over a heated horizontal plate. *Studies in Applied Mathematics, 117*, 265–298.
57. Sengupta, T. K., Ballav, M., & Nijhawan, S. (1994). Generation of Tollmien-Schlichting waves by harmonic excitation. *Physics of Fluids, 6*(3), 1213–1222.

58. Sengupta, T. K., Bhaumik, S., Singh, V., & Shukl, S. (2009). Nonlinear and nonparallel receptivity of zero-pressure gradient boundary layer. *International Journal of Emerging Multidisciplinary Fluid Sciences, 1,* 19–35.
59. Sengupta, T. K., Chattopadhyay, M., Wang, Z. Y., & Yeo, K. S. (2002). By-pass mechanism of transition to turbulence. *Journal of Fluids and Structures, 16,* 15–29.
60. Sengupta, T. K., De, S., & Sarkar, S. (2003). Vortex-induced instability of an incompressible wall-bounded shear layer. *Journal of Fluid Mechanics, 493,* 277–286.
61. Sengupta, T. K., Rao, A. K., & Venkatasubbaiah, K. (2006). Spatiotemporal growing wave fronts in spatially stable boundary layers. *Physical Review Letters, 96*(22), 224504.
62. Sengupta, T. K., Rao, A. K., & Venkatasubbaiah, K. (2006). Spatiotemporal growth of disturbances in a boundary layer and energy based receptivity analysis. *Physics of Fluids, 18,* 094101.
63. Sengupta, T. K., Sircar, S. K., & Dipankar, A. (2006). High accuracy compact schemes for DNS and acoustics. *Journal of Scientific Computing, 26*(2), 151–193.
64. Sengupta, T. K., Unnikrishnnan, S., Bhaumik, S., Singh, P., & Usman, S. (2011). Linear spatial stability analysis of mixed convection boundary layer over a heated plate. *Program in Applied Mathematics, 1*(1), 71–89.
65. Sengupta, T. K., Bhaumik, S., & Bhumkar, Y. (2012). Direct numerical simulation of two-dimensional wall-bounded turbulent flows from receptivity stage. *Physical Review E, 85*(2), 026308.
66. Sengupta, T. K., Bhaumik, S., & Bose, R. (2013). Direct numerical simulation of transitional mixed convection flows: Viscous and inviscid instability mechanisms. *Physics of Fluids, 25,* 094102.
67. Shaukatullah, H., & Gebhart, B. (1978). An experimental investigation of natural convection flow on an inclined surface. *International Journal of Heat and Mass Transfer, 21,* 1481–1490.
68. Sparrow, E. M., & Husar, R. B. (1969). Longitudinal vortices in natural convection flow on inclined plates. *Journal of Fluid Mechanics, 37,* 251–255.
69. Sparrow, E. M., & Minkowycz, W. J. (1962). Buoyancy effects on horizontal boundary-layer flow and heat transfer. *International Journal of Heat and Mass Transfer, 5,* 505–511.
70. Skote, M., Haritonidis, J. H., & Henningson, D. S. (2002). Varicose instabilities in turbulent boundary layers. *Physics of Fluids, 14,* 2309–2323.
71. Tollmien, W. (1931). Über die enstehung der turbulenz. I, English translation. *NACA TM 609*
72. Tumin, A. (2003). The spatial stability of natural convection flow on inclined plates. *ASME Journal of Fluids Engineering, 125,* 428–437.
73. Unnikrishnan, S. (2011). Linear stability analysis and nonlinear receptivity study of mixed convection boundary layer developing over a heated flat plate. *M. Tech. thesis* (I.I.T. Kanpur, 2011)
74. Van der Pol, B., & Bremmer, H. (1959). *Operational calculus based on two-sided Laplace integral.* Cambridge, UK: Cambridge University Press.
75. Wang, X. A. (1982). An experimental study of mixed, forced, and free convection heat transfer from a horizontal flat plate to air. *ASME Journal of Heat Transfer, 104,* 139–144.
76. Wu, R. S., & Cheng, K. C. (1976). Thermal instability of Blasius flow along horizontal plates. *International Journal of Heat and Mass Transfer, 19,* 907–913.
77. Zhang, S. L. (1997). GPBi-CG: Generalized product-type methods based on Bi-CG for solving Non symmetric linear systems. *SIAM Journal on Scientific Computing, 18*(2), 537–551.
78. Zuercher, E. J., Jacobs, J. W., & Chen, C. F. (1998). Experimental study of the stability of boundary-layer flow along a heated inclined plate. *J. Fluid Mech., 367,* 1–25.

Chapter 4
Nonlinear Theoretical and Computational Analysis of Fluid Flows

Morkovin [33] classified transition to turbulence in to two main types: (i) The classical primary instability route whose onset is marked along with the presence of TS waves (as in ZPGBL) and (ii) the bypass routes, which encompass all other possible transition scenarios that do not exhibit TS waves. Unfortunately, this is too simplistic a classification scheme for the following reasons.

It is believed that canonical ZPG flow past a flat plate experience disturbance growth via linear spatial viscous instability theory governed by the OSE. This primary instability is indicated by the appearance of TS wave, as was experimentally verified by Schubauer and Skramstad [47]. Ever since, this has become the centerpiece of instability research for external flows past streamlined bodies. We often loose sight of the fact that this experiment achieved success after repeated attempts, starting with creation of a ultra-quiet wind tunnel, with extremely low noise. In this tunnel, the investigators imposed monochromatic time harmonic excitation inside the boundary layer forming over a ZPG flow to create TS wave. Acoustic disturbances imposed from outside was not at all effective in creating TS wave. Thus, experimental verification of transition process is far removed from the natural transition in many respects. The major distinction comes from the fact that in natural setting, the boundary layer is excited simultaneously by multiple frequencies by vortical, entropic and acoustic disturbances. As a consequence, the response field will be a combination of multiple TS waves convoluting each other. This has been demonstrated by considering multi-periodic excitation in [60] for Blasius boundary layer, where all the constituent TS waves were unstable, yet the composite response field obtained from solution of the OSE did not show the response that could be construed as TS wave for a long time, essentially caused by wave cancellation due to phase shift among the constituents. Thus, TS waves have been noted in extremely controlled experiments. It is noted that in the presence of adverse pressure gradient, such spatial instabilities are stronger in terms of lower critical Reynolds number and higher growth rate.

Flow transitions which do not display TS waves as a marker of instability, are said to follow the bypass route. A similar point of view was also advanced in [6].

© Springer Nature Singapore Pte Ltd. 2019
T. K. Sengupta and S. Bhaumik, *DNS of Wall-Bounded Turbulent Flows*,
https://doi.org/10.1007/978-981-13-0038-7_4

There are many such flows, where transition are caused without the presence of monochromatic unstable TS waves. Prime examples of bypass transition are flows past bluff bodies, Couette and pipes flows, leading edge contamination on swept wings, two-dimensional and three-dimensional roughness effects, etc. In some of these flows, unstable TS waves are not noted; instead disturbances are seen to grow with time due to the presence of inflection point(s) for the velocity profile. Flow past a bluff body is a typical example. In contrast, plane Poisseuille flow displays a critical Reynolds number in excess of 5770, while the flow is noted experimentally to be turbulent at significantly lower Reynolds numbers (around 1000 in [10]). This flow is also not characterized by an inflectional velocity profile, yet it displays subcritical transition. Another example of interest is the flow undergoing transition, right at the leading edge of a swept wing of an aircraft. There are no definitive physical mechanisms identified in the system portrait of Fig. 3.1 in [33] for any of these bypass transition examples. Instead, question marks have been raised related to the possibility of nonlinear, nonparallel or some unknown mechanisms as potential reasons. In this chapter, we focus upon the physical mechanism for a few prototypical examples of flow transition with the help of results from carefully designed receptivity experiments and accurate flow computations. Next, we study cases where the excitation is applied at the free stream without introducing any apparent time scale(s), unlike the multi-periodic cases discussed in the previous chapter.

The idea that a distant vortex can induce a small longitudinal adverse pressure gradient to destabilize a wall-bounded flow was mooted in [76], while studying dependence of critical Reynolds number upon free-stream turbulence (FST). Monin and Yaglom [32] in discussing this work noted that the change in critical Reynolds number by the small longitudinal adverse pressure gradient is due to a sequence of unsteady separations, presumably created by a train of vortices embedded in the FST. The assumption implicit in this scenario is that the effect is connected with the generation of fluctuations of longitudinal pressure gradient by these disturbances, leading to the random formation of individual spots of unstable S-shaped velocity profiles [32].

As mentioned above, that exciting a laminar flow by acoustic excitation from the free stream in [47] was not successful, while vortical disturbances inside the boundary layer triggered transition. In the experiment of Leib et al. [27], authors used vibrating ribbon in the free stream to cause transition. It has also been shown in [55] that free stream excitation can be effective also in the linear viscous instability theory model given by the OSE.

To model the FST effects on boundary layer, experiments have been performed in [28, 54, 56], to study the unit process of a convecting vortex moving at a constant height and speed over a zero pressure gradient boundary layer. In this point of view, FST can be thought of as an assembly of vortices moving aperiodically at arbitrary heights. One of the key features of this experimental observation was that the ensuing instability is subcritical with respect to instability given by the OSE (displaying unstable TS waves) and this vortex-induced instability gives rise to unsteady separation. As there are no linear theories available for studying vortex dominated flows, a nonlinear theory was developed [56], which is explained in the following for a general nonlinear instability for incompressible flows from NSE without making any assumption, based on mechanical energy.

4.1 Nonlinear Instability Theory Based on Mechanical Energy

Landahl and Mollo-Christensen [26] noted that to understand turbulence, one must consider growth of total mechanical energy and not just simply the disturbance turbulent kinetic energy (TKE). According to these authors, "it is possible to understand such behaviour by studying the redistribution of the total mechanical energy of the flow". In contrast to popular practice of characterizing turbulence by discussing about TKE, one must consider the roles of fluctuating pressure, as Morkovin [33] suggested that unsteadiness during bypass transition can be due to shear noise term in the Poisson equation for the static pressure. This prompted the authors in [56] to develop an equation for the total mechanical energy from NSE, for the equilibrium and the disturbance flow fields. Such equations for receptivity must also be capable to explain classical linear theories as special cases. This has been demonstrated in [59, 60] showing the existence of TS waves caused by spatially localized excitation by this energy-based approach. In the following, we show the derivation of the equation for mechanical energy of incompressible fluid flow. NSE is written using vector notation in rotational form, with the vector identity for the convective acceleration term,

$$(\vec{V} \cdot \nabla)\vec{V} = \frac{1}{2}\nabla(|\vec{V}|^2) - \vec{V} \times (\nabla \times \vec{V})$$

to yield the NSE as,

$$\frac{\partial \vec{V}}{\partial t} + \frac{1}{2}\nabla(|\vec{V}|^2) - \vec{V} \times \vec{\omega} = -\frac{\nabla p}{\rho} + \nu\nabla^2\vec{V} \tag{4.1}$$

Further, one can alter the diffusion term with the following identity,

$$\nabla^2\vec{V} = \nabla(\nabla \cdot \vec{V}) - \nabla \times (\nabla \times \vec{V})$$

For incompressible flows, one gets the rotational form of NSE given by,

$$\frac{\partial \vec{V}}{\partial t} - \vec{V} \times \vec{\omega} = -\nabla\left(\frac{p}{\rho} + \frac{|\vec{V}|^2}{2}\right) - \nu(\nabla \times \vec{\omega}) \tag{4.2}$$

Reverting back to Laplacian form of diffusion term and defining the total mechanical energy as

$$E = \frac{p}{\rho} + \frac{|\vec{V}|^2}{2}$$

NSE is written with E as

$$\frac{\partial \vec{V}}{\partial t} - \vec{V} \times \vec{\omega} = -\nabla E + \nu \nabla^2 \vec{V} \tag{4.3}$$

Taking a divergence and making incompressibility assumption for the equation above, one gets the governing Poisson equation for E as,

$$\nabla^2 E = \nabla \cdot (\vec{V} \times \vec{\omega}) \tag{4.4}$$

Further using the identity,

$$\nabla \cdot (\vec{V} \times \vec{\omega}) = \vec{\omega} \cdot (\nabla \times \vec{V}) - \vec{V} \cdot (\nabla \times \vec{\omega})$$

one gets finally the governing equation for the total mechanical energy as

$$\nabla^2 E = \vec{\omega} \cdot \vec{\omega} - \vec{V} \cdot (\nabla \times \vec{\omega}) \tag{4.5}$$

Thus, the distribution of E is directly related to enstrophy (the first term on the right hand side). Without solving this equation, one can comment about the solution from the sign of the right-hand side. It indicates the presence of a source or a sink of E, as given by the property of Poisson equation for energy in [73]. A negative sign signifies a local source. Thus, it may appear that the enstrophy stabilizes E in Eq. (4.5). The development here is for nonlinear evolution of a flow, by tracing destabilization of an equilibrium state. A more subtle picture emerges, if one divides E into an equilibrium and a disturbance part with the help of the subscripts, m and d in, $E = E_m + \varepsilon E_d$. If one is interested in the growth of small perturbation applied to equilibrium state, then the parameter ε represents a small value, and apply same splitting scheme for all the quantities appearing in Eq. (4.5). The disturbance energy equation for E_d is obtained by subtracting the equation for equilibrium quantities from the equation for total quantities as,

$$\nabla^2 E_d = 2\vec{\omega}_m \cdot \vec{\omega}_d + \varepsilon \vec{\omega}_d \cdot \vec{\omega}_d - \vec{V}_m \cdot (\nabla \times \vec{\omega}_d) - \vec{V}_d \cdot (\nabla \times \vec{\omega}_m) - \varepsilon \vec{V}_d \cdot (\nabla \times \vec{\omega}_d) \tag{4.6}$$

which can describe the onset of instability for an equilibrium flow. In [59, 60], existence of TS waves caused by spatially localized excitation for a boundary layer is shown, by taking the equilibrium state as the Blasius profile and solving the linearized version in Eq. (4.6). In the developed theory based on NSE without any assumption, growth of primary disturbances are due to interactions of velocity and vorticity field acting as source terms on the right-hand side of Eqs. (4.5) and (4.6). NSE is a consequence of conservation of translation momentum, and E remains conserved, as viscous term is absent in Eq. (4.5). Any instability that is observed, is caused by growth of disturbance quantity, whose supply must be from the equilibrium flow or by external agencies of boundary condition(s). Thus, the major issue is about how

he energy is initially exchanged from the equilibrium to the disturbance field and
his is clearly brought out by the first term on the right-hand side of Eq. (4.6), which
indicates an interaction between equilibrium and disturbance vorticity fields or the
disturbance enstrophy.

The above equation required no approximation on the equilibrium and disturbance
ields. Even for unsteady equilibrium flows, one can trace nonlinear instantaneous
nstability. For the creation of surface gravity waves and side-band instability of
surface gravity waves by the mechanism in [2], unsteadiness/instability onset is
triggered by the boundary condition.

4.2 Vortex-Induced Instability: Application of a Nonlinear Theory for Total Mechanical Energy

We focus on the topic of bypass transition by studying the unit process caused by a
single convecting vortex. The same physical mechanism is seen in other examples
of unsteady flow separation (as discussed in [12]), near-wall eddy formation in turbulent boundary layer (as studied in [7]) In [43, 72], the authors have also discussed
formation of hairpin vortices in the near-wall region of turbulent flows. Other studies
[14, 39, 40, 60] considered the scenario where a vortex placed above a plane wall
caused the vortex to move and thereby create a thin unsteady boundary layer over
the wall. All these are examples of vortex-induced instability.

4.2.1 An Experimental Observation of Vortex-Induced Instability

In [28, 54], a vortex with finite core size was created experimentally by a rotating
and translating circular cylinder, whose strength is Γ, and which is at a fixed height
from the plate (H), and the sign of circulation can be easily controlled. The main
emphasis of the experiment was to control all the relevant parameters, so that the
resultant bypass transition can be reproduced. In Fig. 4.1, the schematic of the flow
is shown. It was seen that a slowly convecting vortex of anticlockwise circulation,
creates transition/unsteady separation ahead of it [28, 51, 56].

The experiment was performed in a recirculating water tunnel. One of the cases
is reproduced here to highlight the receptivity mechanism of the shear layer to a
convecting vortex in the free stream. In this experiment, the boundary layer was
formed on a flat plate, held vertically on its edge in the tunnel. A coherent bound
vortex is created by rotating a circular cylinder of diameter 15 mm, whose axis
was along the spanwise direction of the plate. The cylinder can be rotated in either
directions and was rotated at $\Omega = 5$ r.p.s. in the anticlockwise direction, for the case
shown in Fig. 4.2. For flow visualization, food dye was released from six dye ports
located 88 mm downstream from the leading edge of the plate, as seen in Fig. 4.1.

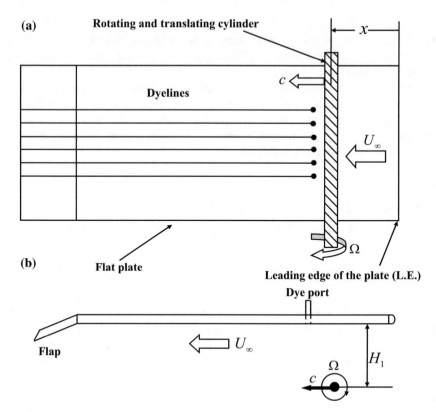

Fig. 4.1 Schematic of the experimental set-up. **a** Side view and **b** top view as seen in the tunnel. Broken line boundary in **b** indicates the computational domain

The Reynolds number based on the diameter of the cylinder and free stream speed ($U_\infty = 162 \, \text{mm s}^{-1}$) was 2600. The cylinder was convected at $c = 0.15U_\infty$, at a height $H = 90 \, \text{mm}$. The ratio of surface speed to the relative speed of the free stream and cylinder was 1.71. The noise level of the water channel is 1% at maximum speed. It is known [1, 58] that a rotating cylinder ceases to shed coherent vortices associated with *Kármán* vortex streets, when the surface speed is more than 1.5 times the free-stream speed. In this experiment, the bound vortex circulation is fixed by controlling the rotation rate of the cylinder, and this controls the dynamics of the flow.

In Fig. 4.2, flow visualization sequences indicate unsteady separation followed by bypass transition in frames ($b - h$). The dye filaments are essentially parallel at the onset (as in frames $a - d$), showing overall two-dimensionality during this stage. In frame (b), the dye filaments released very close to the plate is lifted up due to the imposed disturbances, with negligible spanwise spreading. The location where unsteadiness is seen is laminar in the absence of the convecting vortex, indicating the subcritical nature of the instabilities. Only flow visualization was used, as

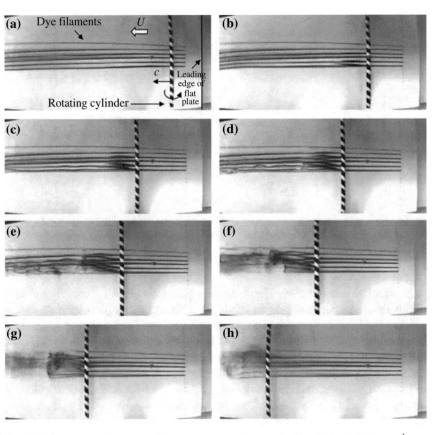

Fig. 4.2 Bypass transition created by a counter-rotating vortex for ($U_\infty = 16.26\,\mathrm{cm\,s}^{-1}$, $c = 0.154$, $H = 9\,\mathrm{cm}$ and $\Omega = +5\,\mathrm{r.p.s.}$) for the experimental arrangement shown in Fig. 4.1

intrusive measurements changed the dynamics drastically. Therefore to quantify the experimental observation, we have undertaken a numerical simulation of the problem described next. As we are interested in the onset stage of the instability, it is sufficient to perform a 2D simulation of NSE.

4.3   Numerical Simulation of Vortex-Induced Instability

Vortex-induced instability problem is used here as a typical case to demonstrate bypass transition caused by a convecting vortex in the free stream. In [56], this problem was solved using (ψ, ω)-formulation due to its inherent advantages. The divergence-free or solenoidality condition for velocity and vorticity is automatically satisfied for 2D flow fields, for this formulation. However, for 3D disturbance computations, direct extension of (ψ, ω)-formulation by the vorticity ($\vec{\omega}$)-vector potential

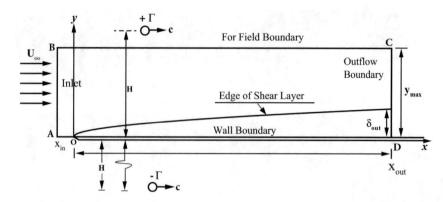

Fig. 4.3 Computational domain used in studying vortex-induced instability by an isolated convecting (at a speed, c) vortex (of circulation, Γ) in the free stream over a flat plate supporting a zero pressure gradient boundary layer

($\vec{\psi}$) formulation is not easy. The authors in [4, 5], developed a velocity-vorticity formulation in a staggered grid, in solving the 3D receptivity problem from excitation stage of laminar flow over a flat plate, to fully developed turbulent state. Here, the same formulation is used for the 2D flow field shown in Fig. 4.3. In this case, receptivity of ZPGBL to convected vortex in the free stream is investigated. This problem requires time-dependent boundary forcing to excite physical instability of the flow. The equilibrium flow is calculated by solving NSE, without the application of external forcing. The origin of the co-ordinate system is placed at the leading edge of the plate, whereas the computational domain starts slightly ahead of it. A counter-clockwise rotating vortex is placed above the plate and this translates from left to right at the constant height of H, with a constant speed, c. Even though this free stream vortex convects far outside the shear layer, the presence of the flat plate requires an image vortex below the plate, to ensure zero wall-normal velocity boundary condition, as shown in the figure.

4.3.1   Velocity-Vorticity Formulations for 2D Flows

Vorticity transport equation is found very relevant for the analysis and solution of viscous incompressible flows [48, 53, 65]. Attendant velocity field can be obtained from the solution of the Poisson equation for the velocity components. In this formulation, the unknowns are velocity components, u, v, and the out of plane vorticity component, ω. For 2D flows, the non-dimensional governing equations for ($\vec{V}$, ω)-formulation in Cartesian co-ordinate are given in [9, 19, 36, 81].

Here, we solve the governing equation in a transformed plane by high accuracy compact scheme for discretization in finite difference framework. More details about the numerical methods, used grids etc. are given in [3, 49]. The governing vorticity transport and velocity Poisson equations in the transformed (ξ, η)-plane are given as

$$\frac{\partial \omega}{\partial t} + \frac{1}{h_1}\frac{\partial}{\partial \xi}(u\omega) + \frac{1}{h_2}\frac{\partial}{\partial \eta}(v\omega) = \frac{1}{Re}\frac{1}{h_1 h_2}\left[\frac{\partial}{\partial \xi}\left(\frac{h_2}{h_1}\frac{\partial \omega}{\partial \xi}\right)\right.$$
$$\left. + \frac{\partial}{\partial \eta}\left(\frac{h_1}{h_2}\frac{\partial \omega}{\partial \eta}\right)\right] \qquad (4.7)$$

$$\left[\frac{\partial}{\partial \xi}\left(\frac{h_2}{h_1}\frac{\partial u}{\partial \xi}\right) + \frac{\partial}{\partial \eta}\left(\frac{h_1}{h_2}\frac{\partial u}{\partial \eta}\right)\right] = -\frac{\partial \omega}{\partial \eta} \qquad (4.8)$$

$$\left[\frac{\partial}{\partial \xi}\left(\frac{h_2}{h_1}\frac{\partial v}{\partial \xi}\right) + \frac{\partial}{\partial \eta}\left(\frac{h_1}{h_2}\frac{\partial v}{\partial \eta}\right)\right] = \frac{\partial \omega}{\partial \xi} \qquad (4.9)$$

where h_1 and h_2 are the scale factors of the transformation given by, $h_1 = \sqrt{x_\xi^2 + y_\xi^2}$ and $h_2 = \sqrt{x_\eta^2 + y_\eta^2}$. The divergence of the velocity field, $D_v = \nabla \cdot \vec{V}$, in the computational (ξ, η)-plane is given as,

$$D_v = \frac{1}{h_1 h_2}\left[\frac{\partial (h_1 v)}{\partial \eta} + \frac{\partial (h_2 u)}{\partial \xi}\right] \qquad (4.10)$$

After solving Eq. (4.8) for u-component of velocity, v-component of velocity is obtained by integrating the right hand side of Eq. (4.10) as equal to zero (solenoidality condition) given as,

$$v(\xi, \eta) = v(\xi, 0) - \frac{1}{h_1}\int_0^\eta \left[\frac{\partial (h_2\, u)}{\partial \xi}\right] d\eta \qquad (4.11)$$

This ensures satisfaction of the divergence-free condition of the velocity vector numerically in the computational plane. For the receptivity of the ZPG flow, boundary condition on v-velocity at the far-field boundary is satisfied asymptotically and hence, Eq. (4.11) is suitable for external flows, which requires only the prescription of v at the bottom boundary segment.

4.3.2 Boundary Conditions

The boundary conditions used in solving this problem are listed as following,

(i) At the inflow (segment AB of Fig. 4.3), uniform inlet velocity U_∞ is imposed along with the contribution imposed by the free stream vortex and its image system.

Corresponding condition is applied on v-velocity component, as computed by the free stream inviscid vortex and its image. The induced stream function created by the finite core translating vortex in conjunction with the uniform flow is given by,

$$\psi_\infty = U_\infty y - (U_\infty - c)\left[\frac{(y-H)(d/2)^2}{\bar{x}^2 + (y-H)^2} + \frac{(y+H)(d/2)^2}{\bar{x}^2 + (y+H)^2}\right] + \frac{\Gamma}{4\pi} Ln \frac{\bar{x}^2 + (y+H)^2}{\bar{x}^2 + (y-H)^2}$$

where d is the core diameter of the convecting vortex of strength, Γ, convecting at a constant height, H. The convection speed of the vortex is given by c, and hence the displacement effect of the finite core vortex is determined by the relative speed, $(U_\infty - c)$, and the circulation effect is given by the last term on the right hand side. Also note that the time dependence of the boundary condition is given by the translated co-ordinate, $\bar{x} = x_0 - ct$, with x_0 indicating the initial co-ordinate of the convecting vortex.

(ii) At the wall, no-slip condition is imposed on both the components of velocity. Wall vorticity is computed based on its kinematic definition.

(iii) In the segment AO of Fig. 4.3, vorticity and the v are prescribed to be zero, by symmetry condition. The boundary condition used on u is also fixed by the condition given by $\partial u/\partial y = 0$.

(iv) At the far-field boundary: $\omega = 0$, and u, v are as calculated from the Biot–Savart law caused by the free stream vortex and its image.

(v) At the outflow, vorticity is calculated using radiative Sommerfeld boundary condition given as

$$\frac{\partial \omega}{\partial t} + U_c \frac{\partial \omega}{\partial x} = 0$$

where the convective speed U_c is chosen to be U_∞. The condition used on u is $\partial^2 u/\partial x^2 = 0$, whereas a condition given by, $\partial v/\partial x = \omega + \partial u/\partial y$ is used for v.

4.3.3   Numerical Methods and Grids

A staggered variable arrangement, as used in [20, 36], is adopted here. Staggered grid arrangement is needed to reduce errors, which is smaller in staggered grid, as compared to non-staggered grid arrangement, shown using this formulation in [24]. In the transformed (ξ, η)-plane, the staggered grid is shown in Fig. 4.4, with contravariant components of velocity marked at the mid-points of control surface over which the component is normal. In Fig. 4.4, the vorticity is placed at the nodes.

In the numerical methods used, time advancement of ω is done by RK_4-scheme, while convective derivatives ($\frac{\partial \omega}{\partial \xi}, \frac{\partial \omega}{\partial \eta}$) are discretized using OUCS3 scheme developed in [57] and second derivatives are discretized using second-order central difference scheme. To evaluate convective derivatives in the staggered grid, it is required to interpolate u- and v-components at the locations of the vorticity; which is carried out by an optimized compact mid-point interpolation scheme of [35].

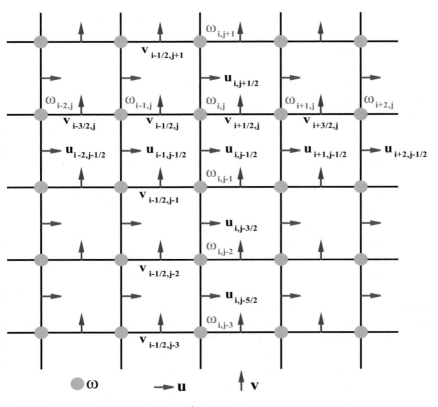

Fig. 4.4 Staggered grid system used in ($\vec{V}$, ω)-formulation for 2D problems. The velocity components are at the mid-point of the elemental surface over which it is normal

A value of $\varepsilon = 10^{-8}$, is used for solving Poisson equations with the Bi-CGSTAB method of [80]. Same numerical methods are used to obtain the equilibrium and disturbance flow, as excited by the free stream vortex. Once the steady equilibrium flow is established, the convecting vortex is activated to excite the flow.

After solving Eq. (4.7) for the vorticity, Eq. (4.8) is solved for the u-component of velocity, while v-component of velocity is obtained by integrating the right hand side of Eq. (4.10), as equal to zero (solenoidality condition). The non-dimensionalized equations have been obtained with a length scale (defined later) and the free-stream speed of the oncoming flow as the velocity scale. From these two scales, the time scale is constructed and all computational results are in non-dimensional units.

To solve the governing equations in the computational domain of Fig. 4.3, the parameters are similar to those in the experiments of [28], except for the strength of the convecting vortex, which cannot be measured and is treated here as the parameter of the problem. In solving the problem computationally, we define a length scale, L, based on which the Reynolds number is 10^5. The computational domain extends over, $-0.05 \leq x^* \leq 20$, i.e., the domain includes the leading edge of the plate. The height of the computational domain is taken as $y_{max}^* = 1.0$, with the wall resolution given by $\Delta y = 3.688 \times 10^{-4}$, and this is half the wall resolution taken in [56]. For the present computations, the strength of the convecting free stream vortex is taken as $\Gamma = 2.0$ (as compared to 9 taken in [56]). This is important for experimental and computational study of flow instability, where we need as small an excitation as possible to study receptivity. The free stream convected vortex moves at constant fixed height of 2, at a fixed speed. The Reynolds number based on displacement thickness at the outflow of the undisturbed flow is 2432.44, and thus the domain is more than five times, that was taken in [56]. The flow computed in the latter reference remained sub-critical over the computational domain, and the present domain is significantly longer.

The authors in [7, 38], have used numerical methods, which are $O(\Delta x \Delta t, \, \Delta y \Delta t)$ accurate. In contrast, the computations in [56] have used OUCS3 schemes for spatial discretization of convection terms, which has more than seven times higher spectral resolution as compared to second-order accurate schemes. The dispersion error of the method is lower for the schemes in [56]. The present formulation and used staggered compact scheme, further improves accuracy by reducing aliasing and dispersion errors, as studied in [61]. It has been firmly established that in computing space-time dependent problems, dispersion error is the largest source of error, as compared to spatial and temporal discretization errors viewed in isolation. In [69], it has been conclusively demonstrated that using a sixth order spatial compact scheme, along with fourth order accurate temporal scheme for model convection equation, the error scales as $O(\Delta x^4)$ in a non-uniform grid. Thus, the role of error dynamics, as described in [49, 61, 75] cannot be overemphasized.

At the inflow and the top-lid of computational domain, one obtains flow variables induced by Biot–Savart law due to the convecting vortex outside the computational domain. At the outflow, the fully developed condition on the wall-normal component of velocity is used. The above conditions are used to derive variables at all boundary segments. The flow is started impulsively, with the initial location of the vortex being ahead of the leading edge. In computing the flow by $(\vec{V}, \omega)$-formulation, computations have been performed with a time step of $\Delta t = 8 \times 10^{-5}$.

4.3.4 Grid Generation

In order to resolve the points near the leading edge (which is the Goldstein singularity), a non-uniform stretched grid is used in the streamwise direction that clusters points at the leading edge. The tangent-hyperbolic function is used for grid clustering as described in [18], which has been shown to cause lower aliasing error [49]. Similar clustering of grid points have been performed in the wall-normal direction as well, to accurately resolve the boundary layer and events very close to the wall for receptivity problem, as one expects high gradients of the flow variables near the wall. This grid-point clustering at the leading-edge and wall, causes the physical problem defined in the Cartesian co-ordinate (x, y) to be transformed to the uniform computational (ξ, η)-coordinate system.

The specific form of grid transformation function in the streamwise direction is given for, $0 \leq \xi \leq \xi_1$ and $x_{in} \leq x \leq x_s$ by,

$$x(\xi) = x_{in} + (x_s - x_{in})\left[1 - \frac{\tanh[\beta_x(1 - \xi)]}{\tanh \beta_x}\right]$$

and for $\xi_1 \leq \xi \leq 1$ (and $x_s \leq x(\xi) \leq x_{out}$)

$$x(\xi) = x_s + (x_s - x_{in})\left[\left(\frac{\beta_x}{\tanh \beta_x}\right)\left(\frac{\xi - \xi_1}{\xi_1}\right)\right] \qquad (4.12)$$

where, $\xi_1 = \frac{1}{1+A_1}$ and

$$A_1 = \left(\frac{x_{out} - x_s}{x_s - x_{in}}\right)\left(\frac{\tanh \beta_x}{\beta_x}\right)$$

Similarly, due to the grid clustering near the wall, the grid transformation function in the wall-normal direction is given as

$$y(\eta) = y_{max}\left[1 - \frac{\tanh[\beta_y(1 - \eta)]}{\tanh \beta_y}\right] \qquad (4.13)$$

where, $0 \leq \eta \leq 1$. Here, β_x and β_y are parameters which control the grid clustering in the streamwise and wall-normal direction, respectively. For most of the cases reported, $\beta_x = \beta_y = 2$, have been used. Because of the transformation of the problem from the physical (x, y)-plane to computational (ξ, η)-plane, one defines the scale factors of transformation h_1 and h_2 by $h_1 = (x_\xi^2 + y_\xi^2)^{\frac{1}{2}}$ and $h_2 = (x_\eta^2 + y_\eta^2)^{\frac{1}{2}}$, respectively. Here, $x_\eta = \frac{\partial x}{\partial \eta} = 0$ and $y_\xi = \frac{\partial y}{\partial \xi} = 0$, and the scale factors are simply given as $h_1 = x_\xi$ and $h_2 = y_\eta$. The above grid transformation in the streamwise direction given by Eq. (4.12) ensures the continuity of both h_1 and $\frac{dh_1}{d\xi}$ at $x = x_s$. The grid used has 1001 points in the streamwise direction and 301 points in the wall-normal direction.

4.3.5 Numerical Results and Discussion

The problem of vortex-induced instability experimentally and analytically studied in [28, 56], has been solved here using $(\overrightarrow{V}, \omega)$-formulation as a 2D problem. As noted, the equilibrium solution is obtained first by solving NSE, without the convecting free stream vortex. Once this is obtained, the free stream vortex is started with constant

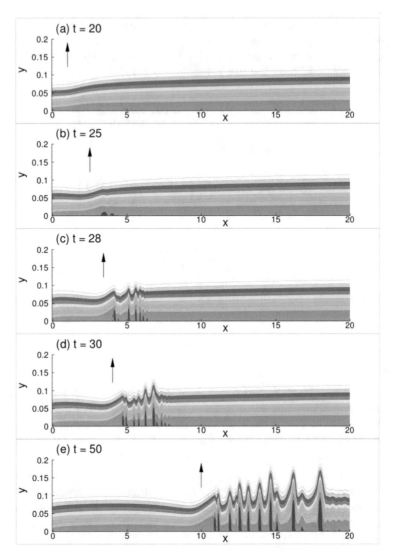

Fig. 4.5 Stream function contours plotted in the computational domain at the indicated times. Arrowheads at the top show the streamwise location of the convecting vortex

convection speed ($c = 0.3$) from the initial location: ($x = -5, y = 2$). Presence of the convecting free stream vortex above the flat plate creates an image vortex, as shown in Fig. 4.3.

In Fig. 4.5, stream function contours are plotted at indicated times chosen to show physically important events. For the frame at $t = 20$, the free stream vortex is just downstream of the leading edge of the plate. As the vortex has circulation in the anti-clockwise direction, the action of the vortex is to lift up the boundary layer ahead of the vortex. At the foot of the vortex (outside the boundary layer), the induced velocity is maximum. Thus upstream of the vortex, one would notice a favourable pressure gradient, while in the downstream direction, the induced velocity reduces with x, i.e., an adverse pressure gradient is created. As the free stream vortex convects at a constant height, its instantaneous stream-wise location is shown by an arrowhead. With passage of time, for farther downstream location of the vortex, the boundary layer experiences sustained adverse pressure gradient and in the frame (b) of Fig. 4.5 one notices formation of unsteady separation bubbles near $x = 4$. Downstream of these bubbles, the flow in the vicinity of the flat plate experiences additional adverse pressure gradient. This is readily seen in frames (c) and (d) at $t = 28$ and 30, respectively, when one notices increased number of unsteady separation bubbles, all of which convect downstream. That such a single convecting free stream vortex, far out in the inviscid part of the flow, can cause bypass transition is noted in frame (e) of Fig. 4.5. Unsteady separation bubbles, as a consequence of bypass transition, on the wall were conjectured in [32] to be caused as a result of buffeting of the boundary layer by FST vortices. The present study is for the unit-process and provides a physical mechanism of a bypass transition process. In the present study no model is required, with the effects governed by NSE. For example in [38], the primary vortex forms as a consequence of unsteady flow evolution and does not move at constant speed, which was modeled as a Batchelor vortex moving at constant speed.

The unsteady separation and vortical structures near the wall are created due to effects of free stream convecting vortex, as shown in Fig. 4.6. These are at the same time instants, as shown for the stream function contours in Fig. 4.5. Despite the fact that free stream acoustic excitation could not trigger disturbances in the experiments of [47], in [27, 56] and here, the coupling between free stream vortical excitation with viscous disturbances inside the boundary layer is established. An explanation of this is provided in [55] with the help of spatial linear instability described by the OSE. It is shown that boundary layers can support disturbances created by sources inside or outside a shear layer by what is referred to as wall- and free stream-modes, respectively. When free stream-modes are excited (as in here), those in turn cause the wall-mode to be excited, by a coupling mechanism that ensures homogeneous boundary conditions at the wall. The growth of the primary bubble and appearance of subsequent separation bubbles are due to an instability, where the disturbance field is enriched from the equilibrium flow. The conditions and mechanism by which these instabilities appear is further discussed next.

4.3.6   The Instability Mechanism

The experimental and accompanying computational results display the existence of a receptivity mechanism inside the shear layer as a consequence of a single vortex migrating in the free stream at a constant speed. The role of various parameters responsible for this instability is by the redistribution of E in the flow. The equation for $E = p/\rho + \frac{1}{2}\overrightarrow{V}^2$, for incompressible flows is the divergence of the rotational form of the NSE given by Eq. (4.5). E depends on the enstrophy and its contribution

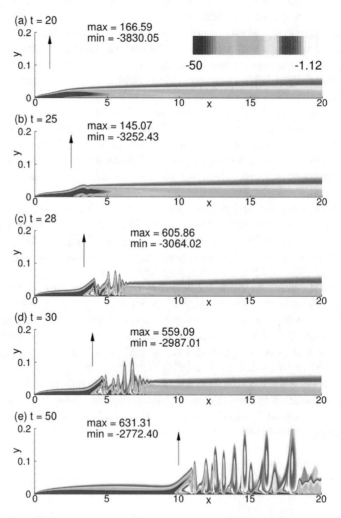

Fig. 4.6 Vorticity contours plotted in the computational domain at the indicated times, as in Fig. 4.5. Same contour values are plotted in all the frames. Arrowheads at the top show the streamwise location of the convecting vortex

on the right hand side appears as positive, i.e., stabilizing. However, as one views the evolution of E_d given in Eq. (4.6), one notices the corresponding term as a product and leads to its growth when the primary and disturbance vorticities are of opposite sign, indicating a transfer of energy from primary to disturbance flow. At the same time, the third and fourth terms on the right hand side of Eq. (4.6) indicates that the spatial variation of the vorticity field can interact with the velocity field to contribute to instability, when this is a negative quantity. Here Eq. (4.6) has been used to explain the complete nonlinear evolution of disturbances.

Here the velocity and vorticity fields at $t = 20$ are taken as representative equilibrium flow. The sign of right hand side being positive or negative, indicates a sink or source, respectively, in Eq. (4.6). Hence for the E_d equation, a negative right hand side anywhere would indicate a source of E_d at that point in the flow field. In Fig. 4.7, these distributed sources are plotted as negative contours. At $t = 20$, there are two sites from where instability originates, one at the leading edge and the other on the plate, downstream of leading edge. It is seen that the leading-edge instability is the weaker of the two and the major one originates near $x = 2.4$, as was also seen in the vorticity contours in Fig. 4.6. Disturbance energy structures from these two regions remain distinct at $t = 25$. But from $t = 28$ onwards, these two sources of E_d interact, as is evident from the bottom three frames in Fig. 4.7, where the spike forming at the downstream site interacts with the vortical structure originating from the leading edge.

In stream function and vorticity contour plots the spike is evident near $x = 4.5$ at $t = 28$ in the form of a secondary bubble. Thus, it is important to include the leading edge in the analysis, otherwise one would compute the unimpeded spike stage, as in [38, 72]. However beyond $t = 28$, the instability originating from the leading edge terminates before the downstream spike and subsequently the distance between the instabilities further increases. The present analysis based on the right-hand side of Eq. (4.6), more clearly reveals the physical nature of the problem compared to the information from stream function and vorticity contours.

In the experimental study it is shown that a vortex with anti-clockwise circulation creates instability ahead of it. The coupling between the convecting vortex outside the shear layer and the generated unsteady vortical field inside is explained theoretically by developing an equation for E_d. This vortex-induced instability is caused by the adverse pressure gradient due to the translating vortex in the free stream. Effects of the finite core translating vortex is shown in Fig. 4.8, indicating the displacement and circulatory components of the input disturbances and associated induced pressure gradient, at the edge of the shear layer. It is noted that unsteady separation occurs at a lower value of adverse pressure gradients, indicated here by plotting the Falkner–Skan parameter, m, plotted as functions of space and time in various frames of Fig. 4.8. This figure should be contrasted with Fig. 4.8a, b in [48], where the induced adverse pressure gradient by the free stream convecting vortex of strengths $\Gamma = 14$ and 30, indicated values of m, which were far in excess of what is needed for steady separation. Thus for vortex-induced instability, it is not merely the strength of adverse pressure gradient that triggers unsteady separation, but the time duration over which the adverse pressure gradient acts matter more.

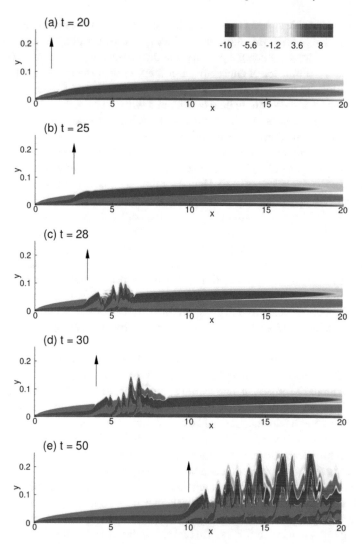

Fig. 4.7 Contours of the right-hand side of the disturbance energy Eq. (4.6). The negative values indicating disturbance energy sources are plotted as dark contours

4.4   Enstrophy Transport Equation: A New Approach to Nonlinear Receptivity Theory

We are already familiar with the nonlinear receptivity/ instability theory based on total mechanical energy. The governing equation for this is derived from NSE without making any assumption, by taking, the divergence of this conservation equation for translational momentum equation. Thus, this provides information about the

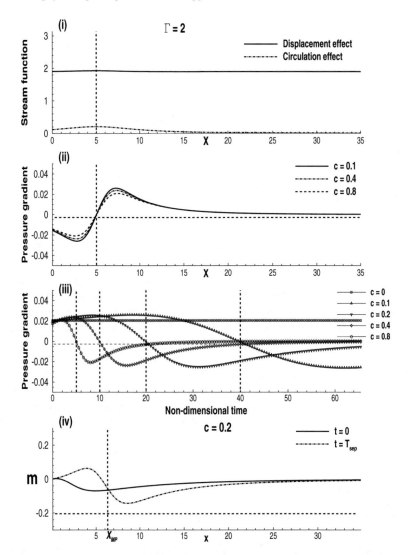

Fig. 4.8 (i) Disturbance stream function versus x caused by circulatory and displacement effects, by a translating and rotating circular cylinder, while the disturbance source is at $x_c = 5$ and $H_1 = 6$ moving at the indicated propagation speed. (ii) Induced pressure gradient for the case of frame (i) for the indicated propagation speed. (iii) The same pressure gradient shown as a function of time for various convecting and non-convecting free stream vortex cases. (iv) Variation of Falkner–Skan parameter m with x shown at two representative times for the case of $c = 0.2$. The time T_{sep} indicates a large time for plotting m

irrotational part of the conservation equation. In fluid mechanics, however, there are no conservation equation for rotational momentum. This is partially addressed by looking at the VTE which is obtained by taking a curl of NSE. Even when vorticity field indirectly provides information about rotationality in a fluid flow, one needs an appropriate measure of it, and one notes that the enstrophy to fulfill this role. Interestingly, the turbulence literature alludes enstrophy to provide a measure of dissipation only, as in [13], who studied the enstrophy transport equation (ETE) for 2D periodic flows and obtained the evolution of integrated enstrophy over the full domain as

$$\frac{d}{dt}\left(\frac{1}{2}||\omega||_2^2\right) = -\nu||\nabla\omega||_2^2 \tag{4.14}$$

Here, the enstrophy is defined over the full periodic domain by $||\omega||_2^2$. Thus, one notes the effects of diffusion to be strictly dissipative for periodic 2D flows viewed globally. This shows that viscous diffusion destructs enstrophy in the full domain, for 2D periodic flows. However for the most generic cases of 3D inhomogeneous flows, authors in [67] have developed a more general ETE, and is the subject of discussion in the following. We also discuss the role of diffusive terms in fluid flows.

Investigation on the true role of diffusion has remained a problem, ever since the time when it was considered to strictly stabilize fluid flow by damping disturbances, an idea attributed to Kelvin, Helmholtz and Rayleigh in [48]. Such heuristic equating of viscous diffusion with dissipation was the main reason in early instability studies, which ignored diffusion, as noted in [17, 48]. Unfortunately, such inviscid studies could not even explain instability of ZPGBL. But the same flow was subsequently investigated by solving the OSE in [21, 46, 78]. We note that the governing equations for E and E_d arises entirely from local and convective acceleration terms (in the absence of body force).

In DNS, one discretizes all terms and numerical solution is obtained without any models. However, diffusion is often equated with dissipation, as in [82], by extrapolating Eq. (4.14), which is valid for homogeneous turbulent flows. However if diffusion is viewed at any instant for any point in the flow, then the effects of diffusion is not strictly dissipative, as has been explained in [67]. When one looks at time-averaged TKE globally, effects of diffusion is again seen as dissipative in [31, 77]. As shown in Eq. (4.34) of [31], time-average of diffusion term in NSE manifests itself, as combination of (i) a strictly dissipation term and (ii) another viscous transfer term. The viscous transfer term does not contribute, when integrated over the flow domain, due to divergence theorem. This term is shown diffusive for homogeneous turbulence [13]. It is further noted that the viscous transfer term is negligible at high Reynolds numbers, except within the thin viscous layers, very near solid surfaces. While on the other hand, the dissipative term is of crucial importance to turbulence energetics everywhere. Similar observations are made in Sect. 3.3 of [77], with respect to time-averaged TKE.

Denoting the instantaneous point property of enstrophy by $\Omega_1 = \vec{\omega} \cdot \vec{\omega}$, Eq. (4.5) can be written as,

$$\nabla^2 E = \Omega_1 - \vec{V} \cdot (\nabla \times \vec{\omega}) \tag{4.15}$$

This equation shows relevance of Ω_1 and the diffusion operator to be central in distributing E. In [82], a similar equation has been written for static pressure, p (see Eq. (1.2) of the reference [82]) which in the present notations is given by

$$\nabla^2 \left(\frac{p}{\rho} \right) = (\Omega_1 - \varepsilon/\nu)/2 \tag{4.16}$$

where $\varepsilon = 2\nu \, s_{ij} \, s_{ij}$ and s_{ij} is the symmetric part of the strain tensor. This equation is valid for both homogeneous and inhomogeneous flows. One notes that the second term on the right hand side of Eq. (4.15) can be written as $\frac{\vec{v}}{\nu} \cdot \nabla \vec{V}$, by drawing analogy with the term ε/ν, on the right hand side of Eq. (4.16), even though the right hand side of Eq. (4.15) purely originates from convection term. This confusion prompted the authors in [15, 16, 82], to equate enstrophy with dissipation. To understand the role of diffusion in creating rotationality for inhomogeneous flows, an evolution equation is developed for enstrophy, as a point property and its higher powers. This explains the roles of diffusion, dissipation and creation of rotationality progressively to smaller scales.

4.4.1 Enstrophy Transport Equation

The role of Ω_1 in rotational fluid flow is similar to kinetic energy describing the translational motion in fluid flow. While vorticity describes rotationality, a measure of it is naturally obtained via enstrophy unambiguously, describing the energy expended by the system in creating and sustaining rotationality. In all flows, physical instabilities take an equilibrium state to another one and in the process, the energy of the system is redistributed into rotational and translation degrees of freedom. Thus, enstrophy is a natural dependent variable to study transitional and turbulent flows. We explain instabilities and pattern formations, with the help of ETE derived from the non-dimensional VTE in tensor notation given for 3D flows by

$$\frac{\partial \omega_i}{\partial t} + u_j \frac{\partial \omega_i}{\partial x_j} = \omega_j \frac{\partial u_i}{\partial x_j} + \frac{1}{Re} \frac{\partial^2 \omega_i}{\partial x_j \partial x_j} \tag{4.17}$$

where subscripts, $i, j = 1, 2$ and 3, represent Cartesian axes and repeated index implies summation. Taking a dot product of Eq. (4.17) with ω_i and using $\Omega_1 = \omega_i \omega_i$ to represent enstrophy, one obtains its transport equation as

$$\frac{\partial \Omega_1}{\partial t} + u_j \frac{\partial \Omega_1}{\partial x_j} - 2\omega_i \omega_j \frac{\partial u_i}{\partial x_j} = \frac{1}{Re} \frac{\partial^2 \Omega_1}{\partial x_j \partial x_j} - \frac{2}{Re} \left(\frac{\partial \omega_i}{\partial x_j} \right) \left(\frac{\partial \omega_i}{\partial x_j} \right) \tag{4.18}$$

The third term on the left hand side (LHS) of the above equation is due to vortex stretching (corresponding to the first term on the right hand side (RHS) of Eq. (4.17)), which is absent for 2D flows. The diffusion of ω_i gives rise to RHS terms in Eq. (4.18). The first term of RHS shows diffusion of Ω_1 and the second term represents strictly a loss or the dissipation term for the transport of Ω_1. The present study views Ω_1 as a point property in the flow and is different from the usage in [13], where the enstrophy is defined over the full domain. The traditional approach utilizes the simplification brought about for problems which are homogeneous and periodic. Focusing on 2D flows with the spanwise component of vorticity (ω), ETE can be written in vector form as,

$$\frac{D\Omega_1}{Dt} = \frac{2}{Re}\left[\frac{1}{2}\nabla^2\Omega_1 - (\nabla\omega)^2\right] \qquad (4.19)$$

The first term on RHS of Eq. (4.19) is missing from Eq. (4.14), due to periodicity of the flow and strictly negative RHS shows the viscous action to dissipate energy globally. In contrast for inhomogeneous flows, the first term on RHS of Eq. (4.19) can be either positive or negative. Therefore, the diffusion term in ETE can create or destroy rotationality, depending upon the sign of RHS in Eq. (4.19). Thus, the diffusion term should not be identified strictly as dissipation for general flows. As $\Omega_1 > 0$, positive RHS indicate the diffusion to cause instability. Negative RHS act as a sink of Ω_1. This provides a mechanism of creating rotationality at different scales by diffusion and is distinctly different form the concept of creating smaller scales by vortex stretching, as the dominant mechanism of generating small eddies for 3D flows. We note that the role of diffusion in creating new length scales is ubiquitous for both 2D and 3D flows.

4.4.2   Enstrophy Cascade for General Inhomogeneous Flows

To further investigate effects of diffusion in Eq. (4.19) at multiple scales, one derives transport equations for higher powers of Ω_1. Multiplying Eq. (4.19) with Ω_1 and defining $\Omega_n = \Omega_1^{2^{n-1}}$, one obtains for Ω_2 the following transport equation,

$$\frac{D\Omega_2}{Dt} = 2Re^{-1}\left[\frac{1}{2}\nabla^2\Omega_2 - (\nabla\Omega_1)^2 - \Omega_1(\nabla\omega)^2\right] \qquad (4.20)$$

Noting further that $\frac{D\Omega_2}{Dt} = 2\Omega_1\frac{D\Omega_1}{Dt}$, one can write Eq. (4.20) as the ETE, i.e., an evolution equation for Ω_1. Multiplying the above equation with Ω_2 and simplifying, one can obtain transport equation for Ω_3, which can be used to write the ETE involving Ω_1, Ω_2 and Ω_3. This process can be generalized to obtain the transport equation for Ω_n as,

$$\frac{D\Omega_n}{Dt} = 2Re^{-1}\left[\frac{1}{2}\nabla^2\Omega_n - (\nabla\Omega_{n-1})^2 - C\right] \qquad (4.21)$$

where,

$$C = \sum_{k=0}^{n-2} 2^{n-k-1} \left(\prod_{j=k+1}^{n-1} \Omega_j \right) (\nabla \Omega_k)^2 \tag{4.22}$$

and $\Omega_0 = \omega$ with $\sum$ indicating summation over all k's and $\prod$ indicating the product of all the jth elements.

Also, the substantive derivative of Ω_n can be written and simplified as

$$\frac{D\Omega_n}{Dt} = 2\Omega_{n-1} \left(\frac{D\Omega_{n-1}}{Dt} \right)$$

$$= 2\Omega_{n-1}(2\Omega_{n-2}) \frac{D\Omega_{n-2}}{Dt} \tag{4.23}$$

$$= 2^{n-1} \left(\prod_{j=1}^{n-1} \Omega_j \right) \frac{D\Omega_1}{Dt}$$

which can be further simplified to yield,

$$\frac{D\Omega_n}{Dt} = 2^{n-1} \Omega_1^{-\beta} \frac{D\Omega_1}{Dt} \tag{4.24}$$

where $\beta = 1 - 2^{n-1}$. Using above relations, one can rewrite Eq. (4.21) as the ETE given by

$$\frac{D\Omega_1}{Dt} = \frac{Re^{-1} \Omega_1^{\beta}}{2^{n-2}} \left[\frac{1}{2} \nabla^2 \Omega_n - (\nabla \Omega_{n-1})^2 - C \right] \tag{4.25}$$

One notes that while writing the transport equation for Ω_n, the diffusion term from the transport equation for Ω_{n-1} contributes two terms; one of which is strictly dissipative (dependent on Ω_{n-2}) and the other as a diffusion term for Ω_{n-1}. The diffusion term involving Ω_{n-1} can be furthermore expressed into two terms involving a strictly dissipative term involving Ω_{n-1} and another diffusive term involving Ω_{n-2}. This process can cascade indefinitely in Eq. (4.25), for increasing n with the leading term as a diffusion term and the rest are strictly dissipative. Higher order moments of Ω_1 will contribute more for higher wavenumbers, implying that the order of even moments of Ω_1 will be restricted by the energy supplied to the flow. For 3D flow as well, the RHS of Eq. (4.25) is present as the forcing term. However in this case, the vortex stretching term is retained. The ETE for 3D flow is same, as given by Eq. (4.18). Following a similar approach as in deriving the transport equation for Ω_n for 2D flows, the transport equation for Ω_n can be derived for 3D flows as,

$$\frac{D\Omega_n}{Dt} = 2Re^{-1}\left[\frac{1}{2}\nabla^2\Omega_n - (\nabla\Omega_{n-1})^2 - C\right] + 2^n\prod_{k=1}^{n-1}\Omega_k\left(\omega_i\omega_j\frac{\partial u_i}{\partial x_j}\right) \qquad (4.26)$$

where expression for C is same as in Eq. (4.22). Using Eq. (4.24), one can rewrite the ETE for 3D flows as

$$\frac{D\Omega_1}{Dt} = \frac{Re^{-1}\Omega_1^\beta}{2^{n-2}}\left[\frac{\nabla^2\Omega_n}{2} - (\nabla\Omega_{n-1})^2 - C\right] + 2\omega_i\omega_j\frac{\partial u_i}{\partial x_j} \qquad (4.27)$$

One also notes that the diffusion term gives rise to enstrophy cascade for both 2D and 3D flows, for which the contribution at higher wavenumbers depends upon the value of n decided by the energy supplied to the fluid dynamical system. However in 3D flows, vortex stretching is also present, which provides an additional mechanism of energy redistribution process. This indicates that in 3D flows, generation of different scales of vorticity is due to enstrophy cascade via the stretching and diffusion terms and the energy cascade is by the vortex stretching implicit in convection process. In 2D flows, it is only the diffusion term, which gives rise to enstrophy (and hence vorticity) at different scales.

It is emphasized that physically the role of diffusion for inhomogeneous flows is not strictly dissipative, as is the case for homogeneous turbulent flows. By developing the ETE, in terms of higher even moments of Ω_1, we identified the index n in this equation, which is fixed from total energy imparted to drive a flow. This view of how smallest scale is fixed is entirely different from the conventional logic used for the dissipation of kinetic energy to heat, even for isothermal flows.

To understand the role of the developed ETE for inhomogeneous flows, in Fig. 4.9, we note the growth rate of total enstrophy, $(D\Omega_1/Dt)$-contours, for the problem of vortex-induced instability, whose results are shown in Figs. 4.5 and 4.6 for the stream function and vorticity contours. Corresponding E_d evolution is shown by contour-plots of the RHS of Eq. (4.6) in Fig. 4.7. The times at which the total enstrophy growth are shown in Fig. 4.9, are identical to the times shown earlier in Figs. 4.5, 4.6 and 4.7. The spatial scales shown in enstrophy growth rates are much more refined, than those seen for E_d. Another aspect that draws our attention is the clear presence of wall- and free stream-modes at early times, as explained in [48, 55]. At $t = 20$, one notices growth of rotationality originating from leading edge of the plate (due to Goldstein singularity), that remains at the edge of boundary layer very prominently. While the wall-mode is in nascent stage, in the region: $1 \le x \le 2$ of the flat plate. Both these modes are noted more prominently in the frame at $t = 25$, and the growing wall-mode causes a bulge in the growth region of the free stream-mode. In a short time interval, one notices severe interaction between these two modes, as noted in the frame for $t = 28$, which is noted more in the vorticity contours of Fig. 4.6, than

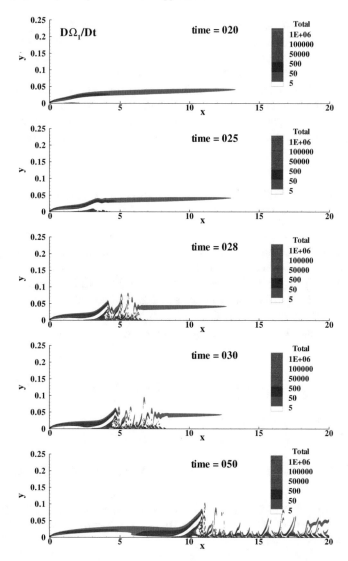

Fig. 4.9 The growth rate of total enstrophy Eq. (4.19) for the vortex-induced problem defined in Fig. 4.3 for which the physical variables are shown in Figs. 4.5 and 4.6 for the identical time instants. The marked regions are for the instability of enstrophy, i.e., where rotationality increases with time at the indicated time instants

in the disturbance mechanical energy of Fig. 4.7. This is apparent that the signature
of vortex-induced instability is more evident in the rotational part of the NSE, i.e.,
in ETE, than in the divergence or irrotational part of NSE for E_d. Also visible is the
presence of much finer wall-normal structures in Fig. 4.9, during early interactions
between wall- and free stream-mode. Such events are equally visible in the frame
at $t = 30$. In the bottom frames of Figs. 4.7 and 4.9, one can notice more energetic
events in E_d, as compared to those in $(D\Omega_1/Dt)$-contours. The latter still grows in
the location over the plate, which is exactly beneath the convecting vortex.

4.5   Theory of Instability for Enstrophy: Creation of Rotationality

Having described the nonlinear growth of E in Sect. 4.1, one can study nonlinear
growth/ decay of rotationality, quantified by Ω_1. We have the equation for the growth
rate of this in Eq. (4.18). If Ω_1 is written as sum of the equilibrium and disturbance
values, $\Omega_1 = \Omega_m + \varepsilon_1\Omega_d$, then the linearized growth rate of Ω_d can be evaluated
retaining the terms of order ε_1. If the primary variables are represented as, $\vec{\omega} = \vec{\omega}_m + \varepsilon_2\vec{\omega}_d$, and $\vec{V} = \vec{V}_m + \varepsilon_2\vec{V}_d$, then the growth or decay rate of Ω_1 can be
written as,

$$\frac{D\Omega_1}{Dt} = \frac{\partial}{\partial t}(\Omega_m + \varepsilon_1\Omega_d) + (\vec{V}_m + \varepsilon_2\vec{V}_d) \cdot \nabla(\Omega_m + \varepsilon_1\Omega_d) = RHS$$

From the above one can rewrite it for

$$\frac{D\Omega_m}{Dt} + \varepsilon_1\frac{\partial\Omega_d}{\partial t} + \varepsilon_1\vec{V}_m \cdot \nabla\Omega_d + \varepsilon_2\vec{V}_d \cdot \nabla\Omega_m + \varepsilon_1\varepsilon_2\vec{V}_d \cdot \nabla\Omega_d = RHS$$

The order ε_1 terms of the left hand side of the above is simply nothing but the
substantive derivative of $(\Omega_m + \varepsilon_1\Omega_d)$ and thus, one can write the evolution equation
for disturbance enstrophy as

$$\frac{D\Omega_d}{Dt} = \left[\left\{2\omega_{im}\omega_{jm}\frac{\partial u_{id}}{\partial x_j} + 2\omega_{im}\omega_{jd}\frac{\partial u_{im}}{\partial x_j} + 2\omega_{id}\omega_{jm}\frac{\partial u_{im}}{\partial x_j}\right\} \right.$$
$$\left. + \frac{1}{Re}\frac{\partial^2\Omega_d}{\partial x_j\partial x_j} - \frac{2}{Re}\left(\frac{\partial\omega_{im}}{\partial x_j}\right)\left(\frac{\partial\omega_{id}}{\partial x_j}\right)\right] \tag{4.28}$$

Thus the growth or decay rate of Ω_d is determined by the vortex stretching terms
given on the RHS of the above equation inside the curly brackets. These will not

be present for 2D disturbance field. The contribution from the diffusion term to the growth rate is obvious. However, the term that is identified as strictly dissipation term in Eq. (4.18) for Ω_1, can contribute to growth of Ω_d, as shown here by the last term on RHS. We also note that unlike Ω_1 (which is strictly positive definite), Ω_d can be either positive or negative, as this is given by $\vec{\omega}_m \cdot \vec{\omega}_d$. It is noted that this Ω_d term also appears in the Poisson equation for E_d, Eq. (4.6). Also as Ω_d can be either positive or negative, its conditions for growth/ decay will be different for different signs of the quantity, i.e., for positive Ω_d, an instability would be indicated when $D\Omega_d/Dt > 0$ and for negative Ω_d, its amplitude will grow when $D\Omega_d/Dt < 0$. These conditions are investigated next during vortex-induced instability.

In Fig. 4.10, this $D\Omega_d/Dt$ is plotted in the domain, at the same indicated times as before. As noted above, we indicate two different conditions for linear growth rate cases, depending upon the signs of Ω_d. The enstrophy contours are drawn by solid (for positive values) and dashed lines (for negative enstrophy) in the frames. Only the growth rate regions are marked by flooded region for both the cases, in the frames of Fig. 4.10. The confluence of the regions are indicative of modulus of Ω_d growing.

Like the $D\Omega_1/Dt$ rate, here also the early-time growth is due to growth of wall- and free stream-modes, with the latter dominating. The wall-mode growth for Ω_d is higher as compared to Ω_1. For the same reason, the wall- and free stream-modes interact earlier and strongly, for the growth of Ω_d. The region of negative Ω_d growth is associated with the wall-mode, as is easily evident in the figure. In Fig. 4.10, one also notices distinct wall-normal streaks, ahead of the convecting free stream vortex from $t = 28$ onwards, which was not the case for Ω_1. These vertical streaks are seen to cover the flat plate ahead of the convecting vortex, near the outflow of the domain. One also notices that for $t \geq 50$, the regions near the leading edge of the plate becomes quieter.

In Fig. 4.11, nonlinear Ω_d growth rate is plotted in the domain, at the same indicated times as before. This is obtained first by noting $\frac{D\Omega_1}{Dt}$ for the total enstrophy at each and every point in the domain as a function of time, and then subtracting $\frac{D\Omega_m}{Dt}$ from it, to obtain the nonlinear growth rate. As before, two different conditions for nonlinear growth rate cases are considered, depending upon the signs of the disturbance enstrophy ($\vec{\omega}_m \cdot \vec{\omega}_d$). The enstrophy contours are drawn by solid (for positive values) and dashed lines (for negative Ω_d) in the frames. Only the growth, rate zones are marked by flooded region for both the cases in the respective frames. The overlap regions are indicative of modulus of Ω_d growing with time. Comparing Figs. 4.10 and 4.11, one notices only very marginal differences between linear and nonlinear growth regions. In nonlinear growth case, the rates are relatively lower and spread over smaller overlap regions.

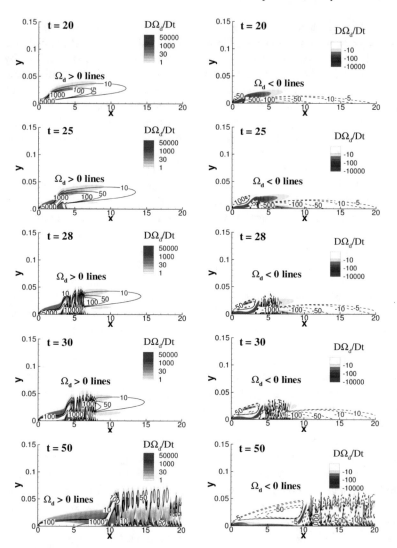

Fig. 4.10 The growth rate of Ω_d Eq. (4.19) for the vortex-induced instability problem defined in Fig. 4.3, for which the physical variables are shown in Figs. 4.5 and 4.6 for the identical time instants. On the left frames shown is the case with positive Ω_d and on the right is the case of negative Ω_d shown at the indicated time instants

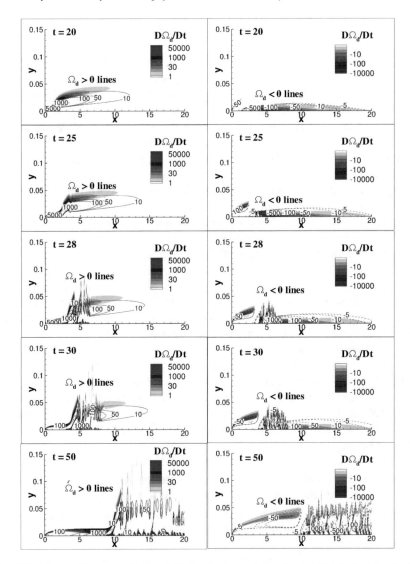

Fig. 4.11 The nonlinear growth rate of Ω_d for the vortex-induced instability defined in Fig. 4.3, for which the physical variables are shown in Figs. 4.5 and 4.6 for the identical time instants. On the left frames shown are the regions where Ω_d is positive and on the right where Ω_d is negative, with overlap regions where rotationality increases with time at the indicated time instants

4.6 Proper Orthogonal Decomposition

For high Reynolds number transitional and turbulent flows, POD analysis [25, 29] can be viewed as providing nonlinear instability portrait, provided we can relate the POD modes with instability modes. In simple terms, POD is nothing but the Galerkin projection of space-time dependent variables describing system dynamics. Such a projection helps describe the space and time dependence of the disturbance field. Principally in the literature, there are two variants of POD. The classical one is based on projection of a random field onto a deterministic basis (eigenfunctions) so that the reduced order model captures maximum energy. While conceptually this is a very appealing approach, but its practice in fluid mechanics is difficult in handling pressure-velocity coupling terms when one is solving NSE, using primitive variable formulation. This has been circumvented by the authors, in developing an enstrophy based POD. This is based on the vorticity transport equation, which does not have the pressure term. The enstrophy-based POD is preferred over those in [11, 23, 30, 37, 41, 71], where kinetic energy is used for POD analysis. In vortex dominated flows, which are neither homogeneous nor periodic, enstrophy is a better descriptor of POD over translational kinetic energy, as highlighted in [48, 67]. This approach has been used in many earlier works [48, 56, 62–64, 68]. Authors in [64], used enstrophy based POD approach to study both external and internal flows to show universality of POD modes in terms of amplitude functions. In [34, 37, 70], the authors devised a new POD mode which was obtained through a Galerkin projection on Reynolds-averaged Navier–Strokes equation. In [8, 44], the authors performed reduced order modeling using Koopman modes. In Sengupta et al. [63], the authors used enstrophy based POD modes to analyze transitional flow past a flat plate. Thus, POD analysis has been shown to be an useful tool in studying both internal and external flows of different kinds.

POD technique was originally introduced by Kosambi [25] for a random field $v_i(\overrightarrow{X}, t)$, where it was projected onto a set of deterministic vectors $\phi_i(\overrightarrow{X})$, so that $\langle |(v_i, \phi_i)|^2 \rangle$ is maximum, where the outer angular brackets signify time-averaging and inner brackets signify an inner product. The computation of $\phi_i(\overrightarrow{X})$ can be posed as an optimization problem in variational calculus,

$$\int \int R_{ij}(\overrightarrow{X}, \overrightarrow{X}') \, \phi_j(\overrightarrow{X}') \, d^2\overrightarrow{X}' = \lambda \, \phi_i(\overrightarrow{X}) \tag{4.29}$$

The kernel of this is the two-point correlation function, $R_{ij} = \langle v_i(\overrightarrow{X})v_j(\overrightarrow{X}') \rangle$ of the random field. It is noted [63] that *classical Hilbert–Schmidt theory applies to flows with finite energy, and, therefore, denumerable infinite orthogonal POD modes can be computed.* Hilbert–Schmidt theory is applicable here because instabilities in flow, derive energy from the equilibrium flow, which itself has finite energy. Disturbance vorticity field is thus, represented in POD formalism as

$$\omega'(\overrightarrow{X}, t) = \Sigma_{m=1}^{\infty} a_m(t) \, \phi_m(\overrightarrow{X}) \tag{4.30}$$

where $a_m(t)$ represents the amplitude function, which describes the spatio-temporal variation of the modal amplitude and $\phi_m(\vec{X})$ is the corresponding spatial eigenfunction. Equation (4.29) is an eigenvalue problem in the integral form, which becomes intractable even for moderate grid resolution. To overcome this difficulty, Sirovich introduced the method of snapshots [71], which has an advantage of dealing with smaller data sets in multiple dimensions. The eigenfunction $\phi_m(\vec{X})$ is expressed as a linear combination of the instantaneous flow fields at distinct instants of time t_i's as,

$$\phi_m(\vec{X}) = \Sigma_{i=1}^{N} q_{mi}\, \omega'(\vec{X}, t_i) \tag{4.31}$$

where N is the number of snapshots used. This together with the expression for R_{ij} reduces Eq. (4.29) to an algebraic eigenvalue problem, $[\bar{C}]\{\mathbf{q}\} = \lambda\{\mathbf{q}\}$. The entries of the matrix $[\bar{C}]$ are obtained by integrating over the domain by using,

$$\bar{C}_{ij} = \frac{1}{N} \int \int \omega'(\vec{X}, t_i)\, \omega'(\vec{X}, t_j)\, d^2\vec{X} \tag{4.32}$$

with $i, j = 1, 2 \ldots N$ defined over the snapshots of the flow. The amplitude functions $a_m(t)$ are obtained from

$$a_m(t) = \frac{\int \int \omega'(\vec{X}, t)\, \phi_m(\vec{X})\, d^2\vec{X}}{\int \int \phi_m^2(\vec{X})\, d^2\vec{X}} \tag{4.33}$$

The eigenvalues λ and eigenvectors $\{\mathbf{q}\}$ of $[\bar{C}]$ are computed using a MATLAB function based on QR decomposition. Once $\{\mathbf{q}\}$ is obtained, Eq. (4.31) is used to find the eigenfunctions $\phi_m(\vec{X})$ and the corresponding amplitude functions, $a_m(t)$ are evaluated using Eq. (4.33). The *kernel* $R(\mathbf{X}, \mathbf{Y})$ is defined as

$$R(\mathbf{X}, \mathbf{Y}) = <\omega'(\mathbf{X}, t), \omega'(\mathbf{Y}, t)> = \frac{1}{T} \int_{t_0}^{t_0+T} \omega'(\mathbf{X}, t)\omega'(\mathbf{Y}, t)dt \tag{4.34}$$

Such a decomposition of $\omega'(\mathbf{X}, \mathbf{t})$ into the various orthogonal modes that satisfies Eq. (4.29) is called its POD. Equation (4.29) represents an eigenvalue problem, where the eigenvalues λ_m represent the fraction of average enstrophy captured by the mth-mode.

4.6.1 Some Useful Mathematical Relations

In this subsection, some useful mathematical relations regarding POD are provided. These are as enumerated below:

1. Multiplying Eq. (4.30) by $\phi_m(\mathbf{X})$, and integrating it over the full computational domain S, by using the orthogonality condition, one obtains Eq. (4.33).
2. Using Eqs. (4.29), (4.30), and orthogonal properties of ϕ_m's and (4.33), it can be shown that

$$\int_{t_0}^{t_0+T} a_m(t)a_n(t)dt = \lambda_m \delta_{mn} \tag{4.35}$$

This implies that that the amplitude functions $a_m(t)$'s are also orthogonal to each other.
3. Multiplying Eq. (4.30) by the amplitude function $a_n(t)$, and integrating over the time interval $(t_0, t_0 + T)$ and using Eq. (4.35), it can be shown that

$$\phi_m(\mathbf{X}) = \int_{t_0}^{t_0+T} \alpha_m(t)\omega'(\mathbf{X}, t)dt \tag{4.36}$$

where $\alpha_m(t) = \frac{a_m(t)}{\lambda_m}$.
4. Inserting the decomposition of $\omega'(\mathbf{X}, t)$ and $\omega'(\mathbf{Y}, t)$, as given by Eq. (4.30) in the definition of $R(\mathbf{X}, \mathbf{Y})$, given in Eq. (4.34), and using Eq. (4.35), it can be shown

$$R(\mathbf{X}, \mathbf{Y}) = \sum_{m=1}^{\infty} \lambda_m \phi_m(\mathbf{X})\phi_m(\mathbf{Y}) \tag{4.37}$$

4. One can define the instantaneous enstrophy $\Omega_1(t)$ and the average enstrophy $\hat{\Omega}_1$ over the time period T as

$$\Omega_1(t) = \int_S \omega'(\mathbf{X}, t)\omega'(\mathbf{X}, t)d^2\mathbf{X} \tag{4.38}$$

$$\hat{\Omega}_1 = \frac{1}{T}\int_{t_0}^{t_0+T} \Omega_1(t)dt \tag{4.39}$$

From Eqs. (4.37), (4.38) and (4.39), one can show that,

$$\hat{\Omega}_1 = \int_S R(\mathbf{X}, \mathbf{X})d^2\mathbf{X}$$

$$= \sum_{m=1}^{\infty} \lambda_m \int_S \phi_m(\mathbf{X})\phi_m(\mathbf{X})d^2\mathbf{X}$$

$$= \sum_{m=1}^{\infty} \lambda_m \tag{4.40}$$

This justifies the statement made above that, the eigenvalues λ_m represent the fraction of average enstrophy $\hat{\Omega}_1$ captured by the mth POD mode.

4.6.2 Method of Snapshots

For a 2D flow field, to calculate POD eigenvalues λ_m, and eigenfunctions $\phi_m(\mathbf{X})$, one has to solve the eigenvalue problem given by Eq. (4.29). If the DNS is carried out using M and N points in ξ- and η-directions and POD is performed over Q time-snapshots at equal time interval of Δt, then $R(\mathbf{X}, \mathbf{Y})$ given by Eq. (4.34) would represent a $L \times L$ square symmetric matrix, where $L = M \times N$. Each element of this symmetric square matrix R is given as

$$R_{mn} = \sum_{q=1}^{Q} \omega_{ij}^{\prime q} \, \omega_{kl}^{\prime q} \, \Delta t$$

where, $m = (j - 1)M + i$ and $n = (l - 1)M + k$. Here, (i, k) and (j, l) denotes the indices along ξ- and η-directions, respectively.

Since, M and N are generally orders of magnitude higher than Q, it is computationally very expensive to evaluate the eigenvalues of the matrix R. To avoid this computational problem of evaluating the POD modes, the *method of snapshots* is used, and described next in brief.

In the *method of snapshots*, one proposes formation of a correlation function $D(t, s)$ as

$$D(t, s) = \int_S \omega'(\mathbf{X}, t)\omega'(\mathbf{X}, s)d^2\mathbf{X} \tag{4.41}$$

It is to be noted that for discrete computations $D(t, s)$ represents a square symmetric matrix of size $Q \times Q$ with elements

$$D_{pq} = \sum_{i=1}^{M} \sum_{j=1}^{N} \omega_{ij}^{\prime p} \, \omega_{ij}^{\prime q} \, h_{1ij} \, h_{2ij} \, \Delta \xi \, \Delta \eta$$

Hence, computing the eigenvalues of $D(t, s)$ will be much more efficient and less time-consuming than the direct method of solving the eigenvalues from Eq. (4.37). This is in essence, the method of snapshots. Once, the eigenvalues λ_m, and amplitudes $a_m(t)$ are obtained, one can calculate the POD spatial eigenfunctions $\phi_m(\mathbf{X})$ from Eq. (4.34).

4.6.3   TS Wave Instability over Zero-Pressure Gradient Boundary Layer

Here, POD analysis is done for a ZPGBL for a semi-infinite flat plate excited by a SBS exciter strip. For the present study, the domain along the streamwise direction is taken as $-0.1 \le x \le 22$; while in the wall-normal direction, it is taken as $0 \le y \le 1$. In the presented simulations, Reynolds number based on reference length scale L is selected again to be $Re = 10^5$. The amplitude and non-dimensional frequency parameter of the excitation are chosen as $\alpha_1 = 0.01$ and $F_f = 1 \times 10^{-4}$, respectively.

For nonlinear receptivity and stability calculations, full NSE is solved for the total quantities (without splitting into equilibrium and disturbance components) similar to the cases discussed earlier. Disturbance quantities are obtained by subtracting the equilibrium solution from the total solution.

In Fig. 4.12, time evolution of disturbance field is shown by plotting the vorticity contours at six different time instants ($t = 10, 15, 20, 30, 50$ and 60). Development of spatio-temporal disturbance field is clearly noticed in these frames. From $t = 10$ onwards, a wave-front can be seen to develop, which is prominently noted at $t = 15$. This wave-front amplifies at a faster rate than the main trailing wave-packet. This

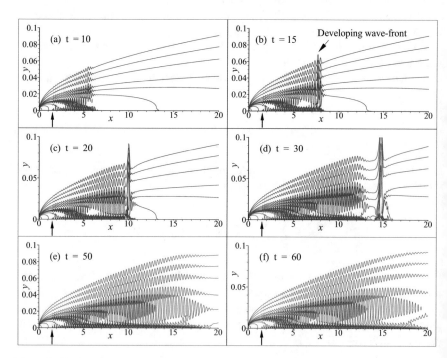

Fig. 4.12 Vorticity contours at indicated times are shown plotted in frames (a) to (b) for ZPG flow past a flat plate excited by SBS strip-exciter. The location of the midpoint of the exciter strip is pointed by an arrow

s further evident from the contours at $t = 20$, when one also notices a vortical eruption associated with the wave-front. The mismatch of speed between the leading wave-front and the following main wave-packet causes these to separate clearly at later times. At $t = 30$, one notes two distinct coherent structures: the main wave-packet and a separated, highly amplified fast moving STWF that after sufficient magnification forms a vertically erupting separation bubble. This erupting bubble perturbs the flow locally, creating secondary new bubbles in its vicinity. The wave-packet after $t = 20$, continues to grow and move forward, but at a relatively lower, exponential rate. The STWF leaves the computational domain at $t = 45$ cleanly due to the application of Sommerfeld radiative boundary condition, as discussed earlier. Growth and expansion of the region perturbed by the wave-packet almost remains invariant after $t = 50$, which is evident from the contour plots at $t = 50$ and 60. Here, for the purpose of POD analysis of the solution incorporating nonlinear and nonparallel effects, the full nonlinear equation is solved without neglecting the growth of the underlying equilibrium flow.

POD of the disturbance vorticity field is performed using the time series taken from $t = 0$ to $t = 60$, by storing flow-snapshots at a time interval of $\Delta t = 0.01$, in a domain whose streamwise and wall-normal extent is $-0.1 \leq x \leq 20$ and $0 \leq y \leq 0.5$, respectively. Input wall excitation imposes a non-dimensional time scale of $T = 0.63$ on this fluid dynamical system, as the chosen non-dimensional circular frequency of excitation is $\beta_L = F_f \times Re = 10$. Hence, the chosen time interval of taking snapshots is more than adequate, as there are 63 snapshots within one fundamental time interval of the perturbed flow-field.

In Fig. 4.13, the cumulative disturbance enstrophy content of the flow obtained by POD analysis is shown, which helps one to identify the dominant modes playing a vital role in determining response of the dynamical system by forcing. One hundred and twenty modes shown in the figure collectively contribute 98.96% of the total enstrophy. The first 30 modes contribute about 90% of the total enstrophy.

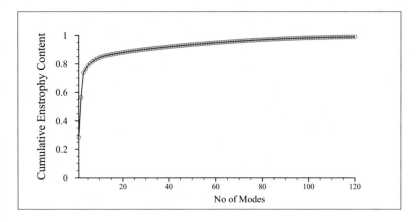

Fig. 4.13 Cumulative enstrophy content are shown plotted against corresponding mode number for zero-pressure gradient flow over semi-infinite flat-plate with wall-excitation

Table 4.1 Eigenvalues and cumulative enstrophy of the POD modes

POD mode index j	Eigenvalues (λ_j)	Cumulative enstrophy content $(e_j = \sum_{k=1}^{k=j} \lambda_k)$
1	0.2829E+00	0.2829
2	0.2808E+00	0.5638
3	0.1731E+00	0.7369
4	–	0.7369
5	0.2546E-01	0.7623
6	0.2532E-01	0.7878
7	0.1663E-01	0.8044
8	–	0.8044
9	0.1228E-01	0.8167
10	0.1227E-01	0.8290
11	0.7578E-02	0.8366
12	0.7534E-02	0.8441
13	0.6547E-02	0.8506
14	–	0.8506
15	0.4622E-02	0.8553
16	0.3907E-02	0.8592
17	0.3517E-02	0.8627
18	0.3189E-02	0.8659

Higher modes individually contribute very less but their collective contributions are significant.

In Table 4.1, first eighteen POD modes with corresponding eigenvalues, and cumulative enstrophy contribution up to that mode, have been listed. The numbering of the modes here, follow the conventions used in [62, 64]. The regular POD modes form pairs with almost equal eigenvalues, and closely resembling eigenfunctions. Constituents of the pairs exhibit an approximate phase shift of 90° between corresponding eigenfunctions, as well as, amplitude functions. Such lead pair of modes have been postulated to be governed by Stuart-Landau equation [11, 30, 37, 74]. But authors in [62] have shown from their DNS results for flow past a circular cylinder, that in addition to these regular modes, solitary modes appear which do not form pair. Noack et al. [37] have shown that the inclusion of these solitary mode effects are necessary to explain the dynamics of flow past a circular cylinder. This *shift* mode in their work was constructed as a 'mean-field correction' using a Gram–Schmidt procedure. Why this is necessary, can be understood by looking at the disturbance velocity components given by

$$u_d(x, y, t) = \int \int [\hat{u}\, e^{i(\alpha x - \omega_0 t)} + \hat{u}^*\, e^{-i(\alpha x - \omega_0 t)}]\, d\alpha\, d\omega_0 \qquad (4.42)$$

$$v_d(x, y, t) = \int \int [\hat{v} \, e^{i(\alpha x - \omega_0 t)} + \hat{v}^* \, e^{-i(\alpha x - \omega_0 t)}] \, d\alpha \, d\omega_0, \qquad (4.43)$$

where quantities with asterisks denote complex conjugates. If one represents the variables as a sum of mean field plus a disturbance and substitute these in NSE, then one obtains the averaged governing equation (indicated by angular brackets) for the mean field quantities to have gradient transport contribution terms of Reynolds stress-like term given by,

$$< u_d(x, y, t)v_d(x, y, t) >= \int \int \left([\hat{u} \, \hat{v}^* + \hat{u}^* \, \hat{v}] + \hat{u} \, \hat{v} e^{2i(\alpha x - \omega_0 t)} \right.$$
$$\left. + \hat{u}^* \, \hat{v}^* \, e^{-2i(\alpha x - \omega_0 t)} \right) \, d\alpha \, d\omega_0 \qquad (4.44)$$

One notes that in Eq. (4.44), the first term on the right hand side is phase independent and should be included in the mean-field equations, irrespective of whether one is talking about time or ensemble averaged equations. In [50] this was called the wave-induced stress term, while studying flow past a wavy wall. This steady streaming is always present in NSE for any flow field, in the presence of oscillatory disturbances. This also explains, why one can have many such terms, coming from different modes, with α and ω_0 related via the dispersion relation of the problem from governing equation and \ or boundary conditions. These type of terms will not be present, when the modes appear in pairs. There is absolutely no restrictions on the number of isolated modes and in the present case, third, seventh and thirteenth are the isolated modes, as shown in Fig. 4.14. In Table 4.1, blanks have been left to indicate the missing mode, next to the isolated modes. Note that these stress terms are at most of order two in NSE, if the mean field equations are of zeroth order and thus represent a correction to the mean field and is a nonlinear term.

In Figs. 4.14 and 4.15, time dependent amplitude functions $a_j(t)$ of the listed eighteen modes in Table 4.1 and the Fast Fourier transform (FFT) are plotted against nondimensional circular frequency ($\beta_L \Delta t$). The pair-forming modes are plotted in the same frame. Among the plotted modes in Fig. 4.15, the third, the seventh and the thirteenth modes are anomalous modes of the first kind (T_1), as defined in [62, 64]. These are similar to the shift modes of [37]. Time variation of $a_1(t)$ and $a_2(t)$ follows the Stuart–Landau–Eckhaus model for instability modes. These type of modes which follow Stuart–Landau–Eckhaus model are the regular POD modes (R) of [62].

There are certain salient features in the time evolution of first and second POD modes. It is seen that for these modes, amplitude function grows exponentially up to approximately $t = 15$. This exponential amplification factor is calculated to be about 0.19. From $t = 15$ to $t = 19$, this growth is saturated. But, thereafter the amplitude again grows at a lower rate up to $t = 43$, and thereafter reaches a saturated value. FFT of these two modes show the presence of only one dominant frequency, which is exactly the frequency of excitation. The variation from $t = 15$ to $t = 19$ can be attributed to the growth and detachment of the STWF from the wave-packet, as can be seen from the vorticity contour plots at these times. The wave-front derives

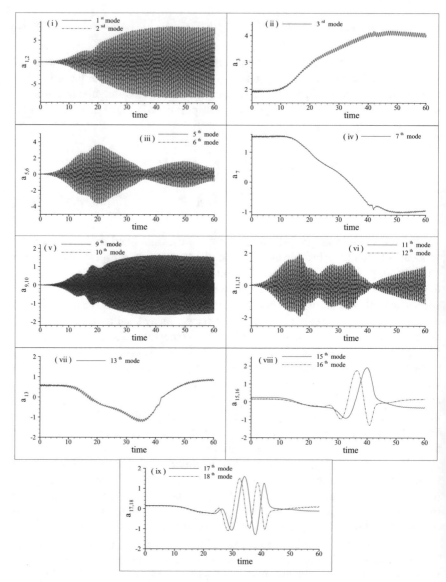

Fig. 4.14 Time dependent amplitude function $a_j(t)$ of first eighteen POD modes for zero-pressure gradient flow over semi-infinite flat-plate plotted as a function of time. The pair-forming modes are shown together in a frame as indicated. The numbering of the modes follow the conventions used in [62, 64]

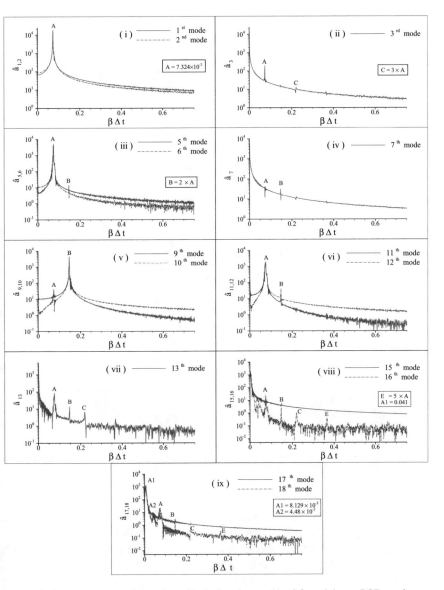

Fig. 4.15 Fast Fourier transform of amplitude functions $a_j(t)$ of first eighteen POD modes are shown plotted as a function of non-dimensional frequency $\beta \Delta t$

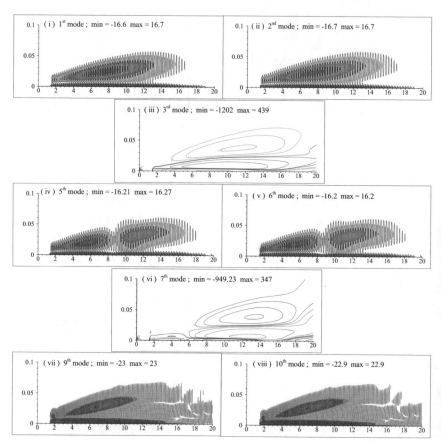

Fig. 4.16 Contours of eigenfunctions $\phi_j(x, y)$ corresponding to the first ten POD modes for zero-pressure gradient flow semi-infinite flat-plate with wall-excitation are shown plotted

substantial energy from the wave-packet for its growth, and subsequent detachment in the mentioned time period causing a saturation of the amplification of these modes.

In Figs. 4.16 and 4.17, eigenfunctions $\phi_j(x, y)$ of the first eighteen POD modes listed in Table 4.1 are plotted. The eigenfunction of first two modes in Fig. 4.16, represent previously described wave-packet, which are closely related to the generated TS waves. The average wavenumber of the streamwise variation of these eigenfunctions, in terms of the displacement thickness at the exciter is 0.16, whereas the same is predicted from the linear parallel theory as 0.17. At the wall, these eigenfunctions show amplification from $Re_{\delta^*} = 835$ to $Re_{\delta^*} = 1598$. This range of amplification though decreases, as one moves away from the wall, which can be attributed to nonparallel, nonlinear effects. Linear theory predicts amplification from $Re_{\delta^*} = 740$ to $Re_{\delta^*} = 1235$, for the chosen non-dimensional circular frequency of excitation $\beta_L = F_f \times Re_L = 10$.

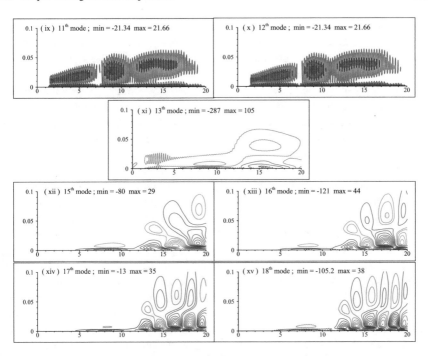

Fig. 4.17   Contours of eigenfunctions $\phi_j(x, y)$ corresponding to the first eighteen POD modes for zero-pressure gradient flow semi-infinite flat-plate with wall-excitation are shown plotted

The third mode is an isolated one, responsible for 17% of total enstrophy of disturbance vorticity field. Figure 4.14 shows that the amplitude of this mode contains low amplitude oscillations, overriding a slower time scale variation, which corresponds to the frequency of excitation, as evident from the corresponding FFT in Fig. 4.15. Up to $t = 10$, mean of this amplitude function has a constant value, after which it grows almost linearly up to $t = 43$, and becomes almost constant thereafter. Amplitude function of the other two isolated modes shown in Fig. 4.14 (the seventh and the thirteenth modes) also show similar kind of time variation, namely an oscillating part, over and above a slow time scale variation, that is constant up to $t = 10$ and varies almost linearly up to $t = 43$ for the seventh mode, and up to $t = 35$, for the thirteenth mode. The oscillatory part can be attribute to the second and third terms in the right hand side of Eq. (4.44). After $t = 43$, mean of the amplitude function for the seventh mode attains a constant value, similar to the third mode. All isolated modes are characterized by the maximum contribution coming from time-independent part.

The eigenfunctions for these isolated modes, as shown in Figs. 4.16 and 4.17, illustrate that these modes basically represent large scale vortices of opposite sign, stacked one above the other. Unlike other modes, eigenfunctions of these modes include the effect of the leading edge, as their maximum and minimum values occur exactly at the leading edge. Eigenfunction of the third mode exhibits a three layered structure, a zone of negative vorticity very near to the wall, covered by a positive vorticity zone, above which there is a negative vorticity layer on top. Eigenfunction of the seventh mode contributes only 1.66% of total enstrophy, and also displays similar structures with vortex centers shifted downstream with opposite signs of vorticity stacked, one over the other.

It has been shown that these isolated modes depict nonlinear interactions by the disturbance field on the mean flow arising due to averaged Reynolds stress terms $< u'v' >$. FFT of the third mode in Fig. 4.15, exhibits a single distinct peak (marked by A) at the exciter frequency β_L. Also a smaller peak at the second super-harmonic $3\beta_L$, is noted. However, dominant frequencies near the origin representing slow time-scale variation, play a significant role for this mode, as explained earlier from Eq. (4.44). This demonstrates two type of nonlinear interactions, the dominant one between the mean flow-field and the disturbance field excited at almost zero frequency (due to first term on right hand side of Eq. (4.44)) and another less dominant one at β_L

In the hierarchy of enstrophy content, the fifth and the sixth modes are the next important ones, which form pair and exhibit wave-packet like features. The amplitude function, as shown plotted in Fig. 4.14, initially grows exponentially at different rates in different time intervals. However beyond $t = 20$, the amplitude starts decaying up to $t = 37$, thereafter it grows and decays again. In [62, 64], this type of modes have been termed as anomalous mode of second kind (T_2), which also do not follow Stuart–Landau–Eckhaus equation. These modes show time variation, that resembles closely like a modulated wave-packet. The eleventh and the twelfth modes also are of this type, as noted in Fig. 4.14. FFT of fifth-sixth and eleventh-twelfth mode pairs show dominance of excitation frequency, though its first super-harmonic is also seen to be present, with an amplitude that is three orders of magnitude lower than the fundamental. For this reason, these mode pairs can be reconstructed by taking a small band of frequency range around the dominant peak, as will be shown shortly. Role of the first super-harmonic is very limited to small departure of oscillations from pure sinusoidal variation.

Pair-forming fifth-sixth and eleventh-twelfth eigenfunctions are shown in Figs. 4.16 and 4.17, which represent same wavenumber, as seen for the first-second mode pair. Eigenfunction of fifth-sixth modes exhibit formation of two side-by-side packets with a node at around $x = 8.5$ ($Re_{\delta^*} = 1585$). Similarly, eleventh-twelfth eigenfunctions show formation of three side-by-side packets with first node at around $x = 7.1$ ($Re_{\delta^*} = 1499$) and a second node at around $x = 11$ ($Re_{\delta^*} = 1803$).

The ninth-tenth mode-pair depicts excitation of the first super-harmonic of β_L, which is evident from the FFT plots in Fig. 4.15. Corresponding amplitude functions in Fig. 4.14, show these modes to be present alongside first-second modes at all times, with a frequency that is significantly higher. Eigenfunctions of these modes in Fig. 4.16, show the corresponding wavenumber to be double the wavenumber of the eigenfunctions of first-second, fifth-sixth and eleventh-twelfth modes. Time variation of amplitude function of these modes in Fig. 4.14, show that they qualitatively follow variations similar to first-second modes. One notes that this variation is different from that is seen for bluff-body flow in [62], where a primary instability is moderated by nonlinear self-interaction. For the cases of bluff body flows, no STWF is formed, and there is a single dominant mode that follows the Stuart–Landau–

Eckhaus equation. In Fig. 4.14, for first-second and ninth-tenth modes, punctuated growth during $t = 0$ to 20 is due to the presence and the dynamics of the wave-front not allowing the following wave-packet to grow, as dictated by the primary linear spatial instability mechanism. However, this mode is similar to the fifth-sixth regular modes in [62], which was the first super-harmonic of the first-second regular mode pair shown in Fig. 4.20 of the reference. From Fig. 4.14, one notices similarity between the amplitude envelope of first-second and ninth-tenth modes, except in the time range $15 \leq t \leq 20$. Difference between the envelopes of first-second and ninth-tenth modes, during $t = 15$ to 20, is due to the generation of additional high wavenumber oscillations near the wave-front, which is a secondary instability.

The fifteenth-sixteenth and seventeenth-eighteenth modes depict the STWF, as seen in Fig. 4.14. Existence of STWF was established using Bromwich contour integral method of receptivity analysis and DNS results in Sengupta et al. [53, 59, 60]. Here, the same is explained in terms of POD modes. Amplitude and FFT of these modes in Figs. 4.14 and 4.15, and eigenfunctions in Fig. 4.17, show that seventeenth-eighteenth modes are the super-harmonic of fifteenth-sixteenth modes. Amplitudes of theses modes are noticed to be almost zero up to $t = 13$, as the wave-front starts developing after this time. Perceptible variation of the modal amplitudes is observed from $t = 25$ to $t = 45$, the time interval when this wave-front forms and eventually separates from the wave-packet leaving the computational domain.

Rempfer and Fasel [41, 42] have tried to relate regular modes with the coherent structures in the flow. The jth coherent structure in [41, 42] have been defined as a sum of $(2j)$th and $(2j - 1)$th pair-forming POD modes as

$$\rho_j(\overrightarrow{X}, t) = a_{2j}(t)\,\phi_{2j}(\overrightarrow{X}) + a_{2j-1}(t)\,\phi_{2j-1}(\overrightarrow{X}) \tag{4.45}$$

Note that this splitting does not show that the pairs are orthogonal to each other. A consistent splitting is given in Eqs. (5.1) and (5.2) in [62]. For example, for unstable flows, vorticity perturbation can be expressed by Galerkin-type expansion in terms of various instability modes as given in [17, 48],

$$\omega'(\overrightarrow{X}, t) = \sum_{j=1}^{\infty}[A_j(t)f_j(\overrightarrow{X}) + A_j^*(t)f_j^*(\overrightarrow{X})] \tag{4.46}$$

where quantities with asterisks again denote complex conjugates. Here, corresponding to the jth instability mode, $A_j(t)$ denote space-time dependent amplitude and $f_j(\overrightarrow{X})$ describes the space-dependent eigenfunction, that satisfies prescribed boundary conditions. As shown in [62], the instability modes are related to POD modes by defining a normalization factor $\varepsilon_j = (\lambda_{2j} + \lambda_{2j-1})/\sum_{k=1}^{N}\lambda_k$. By noting that pair-forming POD modes have a phase shift of $90°$, $A_j(t)$ and $f_j(\overrightarrow{X})$ can be defined in terms of the POD modes as,

$$A_j(t) = \sqrt{\varepsilon_j}[a_{2j-1}(t) + i\ a_{2j}(t)] \tag{4.47}$$

Table 4.2 Normalization factor ε_j and its square root $\sqrt{\varepsilon_j}$ for instability modes

Instability mode index j	ε_j	$\sqrt{\varepsilon_j}$
1	0.563E+00	0.751E+00
2	0.173E+00	0.416E+00
3	0.508E-01	0.225E+00
4	0.166E-01	0.128E+00
5	0.244E-01	0.156E+00
6	0.151E-01	0.123E+00
7	0.650E-02	0.806E-01
8	0.850E-02	0.922E-01
9	0.670E-02	0.818E-01

$$f_j(\overrightarrow{X}) = \frac{1}{2\sqrt{\varepsilon_j}}[\phi_{2j-1}(\overrightarrow{X}) - i \ \phi_{2j}(\overrightarrow{X})] \qquad (4.48)$$

The POD analysis is performed for disturbance enstrophy, and the eigenvalues measure the same. We provide a note of caution that the disturbance enstrophy in the POD context is not the same one as Ω_d, that was used to describe enstrophy-based instability theory. Thus, in representing the disturbance vorticity, the normalization in Eqs. (4.47) and (4.48), is as indicated in terms of the square root of ε_j. In Table 4.2, the normalization factor ε_j and its square root (used to relate instability modes with POD modes) are tabulated for the first nine instability modes. Table 4.2 shows that ε_4 is less than ε_5, and ε_7 is less than both ε_8 and ε_9. Reason behind this is clear from the corresponding eigenvalues listed in Table 4.1. One can see from Table 4.1 that: (1) The seventh and the thirteenth POD modes are isolated, anomalous modes of first kind; (2) λ_7 is less than $(\lambda_9 + \lambda_{10})$ and (3) λ_{13} is less than both $(\lambda_{15} + \lambda_{16})$ and $(\lambda_{17} + \lambda_{18})$. Definition of normalization factors gives $\varepsilon_4 = \lambda_7/\sum \lambda_k$, $\varepsilon_5 = (\lambda_9 + \lambda_{10})/\sum \lambda_k$, $\varepsilon_7 = \lambda_{13}/\sum \lambda_k$, $\varepsilon_8 = (\lambda_{15} + \lambda_{16})/\sum \lambda_k$ and $\varepsilon_9 = (\lambda_{17} + \lambda_{18})/\sum \lambda_k$, which justify the above observations. In fact, it is noted that whenever anomalous mode of first kind appears (barring the first one), normalization factor ε_j of the respective instability mode is less than several subsequent normalization factors.

In Figs. 4.18 and 4.19, modulus of the amplitude function $|A_j(t)|$ and the space dependent function $|f_j(\overrightarrow{X})|$ of the instability modes are plotted. Comparing each $|A_j(t)|$ in Fig. 4.18 with corresponding $a_{2j-1}(t)$ and $a_{2j}(t)$ plotted in Fig. 4.14, one can see that for each instability mode, $|A_j(t)|$ is essentially the upper envelope of $a_{2j-1}(t)$ and $a_{2j}(t)$. Similarly, if the modulus of space dependent functions $|f_j(\overrightarrow{X})|$ and $(\phi_{2j}(\overrightarrow{X}), \phi_{2j-1}(\overrightarrow{X}))$ are compared at any height as a function of x, it is noted that $|f_j(\overrightarrow{X})|$ defines the upper envelope of $(\phi_{2j}(\overrightarrow{X}), \phi_{2j-1}(\overrightarrow{X}))$.

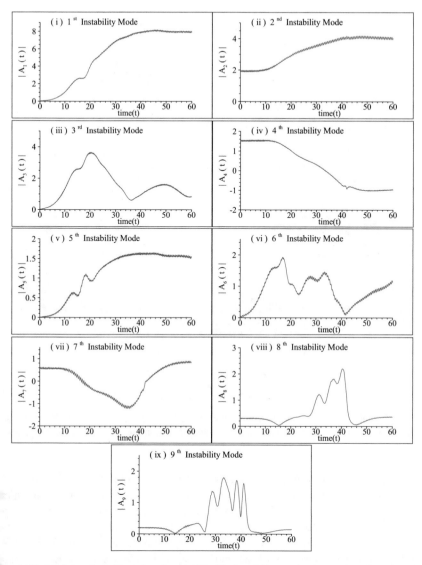

Fig. 4.18 Modulus of amplitude function $|A_j(t)|$ of the instability modes shown plotted as a function of time for zero-pressure gradient flow over semi-infinite flat-plate

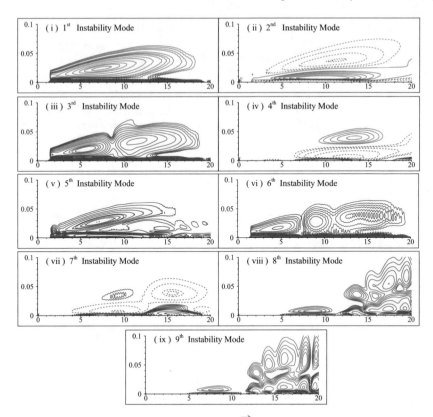

Fig. 4.19 Modulus of space dependent function $|f_j(\vec{X})|$ of the indicated instability modes are shown plotted as a function of time for zero-pressure gradient flow over semi-infinite flat-plate

Here, an attempt is made to reconstruct the amplitude function of anomalous modes of second kind by taking a band of frequencies about the most dominant frequency (F_0) noted in the FFT. The ordinate in Fig. 4.15 is shown in terms of $\beta_L \Delta t$. In Figs. 4.20 and 4.21, this type of reconstruction is shown for the fifth and the eleventh modes, respectively. FFT of amplitude functions performed with 8192 points (hence $\Delta t = (t_1 - t_0)/8192$) indicate $F_0 = 10.158$ for the fifth mode and $F_0 = 10.262$ for the eleventh mode. As mentioned earlier, excitation frequency is $\beta_L = 10$ and thus the dominant frequencies are quite close to β_L. About 170 data samples for each time period corresponding to F_0 have been taken. This shows that the assumption of signal problem used for linear spatial stability theory is strictly not correct and the dynamics should be obtained by spatio-temporal analyses.

Since, POD is performed from $t_0 = 0$ to $t_1 = 60$, the spectrum is obtained with $\Delta F = 2\pi/(t_1 - t_0) = 0.105$. From Figs. 4.20 and 4.21, it is seen that reconstruction with $\pm 12 \Delta F$ ($\Delta \beta_L \Delta t = \pm 0.01841$ as shown in Figs. 4.20 and 4.21) around F_0 reproduces the original, perfectly in the time range $t_s \le t \le t_e$. For the fifth mode, $t_s = 1.8$ and $t_e = 58.6$, whereas for the eleventh mode, $t_s = 3.5$ and $t_e = 56.5$. Mismatch near

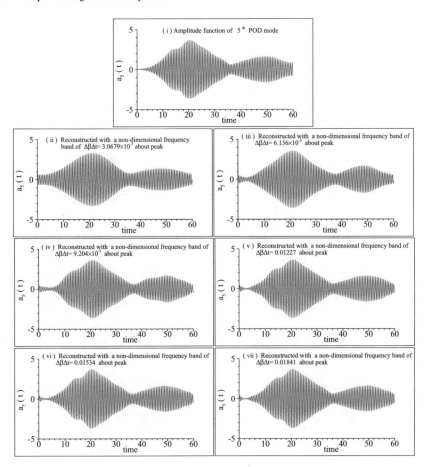

Fig. 4.20   Reconstructed signals of $a_5(t)$ with indicated frequency band about the peak at $\beta \Delta t = 7.234 \times 10^{-2}$ shown in frames (ii) to (vii). The original signal $a_5(t)$ is shown in the top frame

t_0 and t_1 are related to Abel's and Tauber's theorem that relates the unknown in physical and transformed spectral planes. Since calculations are performed with finite resources, we do take a finite spectrum with approximate resolution and thereby having inaccuracies near the origin and the far field. The Tauber's theorem states that signal near $t = 0$ is determined by the spectrum at $\beta_L \to \infty$ (see page 84 of [52] for further details), which we do not have in the full range of $\beta_L \Delta t$. However, the mismatch near t_1 is related to our calculation of FFT by assuming periodic continuation outside the range of consideration in physical plane and this enforced periodicity causes the mismatch at t_1. This type of mismatch near t_1 is typical of any Fourier spectral method.

Reconstruction of the anomalous modes of second kind in Figs. 4.20 and 4.21, emphasizes the fact that the contributions come from the side-band of the main peak noted in the FFT shown in Fig. 4.15. For both the cases, this central peak corresponds

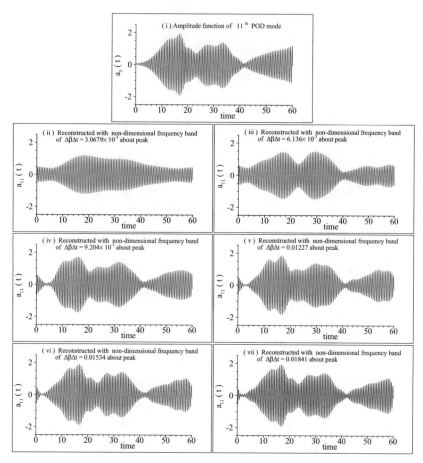

Fig. 4.21 Reconstructed signals of $a_{11}(t)$ with indicated frequency band about the peak at $\beta\Delta t = 7.234 \times 10^{-2}$ are shown plotted in frames (ii) to (vii). The original signal $a_{11}(t)$ is shown in the top frame

to the frequencies mentioned above, those are slightly detuned from the excitation frequency.

A contemporary system reduction strategy for fluid flows is the Koopman mode analysis [44], based on the spectral analysis of the finite-dimensional representation of the Koopman operator using Krylov subspace method, a variant of the commonly used Arnoldi iteration. This linear mapping is equivalent to the dynamic mode decomposition (DMD) technique developed [22, 45, 79]. Koopman analysis results in modes that are orthogonal in time (single frequency modes) while POD gives spatially orthogonal, multi-time periodic modes, which capture the most energetic/enstrophic structures of the flow. The results of Koopman mode analysis and its variants in [8], show that this strategy although successful in the initial linear growth stage, and probably in the limit cycle stage, is not so successful in describing the transient disturbance growth in which the frequency of large scale oscillations is no longer approximately constant. POD modes being free from any restriction on the frequency composition, are effective in capturing transient dynamics, as shown in [62] and here. Also, we have shown here as to how one can relate the POD modes with instability modes. Once this is achieved, one can perform space-time dependent nonlinear analysis of instability.

References

1. Badr, H. M., Coutanceau, M., Dennis, S. C. R., & Menard, C. (1990). Unsteady flow past a rotating circular cylinder at Reynolds numbers 1000 and 10,000. *Journal of Fluid Mechanics*, *220*, 459–484.
2. Benjamin, T. B., & Feir, J. E. (1967). The disintegration of wave trains on deep water. Part 1. Theory. *Journal of Fluid Mechanics*, *27*(3), 417–430.
3. Bhaumik, S., & Sengupta, T. K. (2011). On the divergence-free condition of velocity in two-dimensional velocity-vorticity formulation of incompressible Navier–Stokes equation. In *20th AIAA CFD Conference, 27–30 June, Honululu, Hawaii, USA*
4. Bhaumik, S., & Sengupta, T. K. (2014). Precursor of transition to turbulence: Spatiotemporal wave front. *Physical Review E*, *89*(4), 043018.
5. Bhaumik, S., & Sengupta, T. K. (2015). A new velocity-vorticity formulation for direct numerical simulation of 3D transitional and turbulent flows. *Journal of Computational Physics*, *284*, 230–260.
6. Breuer, K. S., & Kuraishi, T. (1993). Bypass transition in two and three dimensional boundary layers. In *AIAA 93-3050*
7. Brinckman, K. W., & Walker, J. D. A. (2001). Instability in a viscous flow driven by streamwise vortices. *Journal of Fluid Mechanics*, *432*, 127–166.
8. Chen, K., Tu, J. H., & Rowley, C. (2012). Variants of dynamic mode decomposition: Boundary condition, Koopman and Fourier analyses. *Journal of Nonlinear Science*, *22*, 887–915.
9. Daube, O. (1992). Resolution of the 2D Navier–Stokes equations in velocity-vortcity form by means of an influence matrix technique. *Journal of Computational Physics*, *103*, 402–414.
10. Davies, S. J., & White, C. M. (1928). An experimental study of the flow of water in pipes of rectangular section. *Proceedings of the Royal Society of London. Series A*, *119*, 92.
11. Deane, A. E., Kevrekidis, I. G., Karniadakis, G. E., & Orszag, S. A. (1991). Low-dimensional models for complex geometry flows: Application to grooved channels and circular cylinders. *Physics of Fluids*, *3*, 2337–2354.
12. Degani, A. T., Walker, J. D. A., & Smith, F. T. (1998). Unsteady separation past moving surfaces. *Journal of Fluid Mechanics*, *375*, 1–38.
13. Doering, C. R., & Gibbon, J. D. (1995). *Applied analysis of the Navier–Stokes equations*. UK: Cambridge University Press.
14. Doligalski, T. L., Smith, C. R., & Walker, J. D. A. (1994). Vortex interaction with wall. *Annual Review of Fluid Mechanics*, *26*, 573–616.
15. Donzis, D. A., & Yeung, P. K. (2010). Resolution effects and scaling in numerical simulations of passive scalar mixing in turbulence. *Physica D*, *239*, 1278–87.
16. Donzis, D. A., Yeung, P. K., & Sreenivasan, K. R. (2008). Energy dissipation rate and enstrophy in isotropic turbulence: resolution effects and scaling in direct numerical simulations. *Physics of Fluids*, *20*, 045108-1–16.
17. Drazin, P. G., & Reid, W. H. (1981). *Hydrodynamic stability*. UK: Cambridge University Press.
18. Eiseman, P. R. (1985). Grid generation for fluid mechanics computation. *Annual Review of Fluid Mechanics*, *17*, 487–522.

19. Gatski, T. B., Grosch, C. E., & Rose, M. E. (1982). A numerical study of the 2-dimensional Navier–Stokes equations in vorticity-velocity variables. *Journal of Computational Physics, 48,* 1–22.
20. Guj, G., & Stella, F. (1993). A vorticity-velocity method for the numerical solution of 3D incompressible flows. *Journal of Computational Physics, 106,* 286–298.
21. Heisenberg, W. (1924). Über stabilität und turbulenz von flüssigkeitsströmen. *Annalen der Physik, 379,* 577–627. (Translated as 'On stability and turbulence of fluid flows'. NACA Tech. Memo. Wash. No 1291 1951).
22. Hemati, M. S., Williams, M. O., & Rowley, C. W. (2014). Dynamic mode decomposition for large and streaming datasets. *Physics of Fluids, 26*(11), 111701.
23. Holmes, P., Lumley, J. L., & Berkooz, G. (1996). *Coherent structures, dynamical systems and symmetry*. Cambridge, UK: Cambridge University Press.
24. Huang, H., & Li, M. (1997). Finite-difference approximation for the velocity-vorticity formulation on staggered and non-staggered grids. *Computers & Fluids, 26*(1), 59–82.
25. Kosambi, D. D. (1943). Statistics in function space. *Journal of the Indian Mathematical Society, 7,* 76–88.
26. Landahl, M. T., & Mollo-Christensen, E. (1992). *Turbulence and random processes in fluid mech*. New York, USA: Cambridge University Press.
27. Leib, S. J., Wundrow, D. W., & Goldstein, M. E. (1999). Generation and growth of boundary layer disturbances due to free-stream turbulence. In *AIAA Conference Paper, AIAA-99-0408*
28. Lim, T. T., Sengupta, T. K., & Chattopadhyay, M. (2004). A visual study of vortex-induced sub-critical instability on a flat plate laminar boundary layer. *Experiments in Fluids, 37,* 47–55.
29. Loéve, M. (1978). *Probability theory Vol. II* (4th ed., Vol. 46)., Graduate Texts in Mathematics New York, USA: Springer.
30. Ma, X., & Karniadakis, G. E. (2002). A low-dimensional model for simulating three-dimensional cylinder flow. *Journal of Fluid Mechanics, 458,* 181–190.
31. Mathieu, J., & Scott, J. (2000). *An introduction to turbulent flows*. Cambridge, UK: Cambridge University Press.
32. Monin, A. S., & Yaglom, A. M. (1971). *Statistical fluid mechanics: mechanics of turbulence*. Cambridge, MA, USA: The MIT Press.
33. Morkovin, M. V. (1991). Panoramic view of changes in vorticity distribution in transition, instabilities and turbulence. In D. C. Reda, H. L. Reed, & R. Kobyashi (Eds.), *Transition to turbulence* (Vol. 114, pp. 1–12). USA: ASME FED Publication.
34. Morzynski, M., Afanasiev, K., & Thiele, F. (1999). Solution of the eigenvalue problems resulting from global nonparallel flow stability analysis. *Computer Methods in Applied Mechanics and Engineering, 169,* 161–176.
35. Nagarajan, S., Lele, S. K., & Ferziger, J. H. (2003). A robust high-order compact method for large eddy simulation. *Journal of Computational Physics, 19,* 392–419.
36. Napolitano, M., & Pascazio, G. (1991). A numerical method for the vorticity-velocity Navier–Stokes equations in two and three dimensions. *Computers & Fluids, 19,* 489–495.
37. Noack, B. R., Afanasiev, K., Morzynski, M., Tadmor, G., & Thiele, F. (2003). A hierarchy of low-dimensional models for the transient and post-transient cylinder wake. *Journal of Fluid Mechanics, 497,* 335–363.
38. Obabko, A. V., & Cassel, K. W. (2002). Navier–Stokes solutions of unsteady separation induced by a vortex. *Journal of Fluid Mechanics, 465,* 99–130.
39. Peridier, V. J., Smith, F. T., & Walker, J. D. A. (1991). Vortex-induced boundary-layer separation. Part 1. The unsteady limit problem. $Re \to \infty$.. *Journal of Fluid Mechanics, 232,* 99–131.
40. Peridier, V. J., Smith, F. T., & Walker, J. D. A. (1991). Vortex-induced boundary-layer separation. Part 2. Unsteady interacting boundary-layer theory. *Journal of Fluid Mechanics, 232,* 133–165.
41. Rempfer, D., & Fasel, H. (1994). Evolution of three-dimensional coherent structures in a flat-plate boundary layer. *Journal of Fluid Mechanics, 260,* 351–375.
42. Rempfer, D., & Fasel, H. (1994). Dynamics of three-dimensional coherent structures in a flat-plate boundary layer. *Journal of Fluid Mechanics, 275,* 257–283.

43. Robinson, S. K. (1991). Coherent motions in the turbulent boundary layer. *Annual Review of Fluid Mechanics*, *23*, 601–639.
44. Rowley, C., Mezi, I., Bagheri, S., Schlatter, P., & Henningson, D. S. (2009). Spectral analysis of nonlinear flows. *Journal of Fluid Mechanics*, *641*, 1–13.
45. Schmid, P. J. (2010). Dynamic mode decomposition of numerical and experimental data. *Journal of Fluid Mechanics*, *656*, 5–28.
46. Schlichting, H. (1933). Zur entstehung der turbulenz bei der plattenströmung. *Nach. Gesell. d. Wiss. z. Gött., MPK*, *42*, 181–208.
47. Schubauer, G. B., & Skramstad, H. K. (1947). Laminar boundary layer oscillations and the stability of laminar flow. *Journal of Aerosol Science*, *14*, 69–78.
48. Sengupta, T. K. (2012). *Instabilities of flows and transition to turbulence CRC press*. Florida, USA: Taylor and Francis Group.
49. Sengupta, T. K. (2013). *High accuracy computing methods: Fluid flows and wave phenomenon*. USA: Cambridge University Press.
50. Sengupta, T. K., & Lekoudis, S. G. (1985). Calculation of turbulent boundary layer over moving wavy surface. *AIAA Journal*, *23*(4), 530–536.
51. Sengupta, T. K., & Dipankar, A. (2005). Subcritical instability on the attachment-line of an infinite swept wing. *Journal of Fluid Mechanics*, *529*, 147–171.
52. Sengupta, T. K., & Poinsot, T. (2010). *Instabilities of flows: With and without heat transfer and chemical reaction*. Wien, New York: Springer.
53. Sengupta, T. K., & Bhaumik, S. (2011). Onset of turbulence from the receptivity stage of fluid flows. *Physical Review Letters*, *154501*, 1–5.
54. Sengupta, T. K., De, S., & Gupta, K. (2001). Effect of free-stream turbulence on flow over airfoil at high incidences. *Journal of Fluids and Structures*, *15*(5), 671–690.
55. Sengupta, T. K., Chattopadhyay, M., Wang, Z. Y., & Yeo, K. S. (2002). By-pass mechanism of transition to turbulence. *Journal of Fluids and Structures*, *16*, 15–29.
56. Sengupta, T. K., De, S., & Sarkar, S. (2003). Vortex-induced instability of an incompressible wall-bounded shear layer. *Journal of Fluid Mechanics*, *493*, 277–286.
57. Sengupta, T. K., Ganeriwal, G., & De, S. (2003). Analysis of central and upwind compact schemes. *Journal of Computational Physics*, *192*(2), 677–694.
58. Sengupta, T. K., Kasliwal, A., De, S., & Nair, M. (2003). Temporal flow instability for Magnus–Robins effect at high rotation rates. *Journal of Fluids and Structures*, *17*, 941–953.
59. Sengupta, T. K., Rao, A. K., & Venkatasubbaiah, K. (2006). Spatiotemporal growing wave fronts in spatially stable boundary layers. *Physical Review Letters*, *96*(22), 224504.
60. Sengupta, T. K., Rao, A. K., & Venkatasubbaiah, K. (2006). Spatiotemporal growth of disturbances in a boundary layer and energy based receptivity analysis. *Physics of Fluids*, *18*, 094101.
61. Sengupta, T. K., Dipankar, A., & Sagaut, P. (2007). Error dynamics: Beyond von Neumann analysis. *Journal of Computational Physics*, *226*(2), 1211–1218.
62. Sengupta, T. K., Singh, N., & Suman, V. K. (2010). Dynamical system approach to instability of flow past a circular cylinder. *Journal of Fluid Mechanics*, *656*, 82–115.
63. Sengupta, T. K., Bhaumik, S., & Bhumkar, Y. G. (2011). Nonlinear receptivity and instability studies by POD. In *AIAA Conference on Theoretical Fluid Mechanics, AIAA Paper No. 2011-3293*
64. Sengupta, T. K., Singh, N., & Vijay, V. V. S. N. (2011). Universal instability modes in internal and external flows. *Computers & Fluids*, *40*, 221–235.
65. Sengupta, T. K., Bhaumik, S., & Bhumkar, Y. (2012). Direct numerical simulation of two-dimensional wall-bounded turbulent flows from receptivity stage. *Physical Review E*, *85*(2), 026308.
66. Sengupta, T. K., Bhumkar, Y. G., & Sengupta, S. (2012). Dynamics and instability of a shielded vortex in close proximity of a wall. *Computers & Fluids*, *70*, 166–175.
67. Sengupta, T. K., Singh, H., Bhaumik, S., & Chowdhury, R. R. (2013). Diffusion in inhomogeneous flows: Unique equilibrium state in an internal flow. *Computers & Fluids*, *88*, 440–451.

68. Sengupta, T. K., Haider, S. I., Parvathi, M. K., & Pallavi, G. (2015). Enstrophy-based proper orthogonal decomposition for reduced-order modeling of flow a past cylinder. *Physical Review E, 91*(4), 043303.
69. Sharma, N., Sengupta, A., Rajpoot, M., Samuel, R. J., & Sengupta, T. K. (2017). Hybrid sixth order spatial discretization scheme for non-uniform Cartesian grids. *Computers & Fluids, 157*(3), 208–231.
70. Siegel, S. G., Seidel, J., Fagley, C., Luchtenburg, D. M., Cohen, K., & Mclaughlin, T. (2008). Low-dimensional modelling of a transient cylinder wake using double proper orthogonal decomposition. *Journal of Fluid Mechanics, 610*, 1–42.
71. Sirovich, L. (1987). Turbulence and the dynamics of coherent structures. Part (I) coherent structures, part (II) symmetries and trans-formations and part (III) dynamics and scaling. *Quarterly of Applied Mathematics, 45*(3), 561–590.
72. Smith, C. R., Walker, J. D. A., Haidari, A. H., & Soburn, U. (1991). On the dynamics of near-wall turbulence. *Philosophical Transactions of the Royal Society of London A, 336*, 131–175.
73. Sommerfeld, A. (1949). *Partial differential equation in physics*. New York: Academic Press.
74. Stuart, J. T. (1960). On the nonlinear mechanics of wave disturbances in stable and unstable parallel flows. Part 1. The basic behaviour in plane Poiseuille flow. *Journal of Fluid Mechanics, 9*, 353–370.
75. Suman, V. K., Sengupta, T. K., Durga Prasad, C. J., Mohan, K. S., & Sanwalia, D. (2017). Spectral analysis of finite difference schemes for convection diffusion equation. *Computers & Fluids, 150*, 95–114.
76. Taylor, G. I. (1936). Statistical theory of turbulence. V. Effects of turbulence on boundary layer. *Proceedings of the Royal Society A, 156*(888), 307–317.
77. Tennekes, H., & Lumley, J. L. (1971). *First course in turbulence*. Cambridge, MA: MIT Press.
78. Tollmien, W. (1931). The Production of Turbulence. NACA Report-TM 609.
79. Tu, J. H., Rowley, C. W., Luchtenburg, D. M., Brunton, S. L., & Kutz, J. N. (2014). On dynamic mode decomposition: Theory and applications. *Journal of Computational Dynamics, 1*(2), 391–421.
80. Van der Vorst, H. A. (1992). Bi-CGSTAB: A fast and smoothly converging variant of Bi-CG for the solution of non-symmetric linear systems. *SIAM Journal on Scientific and Statistical Computing, 12*, 631–644.
81. Wu, X. H., Wu, J. Z., & Wu, J. M. (1995). Effective vorticity-velocity formulations for 3D incompressible viscous flows. *Journal of Computational Physics, 122*, 68–82.
82. Yeung, P. K., Donzis, D. A., & Sreenivasan, K. R. (2012). Dissipation, enstrophy and pressure statistics in turbulence simulations at high Reynolds numbers. *Journal of Fluid Mechanics, 700*, 5–15.

Chapter 5
Dynamics of the Spatio-Temporal Wave-Front in 2D Framework

5.1 Introduction

A general consensus among fluid dynamicists is that flow transition from laminar to turbulent state occurs due to its instability by imposed and/or background omnipresent disturbances [7, 17]. These ideas have prompted researchers to study the problem of flow transition from the perspective of the stability or receptivity of equilibrium flows. For a ZPGBL, first attempts include analyses by linearized inviscid and viscous instability theories. For a parallel boundary layer the latter approach gives rise to the OSE [7, 17]. As discussed in the previous chapters, the solution of the OSE predicts the existence of spatially modulated wavy solutions, known as TS waves [7]. First experimental detection of spatially evolving TS wave-packets were reported by Schubauer and Skramstad [16], who essentially perturbed the ZPGBL by vibrating a ribbon at a fixed frequency inside it. Following mathematical physics, disturbance evolution in any continuum medium can occurs following temporal or spatial or spatio-temporal routes. For example, there has been an effort [4], where wave propagation problem in electromagnetic medium has been considered with the proviso that the wave-train is preceded by a spatio-temporal wave-packet or STWF. The success of the experiments by Schubauer and Skramstad [16] in detecting TS waves, prompted researchers to predominantly consider the possibility that flow transition in fluid flows (specifically for wall-bounded flows) follow spatial route following the growth of TS waves only. Thus, in fluid mechanics no effects have been made to find STWF for a long time.

The linear spatial theory does not satisfactorily explain all the aspects of the experiment [16], and this has led to search for alternate explanations for route of transition following spatio-temporal growth of disturbances. Wave-front dynamics following spatio-temporal route have been investigated, as a consequence of convective/absolute instability [11], adopting inviscid approach for a pulse excitation. Considering non-localized finite amplitude excitation, Chomaz and co-authors [5] have illustrated spatio-temporal front, as a consequence of initial perturbations of finite extent and amplitude, which separates from undisturbed state. In [23,

© Springer Nature Singapore Pte Ltd. 2019
T. K. Sengupta and S. Bhaumik, *DNS of Wall-Bounded Turbulent Flows*,
https://doi.org/10.1007/978-981-13-0038-7_5

24], spatio-temporal receptivity analysis was presented for the OSE via Bromwich contour integral method (BCIM). These aspects have already been discussed in the previous chapters. In [23, 24], localized time-harmonic excitation with impulsive onset was considered. These approaches of studying spatio-temporal evolution of perturbation are different from those described in [5]. The finite start-up time excites a large band of frequencies, even for the time-harmonic excitation case, which causes the generation of STWF. The resultant STWF obtained by BCIM is distinctly different from the one obtained in various experimental and theoretical approaches [10–14], where wave-packets are created by pulse excitation and not by continuous excitation, as in [16]. Reported results in [23, 24] show that STWF can grow even for excitation parameters, which are stable according to spatial linear theory. Indeed, the dynamics of STWF is distinctly different from the spatially modulating TS wave packet. Similar conclusions were also noted in [9] based on the obtained numerical results, where the authors pointed out that *"the nonlinear behavior of the leading wave-packet would be very different from that of pure Tollmien-Schlichting waves"*. This chapter attempts to bridge this gap by providing analysis of the spatio-temporal evolution of disturbances following localized excitations of various types.

5.2 From Linear Theory to Turbulence via Deterministic Routes

The nonlinear receptivity results of the 2D ZPGBL is considered in the present chapter, with the specific emphasis fixed on understanding the dynamics of the STWF. As discussed in Chap. 3, when the wall bounded shear layer is excited harmonically from the wall, obtained solution displays (i) a STWF, (ii) a local solution and (iii) the wave-packet composed of Tollmien-Schlichting (TS) wave. If the boundary layer is excited by a simultaneous blowing-suction strip at the wall, as in Fig. 5.1, so that there is zero mass transfer at any instant of time, then the peak amplitude, frequency of excitation and exciter width play major roles in deciding the response.

Different routes of disturbance evolution can be noted when 2D ZPG is excited harmonically from the wall. Some of these distinct routes of disturbance evolution are as follows:

1. Following this disturbance evolution route, the STWF is seen to be solely responsible for transition to turbulence which starts off via a linear mechanism, as given in [23], and its subsequent nonlinear growth triggers transition. These features are noted for lower amplitude excitation, at moderate to high frequency cases.
2. If the excitation amplitude is increased for those frequency cases, where disturbance evolution follows previous route, the STWF is noted to directly enter into non-linear phase of amplification, while inducing unsteady separations on the wall to cause bypass transition.

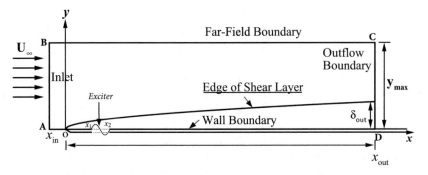

Fig. 5.1 Schematic of 2D receptivity to SBS wall excitation

3. At lower frequencies, distinct presence of a spatially modulated TS wave-packet is not noted. Instead, a spatio-temporally evolving wave-packet is noted, which at subsequent time-instants intermittently issue secondary wave-packets, inducing flow transition.
4. Depending on exciter location or amplitude, another route is noted which is intermediate between the previous and the first route. Following this route, the spatio-temporal wave-packet also interacts with the primary STWF at later times by issuing subsequent secondary STWFs to cause transition of the flow. These different disturbance evolution routes are described in details, in subsequent section.

5.2.1 Governing Equations, Numerical Schemes and Simulation Parameters

The schematic diagram of the receptivity problem is shown in Fig. 5.1. Here, all the reported simulation results use, 4501 and 401 grid points along streamwise and wall-normal directions, respectively, while the computational domain is from $x_{in} = -0.05L$ to $x_{out} = 120L$ in the streamwise direction, and up to $y_{max} = 1.5L$ in the wall-normal direction. Numerical simulations performed with larger y_{max} show it to have virtually no effects on the simulation results.

The Reynolds number based on the reference velocity U_∞ (free-stream speed) and length L is $Re_L = 10^5$, as also used in the previous chapters. A time-step of 8×10^{-5} is used to integrate in time, using optimized three-stage Runge–Kutta (ORK_3) method of [15]. As illustrated in Chap. 2, ORK_3 minimizes the dispersion, phase and attenuation errors and allows one to adopt larger time-steps than the conventional RK_4 method.

Grid clustering in the streamwise and wall-normal directions are performed as given in Sect. 4.3.4. The minimum and maximum resolution along the stream-wise direction are $\Delta x_{min} = 2 \times 10^{-3}$ and $\Delta x_{max} = 2.7 \times 10^{-2}$, respectively. The wall-normal resolution is given, as $\Delta y_{min} = 5.5 \times 10^{-4}$, which stretches up to $\Delta y_{max} = 7.77 \times 10^{-3}$, at the far-field boundary. The Reynolds number based on the displacement thickness δ^* at the outflow of the computational domain (i.e., at $x = x_{out}$) is $Re_{\delta^*}(x_{out}) = 5958$. Simulations are carried out in such a long domain in the streamwise direction, as in [18, 19]. In [8] the longest simulations were reported only up to $Re_{\delta^*} = 450$. The width of the computational domain along the wall-normal direction is 25 times the local displacement thickness at the outflow, in [18, 19].

Stream function and vorticity (ψ, ω)-formulation of 2D NSE is solved here in the (ξ, η)-plane. These equations have been described previously in Chap. 3. These are adopted to simulate the receptivity problem with better accuracy, and the ability to satisfy the mass conservation exactly at all time instants. The convection terms in the vorticity transport equation are discretized with OUCS3 scheme, while the self-adjoint diffusion terms are discretized with second-order central difference scheme. Simulations are performed in MPI parallelization framework using thirty-two cores, using Schwarz domain decomposition technique. The computational domain is decomposed along the streamwise direction only, with thirty points overlap between successive domains. Such a large overlap among each successive domain is taken to drastically minimize reflection and spurious wave generation from the inter-domain boundaries.

One first computes the equilibrium flow governed by NSE in the whole domain before switching on the excitation. At the inflow of the domain (AB) and the far-field (BC) boundary of the domain, as shown in Fig. 5.1, the free stream conditions are specified for both perturbed and unperturbed flow as

$$\frac{\partial \psi}{\partial \eta} = h_2 \tag{5.1}$$

$$\omega = 0 \tag{5.2}$$

This has to be emphasized that, the far-field boundary is practically very far away from the wall, so the prescription of free-stream condition there does not affect the receptivity solutions at all. This has also been confirmed numerically for two different computational domains with different values of y_{max} for some representative cases.

For the equilibrium flow, the no-slip and zero-normal velocity conditions are prescribed at the wall, which specifies the necessary boundary conditions for ψ and ω on the wall by,

$$\psi_w = \psi_0 = \text{Constant} \tag{5.3}$$

$$\omega_w = -\frac{1}{h_2^2} \frac{\partial^2 \psi}{\partial \eta^2} \tag{5.4}$$

For the excited flow, the no-slip condition is still applicable on the wall. However, the wall-normal velocity assumes a prescribed value $v_w(x, t)$, as fixed by the corresponding specific exciter. The corresponding wall-boundary conditions on ψ and ω are specified as

$$\psi_w = \psi_0 + \psi_{wp} \tag{5.5}$$

$$\omega_w = -\frac{1}{h_1 h_2} \frac{\partial}{\partial \xi} \left(\frac{h_2}{h_1} \frac{\partial \psi}{\partial \xi} \right) - \frac{1}{h_2^2} \frac{\partial^2 \psi}{\partial \eta^2} \tag{5.6}$$

where, ψ_{wp} is the perturbation stream-function given as

$$\psi_{wp}(x, t) = \int_0^x v_w(\hat{x}, t) d\hat{x} \tag{5.7}$$

At the bottom boundary, ahead of the leading edge (the segment AO, as shown in Fig. 5.1), the symmetry condition is specified for both perturbed as well as unperturbed flow. This specifies the relevant boundary conditions for ψ and ω as

$$\psi = \psi_0 = \text{Constant} \tag{5.8}$$

$$\omega = 0 \tag{5.9}$$

At the outflow, vorticity is calculated from the convective Sommerfeld boundary condition given by,

$$\frac{\partial \omega}{\partial t} + U_c \frac{\partial \omega}{\partial x} = 0, \tag{5.10}$$

which is time-advanced by using the same time-stepping method used for vorticity-transport equation. The prescription of the radiative Sommerfeld boundary condition allows the disturbances to smoothly convect out of the outflow boundary. Convective speed of disturbances through the outflow, U_c, in the Sommerfeld radiative outflow condition in Eq. (5.10), is taken as the free stream speed U_∞. However, it is seen and verified by some numerical simulations that change in the value of U_c, does not alter the solution in the interior of the computational domain. Only the solution at a very narrow strip near the outflow boundary, is affected in such cases. Both mean and perturbed stream-function ψ, at the outflow boundary is calculated from the following condition on wall-normal component of velocity given as

$$\frac{\partial v}{\partial x} = 0 \tag{5.11}$$

This is a *so-called* soft boundary condition that fixes the stream-function value. Note that the computational domain is so large, that in the streamwise direction this outflow boundary condition is equivalent to evaluating the vorticity at the outflow from $\omega = -\frac{\partial u}{\partial y}$, an equivalent boundary layer approximation.

Other forms of boundary condition on ψ such as the one derived from

$$\frac{\partial u}{\partial t} + U_c \frac{\partial u}{\partial x} = 0$$

have also been tested, and found to produce identical results for both mean and perturbation quantities, except in a very narrow zone, near the outflow boundary. This boundary condition is not used for further 2D simulations.

As for the perturbed flow, the equilibrium solution is used as the initial condition, as mentioned earlier. Equilibrium flow is computed till a steady flow-field is obtained inside the entire domain. This is ensured by checking the time history of $\partial \omega / \partial t$. In [2], the similarity functions $f'(\bar{\eta})$, $f''(\bar{\eta})$ and $f'''(\bar{\eta})$ obtained from the solution of the Blasius equation are compared with the corresponding obtained equilibrium solution at several streamwise locations, varying from very close to the leading edge to x_{out}. The match is good except at streamwise stations, which are close to the leading edge of the plate. Next, the receptivity of the computed solution to monochromatic deterministic excitation frequency is described.

5.3   Small Amplitude Disturbance at Moderate Frequency

Obtained equilibrium flow is perturbed time harmonically by the SBS exciter. The amplitude function of the SBS exciter is defined in Chap. 3 (see Eq. (3.102)). Here, the receptivity results of the ZPG boundary layer is discussed, where the amplitude control parameter of the SBS strip exciter is $\alpha_1 = 0.002$ (see Eq. (3.101)), while the non-dimensional frequency of the excitation is $F = 1.0 \times 10^{-4}$. The nondimensional frequency F, is related to the physical frequency f as $F = 2\pi \nu f / U_\infty^2$, where ν and U_∞ are kinematic viscosity and free-stream velocity, respectively. The amplitude of the excitation is kept at 0.2% of the free-stream speed, U_∞. The exciter is placed at $x_{ex} = 1.5$ (with corresponding $Re_{\delta^*}(x_{ex}) = 666.13$) and has a width of $w_{ex} = 0.09$. One can represent the excitation frequency $F = 1.0 \times 10^{-4}$, by a straight line in the (Re_{δ^*}, β_0)-plane (see Fig. 3.18). This line intersects the lower and upper branches of the neutral curve at $Re_{\delta^*} = 728.7$ ($x = 1.79$) and $Re_{\delta^*} = 1233.88$ ($x = 5.15$), respectively. Therefore, the exciter location is stable according to linear spatial stability theory. As the exciter is switched on impulsively at $t = 0$, a STWF pierces out of the TS wave-packet. For moderate frequency cases, although the TS wave-packet does not play any role in triggering transition of the flow, this STWF is responsible for transition, whose feature is explained next by studying its dynamics.

In Fig. 5.2, the evolution of disturbance at $y = 0.0057$ (which is the tenth line from the wall) is shown by plotting streamwise disturbance velocity u_d as a function of x for the indicated time instants. For this case, STWF is marked in the top frame at $t = 20$. It is also noted in Fig. 5.2, that the amplitude of the TS wave packet changes slightly with time. The TS wave-packet is essentially a progressive wave, which attenuates due to the spatial stability properties (modified by nonlinear, nonparallel

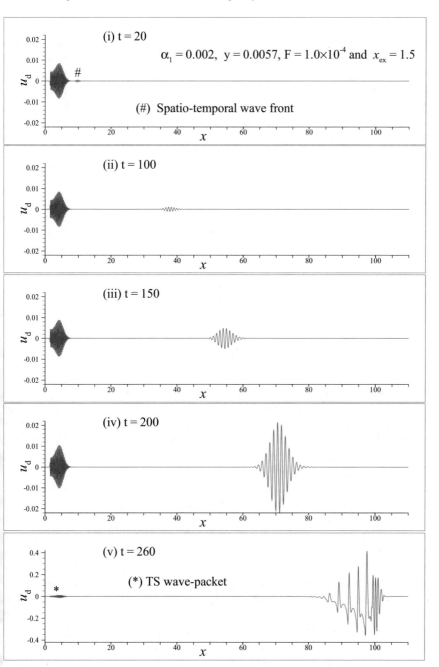

Fig. 5.2 u_d plotted as a function of x at $y = 0.0057$ and indicated times for $\alpha_1 = 0.002$, $F = 1.0 \times 10^{-4}$ and $x_{ex} = 1.5$. The spatio-temporal wave front in the top frame and the TS wave-packet in the last frame is indicated by # and asterisk symbol, respectively

effects), and remains virtually localized in space. In contrast, the STWF propagates in space and time, while growing significantly to a larger amplitude, as compared to the TS wave-packet. In [23, 24], it has been shown that the origin of the first STWF is due to a linear mechanism, whose onset can be captured by the solution of the OSE by BCIM. At $t = 20$ and 100, the wave-front is significantly smaller than the TS wave-packet. At $t = 150$, maximum amplitude of STWF is of the same order as compared to the TS wave-packet. By $t = 200$, the amplitude of the STWF is approximately four times the amplitude of the TS wave-packet, and due to this large amplitude of the wave-front, the nonlinear effects play a dominant role thereafter. With passage of time, nonlinear effects distort the fore-aft symmetry of STWF due to induction of unsteady separation on the wall below it (see the frame at $t = 260$). At $t = 260$, the maximum peak-to-peak variation of the wave-front is of the order of the free-stream speed. Thus, it is apparent that for this moderate frequency excitation case, the disturbance evolution is more dominated by the STWF, rather than the TS wave-packet, contrary to the classical view-point of transition. According to the classical view, the flow transition caused by low amplitude, moderate frequency inputs are solely dominated by TS wave-packets, which in this case is found to be stationary and does not significantly grow with time, as well.

5.3.1 Growth and Speed of STWF

The maximum amplitude of the STWF for streamwise disturbance velocity is traced in Fig. 5.3a, where the variation of maximum amplitude (u_{dm}) with time is shown for the case of Fig. 5.2. Based on the physical processes of growth for this distur-bance, six different stages can be identified (marked in the figure). The origin of STWF can be traced to the linear mechanism governed by the OSE [23]. The stage I denotes the initial spectral redistribution, which causes the maximum amplitude of STWF to decrease due to dispersion of the wave-front, resulting in the increase of the band-width of wavenumbers of STWF. Similar initial decrease in amplitude of the wave-packet, due to dispersion, was also noted for pulse excitation case in [3]. During stage II, one observes growth of STWF, where the amplitude of the front-grows exponentially with time. One notes that despite the monochromatic temporal excitation at the wall, STWF is not monochromatic, with respect to both space and time. Strong nonlinear effects start to affect its growth from stage III onwards, where the amplitude grows at a reduced rate, due to self and multi-modal interaction. In the context of flow past a cylinder [25, 26], self-and multi-modal interactions were accounted for following Stuart–Landau–Eckhaus equation. Higher order non-linear effects, secondary and higher order instabilities due to mean flow distortion by wave-induced stresses [27] become dominant in stage IV, resulting in a super-exponential growth, as seen in Fig. 5.3a. Eventually nonlinear saturation of amplitude is achieved in stage V, where u_{dm} is of the order of the free-stream speed. This subsequently results in the creation of fully developed 2D turbulence in stage VI.

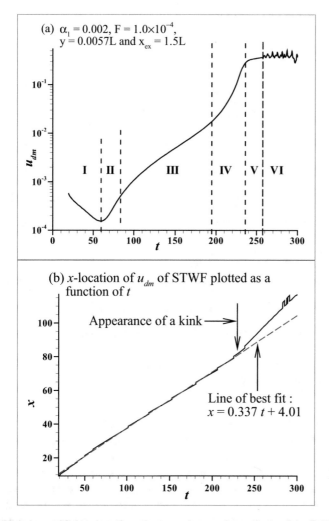

Fig. 5.3 **a** Variation and **b** location of maximum peak-to-peak amplitude of the STWF for u_d at $= 0.0057$ with time $\alpha_1 = 0.002$, $F = 1.0 \times 10^{-4}$ and $x_{ex} = 1.5$. Six stages of the growth of the wave front are identified and marked in frame (**a**)

In Fig. 5.3b, the streamwise location of u_{dm} is shown as a function of time for the height, $y = 0.0057$. One notes that STWF propagates roughly at a speed of $0.337U_\infty$, up to a time when its growth is significantly affected by the distortion of the mean flow in the leading order. This deviation appears as a kink, as marked in Fig. 5.3b by an arrowhead. The respective time instant lies at the end of stage IV of the growth of STWF.

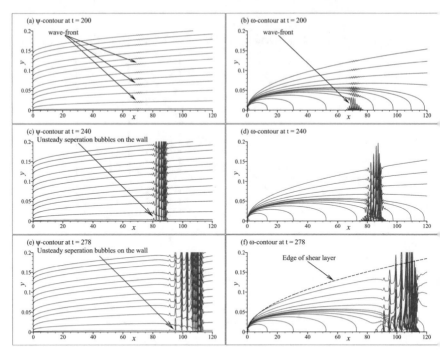

Fig. 5.4 The stream-function ψ- (left column) and vorticity ω-contours (right column) are plotted at indicated times for $\alpha_1 = 0.002$, $F = 1.0 \times 10^{-4}$ and $x_{ex} = 1.5$

On a closer look, it is revealed that in stages V and VI, unsteady separation bubbles form on the wall. This feature is shown in Fig. 5.4, with the help of ψ- and ω-contours at $t = 200$ (in stage III), 240 (at the beginning of stage IV) and 278 (in stage VI, after the creation of a fully developed turbulence). Each bubble creates adverse pressure gradient upstream of it, which in turn creates another bubble and this cascading effect is responsible for the rapid widening of the perturbed zone and the eventual transition. However as bubbles keep appearing, these also convect downstream, creating a dynamical equilibrium, whereby no further upstream penetration of disturbances takes place. This type of saturation of growth rate for the STWF due to nonlinear interaction in stage VI, is shown to be universal, for all input amplitude cases considered and is discussed later. The unsteady separation bubble formation leading to transition is similar to the bypass route shown by free stream convecting vortex discussed in Chap. 4. Once the bubbles form, u_d amplitude saturates to the level of free stream speed, as noted in Fig. 5.3. From the vorticity contours in the right column, one notes that these unsteady separations on the wall cause highly unsteady vortical eruptions, which pierces through the shear layer (see Fig. 5.4f).

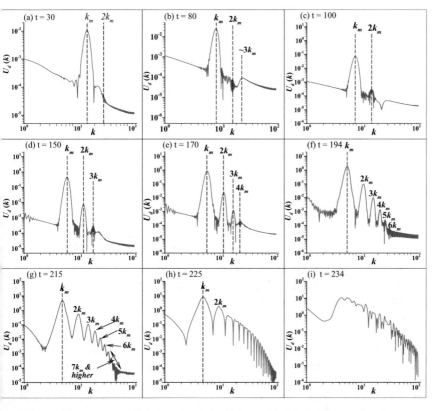

Fig. 5.5 The spatial Laplace transform of the STWF at $y = 0.0057$ shown for indicated times. The excitation parameters are $F_f = 10^{-4}$, $x_{ex} = 1.5$ and $\alpha_1 = 0.002$

5.3.2 Spatial Spectrum and Scale Selection of STWF

Scale selection and spectrum during the downstream propagation of STWF is illustrated in Fig. 5.5, in a log-log plot showing spatial Fourier–Laplace transform of u_d as a function of wavenumber k, plotted at indicated times instants. Frames (a–c) show that as STWF propagates downstream, the dominant wavenumber k_m comes down with the narrowing of the band of wavenumbers around it. The value of k_m decreases from 18.01 at $t = 20$ to 6.5 at $t = 194$. At initial stages of the evolution of the STWF, one also notes that there is a presence of the first superharmonic of the dominant wavenumber k_m, but its amplitude is more than two orders of magnitude lower than k_m (see frames (a–c)). This effectively rules out any nonlinear interaction between these modes.

With progress in time, the amplitude of the first superharmonic also displays growth. During stage *III* of the evolution of STWF, strong nonlinearity starts to interfere with the growth of STWF, as evident from frames (d–f). One notes that not

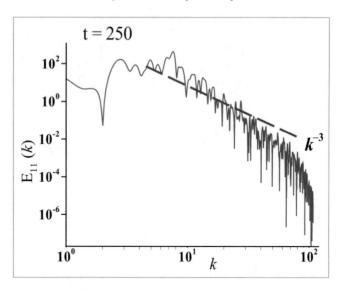

Fig. 5.6 E_{11} corresponding to the STWF plotted as a function of k, for SBS excitation case with $\alpha_1 = 0.002$, $F_f = 10^{-4}$ and $x_{ex} = 1.5$

only the growth rate of the amplitude at k_m increases rapidly, but also more and more higher harmonics appear and at $t = 194$, presence of as many as six super-harmonics of k_m can be noted. The rapid growth and induction of higher harmonics in stage *III* has two obvious effects: first, it reduces the growth rate of the primary mode at k_m, and secondly, it distort the mean flow by wave-induced stresses [27]. The distortion of the mean flow initiates secondary and higher order instabilities of STWF, causing rapid growth in its overall amplitude. Growth of amplitude of the higher harmonics are also accompanied by increase in the band-width resulting in a continuous spectra at $t = 234$. In Fig. 5.6, energy spectra is plotted at $t = 250$, i.e., after the complete breakdown of STWF, and there exists a range of wavenumber band, where the spectra falls off following a k^{-3}-type variation which is typical of 2D turbulence [1, 6, 12, 13]. Here, E_{11} is obtained as

$$E_{11} = |U_d(k)|^2$$

where $U_d(k)$ is the Fourier–Laplace transform of u_d with respect to the streamwise wavenumber k.

5.3.3   Wall-Normal Variation of the Disturbance Velocity

In Fig. 5.7a–f, u_d is plotted as a function of y/δ^*, at indicated times and streamwise locations. Here, δ^* is the local displacement thickness. The indicated streamwise

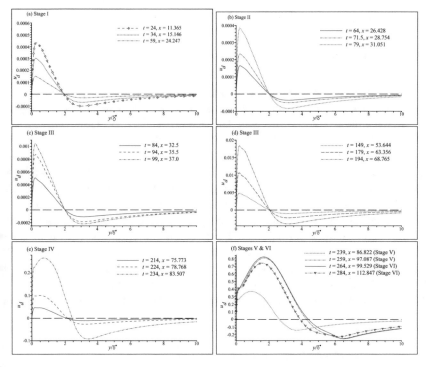

Fig. 5.7 u_d corresponding to the STWF plotted as a function of y/δ^* at indicated time and location. Here, δ^* is the local displacement thickness

locations correspond to where u_d is maximum. Corresponding growth stages are also noted in the respective frames. Figure 5.7 shows that during stages I, II and III of the STWF, the y-variation of u_d roughly follows the variation of the real part of the eigenfunction ϕ' (see Chap. 3). Figure 5.7a–d also shows that u_d becomes zero at $y/\delta^* \simeq 2$. The inner and outer maxima of the wall-normal profile of u_d also appears at identical locations of y/δ^* for stages I and II, which starts to vary slowly from stage III onwards. One observes considerable variations in u_d during stages IV, V and VI of the growth of STWF. Particularly significant is the presence of considerable high perturbation shear stress $\tau_d = \partial u_d/\partial y$ at the wall during these stages. Such a high value of τ_d at the wall is a signature of the presence of high intensity fluctuations inside the boundary layer. One also observes that during stage VI, and later part of stage V ($t = 239$ onwards), u_d is of the order of the free-stream speed from $y/\delta^* \simeq 0.08$ (very close to the wall) to 2.5 (near the shear-layer edge). This kind of very high streamwise disturbance velocity inside the shear-layer promotes rapid transport of momentum and is a hallmark of turbulent flows.

5.4 Dynamics of STWF for High Amplitude Wall-Excitation and Turbulent Spot Regeneration Mechanism

Dynamics of the STWF is different, when the excitation amplitude is higher. For high amplitude wall-excitation cases, unsteady separation bubbles are induced on the wall and the scenario is different than that is described in the previous section, for small excitation amplitude cases. Here, the amplitude of excitation cases with $\alpha_1 = 0.01$ and 0.05 are discussed. In Fig. 5.8, variation of the maximum amplitude u_{dm} of the front is shown. For $\alpha_1 = 0.01$, the time duration of stages I (shown in the inset), II and III are seen to be very small as compared to the $\alpha_1 = 0.002$ case. The stage IV (the stage of secondary and higher order instabilities due to the distortion of the mean flow by wave-induced stresses) for $\alpha_1 = 0.01$, takes over very rapidly, as u_d saturates nonlinearly. The STWF for the case of $\alpha_1 = 0.05$ does not exhibit the six stages of evolution as mentioned earlier, and induces bypass transition right at the beginning by forming unsteady separation bubbles on the plate surface, which causes unsteady vortical eruptions. This feature is shown in Fig. 5.9 with the help of stream function ψ- and vorticity ω-contours at $t = 24$ after the onset of excitation.

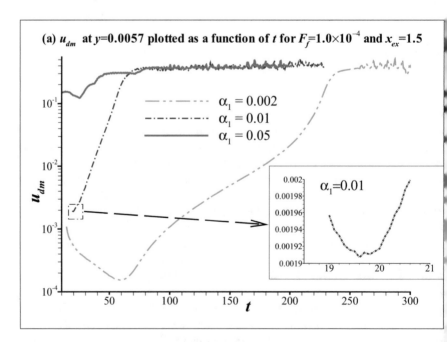

Fig. 5.8 a Variation of the maximum peak-to-peak amplitude u_{dm} corresponding to the STWF at $y = 0.0057$ with time shown for indicated values of α_1. In the inset shown in frame **a**, the variation of u_{dm} for $\alpha_1 = 0.01$ is shown for initial times

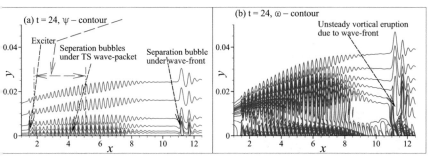

ïg. 5.9 ψ- and ω-contours plotted at indicated times for $\alpha_1 = 0.05$ with $F = 1.0 \times 10^{-4}$ and $_{ex} = 1.5$

)ne notices formation of separation bubbles not only at the location of the front, but ɪlso due to the TS wave-packet, for this amplitude of excitation case.

In the previous section, we demonstrated that turbulence is a deterministic con-;equence of a single STWF convecting downstream, i.e., once STWF is created, ɪts growth and nonlinear saturation leads to small scale unsteady separations on the ᴡall, which culminate into intermittent nature of the wall bounded flow. This could ᴇad one to conclude that this turbulence creation is a buffeting problem, i.e., sus-ᴇnance and creation of turbulence would require subsequent induction of another ;TWF. It is relevant to ask, whether this latter process is due to intrinsic or extrinsic ɪynamics. This is investigated here for $\alpha_1 = 0.01$ and 0.05. In Figs. 5.10 and 5.11, ᴛhe streamwise disturbance velocity component u_d is shown plotted as a function)f x at $y = 0.0057$ for $\alpha_1 = 0.01$ and 0.05, respectively. The exciter is located at ;$_{ex} = 1.5$ $(Re_{\delta^*}(x_{ex}) = 666.67)$, similar to $\alpha_1 = 0.002$ case discussed in the previous ;ection. In the top frame of Fig. 5.10 at $t = 94$, a single STWF is marked as A. In the ﬁollowing frame at $t = 164$, nonlinear saturation of this STWF is noted. At $t = 164$ ɪnsteady separations are created on the wall, underneath this front during this stage)f disturbance evolution, which not only widens the disturbance packet, but also ᴍakes the flow intermittent.

The most important aspect of this propagating STWF during this stage is the ɪnduction of another STWF, marked as B in the frame. The fact that this front is ɪreated upstream of A is very significant, as this event negates the assumption used n parabolized stability equation (PSE) approach, often used to study flow instability. Ⱥs time progresses, the amplitude of the front B also grows rapidly and its nonlinear ᴅistortion is more rapid due to the presence of A, while the gap between the two ;TWFs reduces with time, as shown in the frames at $t = 194$ and 244. Also in the ïrame at $t = 244$, the induction of another STWF is noted, marked as C. In the ﬁollowing frame at $t = 324$, one notices amalgamation of the STWFs marked as ʌ, B and C, while two new STWFs, marked as D and E, which are formed upstream)f the leading packet. These trailing packets grow with time, as the leading part of ᴛhe first packet leaves the computational domain. By $t = 344$, as shown in Fig. 5.10f, ᴛhe trailing fronts D and E grow and become part of the leading packet which finally

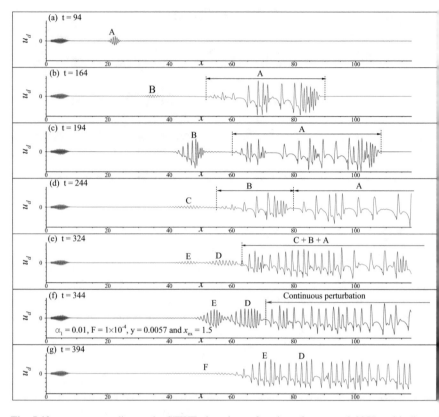

Fig. 5.10 u_d corresponding to the STWF plotted as a function of x at $y = 0.0057$ and indicated times for $\alpha_1 = 0.01$, $F = 1.0 \times 10^{-4}$ and $x_{ex} = 0.5$ in frames (a) to (g)

combines by $t = 394$. The upstream point of the turbulent zone remains in the vicinity of $x \simeq 40$ for this amplitude case at $t = 194, 244, 324$. However, this is seen to move forward up to $x \simeq 60$ at $t = 394$. Another wave-packet F is seen to slowly emerge beyond $t \simeq 360$ (see Fig. 5.10g), which also eventually grows (results not shown here). This process will go on, and will make the flow intermittent in nature. Similar picture is also noted for even higher amplitude of excitation shown in Fig. 5.11, for $\alpha_1 = 0.05$. Thus, the amplitude of excitation causes earlier intermittency of the flow, which also penetrates upstream towards the TS wave-packet. Interestingly the TS wave-packet, for the cases shown in Figs. 5.10 and 5.11, remains rooted at the same spatial location, without affecting the transition process. These features show that a time-harmonic excitation, at a fixed location, gives rise to self-regenerating sequence of STWFs, which are responsible for the generation of turbulent flow field.

For all the amplitude cases of $F_f = 10^{-4}$ and $x_{ex} = 1.5$, the TS wave-packet does not play any role in determining the transition process, that is solely dictated by the growth and subsequent breakdown of the STWFs. However, the TS wave-packet also suffers marginal instability for this moderate frequency of excitation case. This

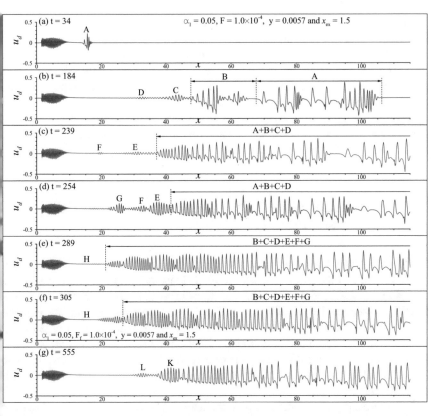

Fig. 5.11 u_d corresponding to the STWF plotted as a function of x at $y = 0.0057$ and indicated times for $\alpha_1 = 0.05$, $F = 1.0 \times 10^{-4}$ and $x_{ex} = 0.5$ in frames (*a*) to (*g*)

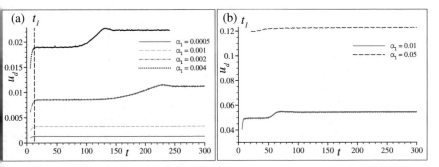

Fig. 5.12 Maximum amplitude of u_d at $y = 0.0057$ corresponding to the TS wave-packet plotted as a function of time for indicated values of α_1 with $F = 1.0 \times 10^{-4}$ and $x_{ex} = 1.5$

feature is noted in Fig. 5.12, where the TS wave-packet amplitude is plotted as a function of time for indicated excitation amplitude cases. For $\alpha_1 = 0.002$, 0.004 and 0.01, two growth stages of the TS wave-packet are noted, *i.e.*, a primary stage (which ends at $t = t_1$ as noted in Fig. 5.12) and a secondary stage which brings the TS wave-packet to a higher saturated level of amplitude. The onset time of the secondary growth stage is dependent on α_1. However, no such secondary growth is observed for $\alpha_1 = 0.05$ case, as the TS wave-packet is nonlinearly distorted displaying a non-zero mean, right from its onset, due to the induction of the unsteady separation bubbles on the wall, as shown in Fig. 5.12b.

5.5   Low Frequency Excitation: Interaction of Near-Field Solution and Primary STWF

For low frequency excitation cases, the evolution of disturbances follow a significantly different route from moderate to high-frequency wall-excitation cases, discussed in previous sections. For latter cases, disturbances evolve such that the STWF and the TS wave-packet remain distinct. The STWF grows in space and time whereas the TS wave-packet remains, more or less, localized in space (close to the exciter location) displaying no spatio-temporal evolution at later times. In contrast, when the excitation frequency is lowered, the near-field solution evolves in both space and time. Here, *the near-field solution*, refers to the part of the solution which corresponds to the TS wave-packet.

For low-frequency excitation cases, the near-field solution evolves in space and time, spawning additional subsequent wave-fronts. These fronts, interact with the primary STWF at later stages, showing an entirely different mechanism of transition. In Fig. 5.13, u_d is plotted at $y = 0.0057$ for $\alpha_1 = 0.002$ and $F_f = 0.5 \times 10^{-4}$, with the exciter located at $x_{ex} = 1.5$. Initially, the primary STWF grows inducing elongated fronts upstream, as noted at $t = 230$ and 250. At $t = 250$, the nonlinear saturation of the primary STWF is noted. The incipient turbulent packet is marked as A, whereas the perturbation wave-packet, which is induced upstream of it, is marked as B. At later times, B also grows and saturate nonlinearly, which in turn induce the disturbance wave-packet C upstream of it, at $t = 290$. At $t = 340$, A convects out of the computational domain with C saturating nonlinearly. These type of spot regeneration mechanism, due to the nonlinear saturation of the primary STWF have also been discussed in the previous section. Between $t = 250$ and $t = 300$, the near-field solution evolves rapidly, which amplifies and extends downstream. The upstream part of the near-field solution is rooted at the location of the exciter. It first shows a streamwise bias, which amplifies and evolves into a sharp wave-front, as noted at $t = 290$. This results in the subsequent emergence of a secondary wave-front (marked by D), which is seen to convect downstream.

Nonlinear saturation of the amplitude of this secondary wave-front happens around $t = 310$, for this case. Subsequently, it attempts to detach from the

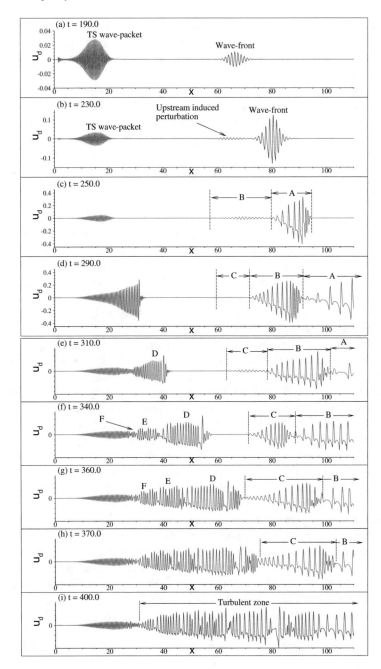

Fig. 5.13 u_d plotted as a function of x at $y = 0.0057$ and indicated times for $\alpha_1 = 0.002$, $F = 0.5 \times 10^{-4}$ and $x_{ex} = 1.5$

parent near-field solution, as it convects and elongates downstream. However, due to the induction of new wave-packets in the intermediate zone, in between the front L and the near-field solution, these are never fully separated from each other. One such induced intermediate wave-packet is shown marked at $t = 340$ as E. The wave-packet E, also grows rapidly, and saturates nonlinearly, while elongating, and eventually undergoing transition, and subsequently merging with D. The nonlinearly saturated wave-packet E, also in turn, induces new perturbation wave-packets (marked as F at $t = 360$) in the intermediate zone, between it and the near-field solution. These newer wave-packets also undergo similar cycle, as its earlier predecessors. The train of disturbance wave-packets, generated due to the spatio-temporal evolution of the near-field solution, finally merges with the rear part of the disturbance packet C (generated due to the cascading effects of the instability of the primary STWF). The difference in length scale between these disturbances are clearly visible at $t = 360$. At later times, it is difficult to distinguish between the perturbations generated from these two different sources, and a continuous highly perturbed zone is obtained, right from the near-field solution. In the present mechanism, more and more new wave-packets keep on appearing in the front part of the TS wave-packet and sustains the extent of the highly unsteady zone, which spans from $x \simeq 30$ to the end of the computational domain. This route of disturbance evolution is completely different from the moderate and high-frequency excitation cases.

Figure 5.14 shows the variation of the maximum peak-to-peak amplitude, u_{dm} (corresponding to the near-field solution) for $y = 0.0057$ is plotted, as a function of time, for the case shown in Fig. 5.13. For this case, four distinct stages of the disturbance evolution can be noted. The first zone marked as S_1 is the zone, where the initial evolution of the near-field solution takes place, with its amplitude increasing by linear mechanism. During S_2, the amplitude attains an almost constant value. Nonlinear stage of growth S_3 is noted subsequently, which brings the amplitude of the near-field solution to the level of the free-stream velocity, leading to continuous spawning of newer secondary wave-fronts, as discussed with respect to Fig. 5.13.

In Fig. 5.15, the spatial Laplace transform of u_d corresponding to the near-field solution ($2.5 \leq x \leq 50$) is plotted for the case of Fig. 5.13, as it evolves spatio-temporally, initiating spawning secondary wave-fronts. This plot highlights the evolution of different peaks and sub-peaks and the nonlinear instability mechanisms. Existence of multiple peaks located at the wavenumbers k_{10} to k_{50}, are noted in Fig. 5.15a. The peaks k_{20} to k_{50} are higher harmonics of the fundamental peak k_{10}. One notes the existence of lower and higher side-band wavenumber components in the neighbourhood of each peak marked in Fig. 5.15a. With progress in time, the principal peak at k_{10} is seen to shift towards lower wavenumber, with additional sub-peaks at higher wavenumbers appearing just adjacent to it, as marked in Fig. 5.15b. These peaks are further traced in Fig. 5.16, where the amplitude of k_{10} and its sub-peaks are plotted as functions of time. One notes form Fig. 5.16, that with increase in time from $t = 250$ to 290, the amplitude of $U_d(k)$ at k_{10} increases five times, while the value of k_{10} decreases from 13.87 to 10.20. The additional sub-peaks k_{11}, k_{12}, k_{13} and k_{14} appear at $t \simeq 270$, which grows till $t \simeq 283$. The sub-peaks at higher wavenumbers, such as k_{13} and k_{14}, grow at a significantly higher rate. For example,

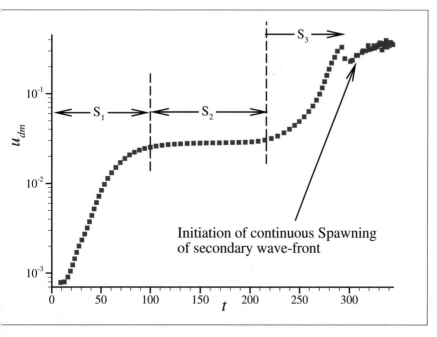

Fig. 5.14 Variation of the maximum peak-to-peak amplitude of u_d corresponding to the TS wave-packet at $y = 0.0057$ plotted as a function of time for $F_f = 0.5 \times 10^{-4}$, $\alpha_1 = 0.002$ and $x_{ex} = 1.5$

the sub-peak at k_{13} grows more than 25 times from its original value at $t = 270$ to 284, whereas the sub-peak at k_{14} grows more than 3000 times during $t = 270$ to 286. From $t \simeq 285$ to 310, the sub-peaks at k_{12}, k_{13} and k_{14} display almost constant amplitude, while the amplitude of the sub-peak at k_{11}, first decreases mildly from $t = 286$ to 296, followed by an increase up to $t = 310$, when it ceases to exist. The amplitude and wavenumber of the peak at k_{11} register rapid drop at $t = 294$, which continues to decrease further up to $t = 310$. These events cause the energy to transfer to higher wavenumber sub-peaks, as these appear and amplify subsequently. The amplitude of such a sub-peak at k_{17} (marked in Fig. 5.15b) is traced in Fig. 5.16a. The sub-peak at k_{17} appears at $t \simeq 282$ and continues to grow up to $t \simeq 310$. In Fig. 5.15c at $t = 309$, one notes the sub-peaks located from k_{11} to k_{17}, to have amplitudes, which are almost similar to that of the principal peak at k_{10}. Appearance of additional sub-peaks adjacent to k_{17} is also noted in this figure. This causes all the main peaks shown in Fig. 5.15c to connect with each other via an array of increasing sub-peaks for $t > 310$, causing the evolution of a continuous spectrum originating from $k = k_{10}$. This initiates the spawning of secondary wave-fronts, as noted earlier in Fig. 5.13, for the low-frequency excitation case.

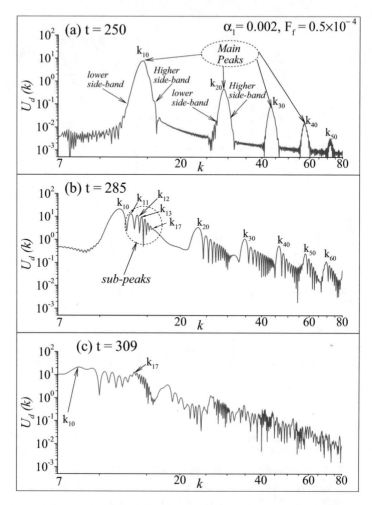

Fig. 5.15 Spatial Laplace transform of u_d within the range $2.5 \le x \le 50$ at $y = 0.0057$ plotted at indicated times with $F_f = 0.5 \times 10^{-4}$ and $\alpha_1 = 0.002$. The exciter is located at $x_{ex} = 1.5$

5.5.1 Low Frequency Excitation: Dominant Role of the Near-Field Solution

Another route of disturbance evolution corresponding to the low-frequency excitation is noted, when amplitude is marginally increased, or the exciter location is changed. This is elaborated here for the case, when following parameters for time-harmonic wall-excitation is used: $F_f = 0.5 \times 10^{-4}$, $x_{ex} = 1.5$ and $\alpha_1 = 0.003$. Therefore, in comparison to the case described in Figs. 5.13, 5.14, 5.15 and 5.16, the current case has slightly higher amplitude of excitation, while the frequency and the exciter

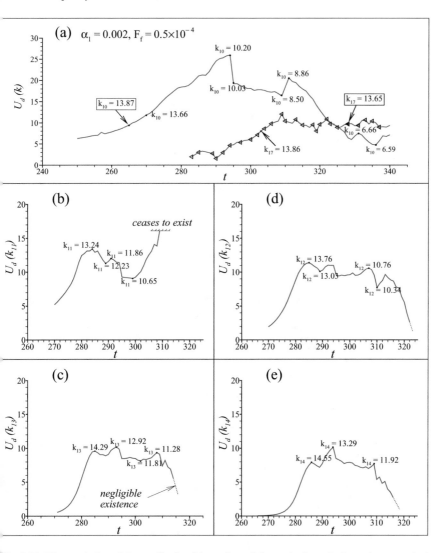

Fig. 5.16 Time variation of the amplitude of the main peak k_{10} and sub-peaks k_{11} to k_{17} as marked in Fig. 5.14 are plotted

location is identical. The evolution of disturbance for this case is shown in Fig. 5.17, where u_d for $y = 0.0057$ is plotted, at indicated time instants. For this case, the amplitude of the near-field solution grows continuously, right from the onset of excitation. The primary STWF is seen initially to come out of the near-field solution, as shown at $t = 60, 90$ and 110. At $t = 90$, the near-field solution is noted to amplify and issue a secondary wave-front, with a sharp leading edge, similar to the case described before. The secondary wave-front convects downstream faster than the

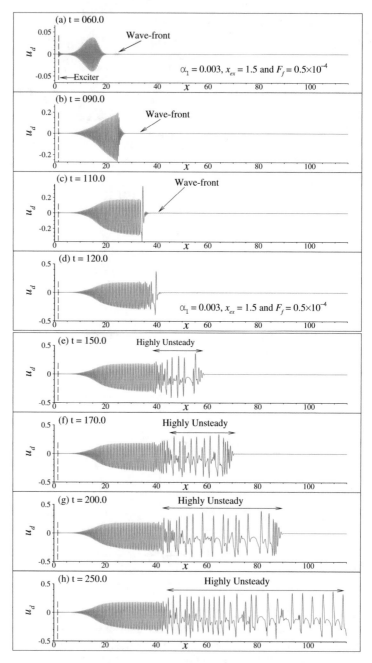

Fig. 5.17 u_d at $y = 0.0057$ plotted as a function of x at indicated times for $\alpha_1 = 0.003$, $x_{ex} = 1.5$ and $F_f = 0.5 \times 10^{-4}$

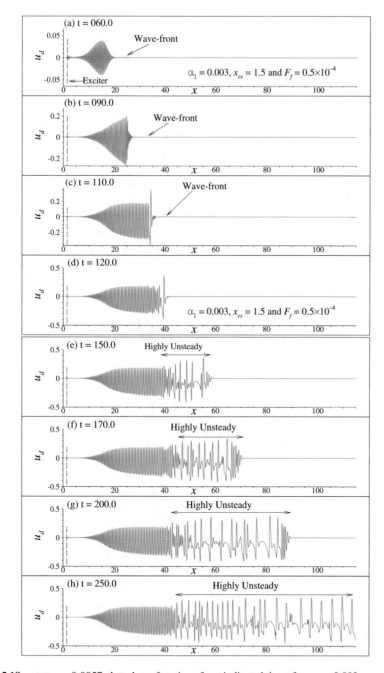

Fig. 5.18 u_d at $y = 0.0057$ plotted as a function of x at indicated times for $\alpha_1 = 0.003, x_{ex} = 1.75$ and $F_f = 0.5 \times 10^{-4}$

corresponding primary STWF, and by $t = 125$, overtakes it. The secondary wave-front grows very rapidly, and as a consequence, one observes a highly unsteady and chaotic zone to form, out of the secondary wave-front, at $t = 150$. This zone is marked by a horizontal arrow in subsequent frames of Fig. 5.17. The front of this zone rapidly progresses downstream, while the upstream end remains fixed at $x \simeq 40$. The rapidly advancing front of this zone, reaches the outflow of the computational domain x_{out} by $t = 250$ (see Fig. 5.17). At this time instant, the entire zone from $x \simeq 40$ to x_{out} is unsteady and chaotic, displaying features of 2D inhomogeneous turbulent flow. However, the perturbation from the location of the exciter to $x \simeq 40$ is composed of traveling waves with a definitive fundamental wavenumber. A similar sequence of disturbance evolution is also noted, when the exciter location is changed to $x_{ex} = 1.75$ from the previous location of 1.5, while keeping the excitation amplitude at $\alpha_1 = 0.003$. This is shown in Fig. 5.18. Figure 5.18 shows that for this case also, the primary STWF does not play any role in determining the dynamics of the disturbance evolution, while the spatio-temporal evolution of the near-field solution, and subsequent continuous spawning of wave-fronts are responsible for eventual flow transition.

5.6 Dynamics of the STWF for Excited Flow Over an Airfoil

To show the relevance of STWF, even for cases with varying pressure gradient, next we study flow past the SHM1 airfoil, by solving NSE. The Reynolds number based on the chord length is $Re = 10.3 \times 10^6$, which corresponds to the cruise condition for Honda-jet aircraft, for which the airfoil is designed. The chosen airfoil is designed as a natural laminar flow (NLF) airfoil, which keeps the flow laminar over a wider range of streamwise stretch. More details of this airfoil and characteristics of flows over it can be found in [20]. The design prediction of the transition location, for both the top and bottom surfaces of this airfoil was performed based on the spatial growth of the TS waves. Here, we compute the flow over SHM1 airfoil without any models for either transition or turbulence. Two-dimensional NSE in (ψ, ω)-formulation using orthogonal coordinate is solved here, similar to the previous flat-plate simulation cases. However, for the present simulations the scale-factors are functions of both azimuthal and wall-normal coordinates.

Figure 5.19 shows the stream function plot for the simulated results at zero degree angle of attack. This figure shows that at zero angle of attack, the flow will accelerate on the top surface almost up to the maximum thickness position (located aft of $0.30c$). For $Re = 10.3 \times 10^6$, the boundary layer formed over this airfoil is very thin, indicating flow separation over the airfoil at very small scales. To note separation bubbles, zoomed view of the flow-field near the trailing edge portion is shown in the bottom two frames of Fig. 5.19. These frames show the formation and downstream convection of micro-bubbles, which are physical in origin. Satisfactorily capturing these micro-bubbles are possible, because of the use of high-accuracy DRP schemes, while solving the unsteady NSE. The presence of unsteady separation bubble on the

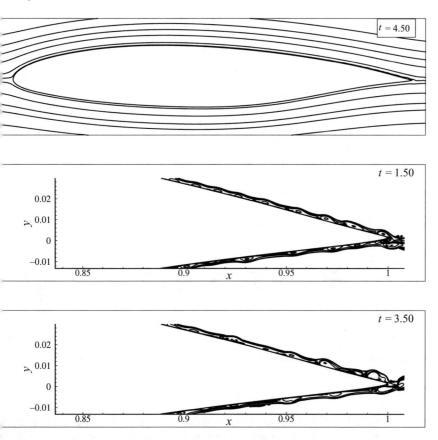

'ig. 5.19 Stream function contours for flow past SHM1 airfoil shown at the indicated time instants
or $Re = 10.3 \times 10^6$, when the airfoil is kept at zero angle of attack (AOA). The top frame shows
he flow field around the full airfoil, while the bottom frames show the zoomed view of the flow
ield near the trailing edge, showing the presence of small bubbles indicative of bypass transition

vall is characteristic of bypass transition, as also noted previously while describing
igh-amplitude excitation cases.

In [17], Falkner–Skan pressure gradient parameter, $m = \frac{x}{U_e}\frac{\partial U_e}{\partial x}$ is plotted for this
low field. For steady boundary layers when m goes below -0.0904, steady flow
eparation is indicated. However, unsteady flows can sustain a higher adverse pressure
gradient (APG), without showing separation. It was noted for the flow past SHM1
irfoil [17], that it exhibits significantly larger swings for m, starting from $x/c = 0.60$,
vhile unsteady separations are only noted after $x/c = 0.75$ (see also Fig. 5.20). Flow-
ield shown in Fig. 5.19, indicates that the flow undergoes bypass transition near the
railing edge of the airfoil, where it encounters high APG. Such bypass transition is
ot due to any explicit excitation on the airfoil surface. In contrast, ZPG boundary
ayer discussed in the previous sections required definitive excitation to trigger flow-
ransition. For flows experiencing an APG, flow is very susceptible to numerical

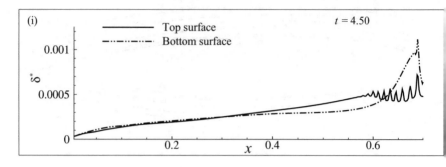

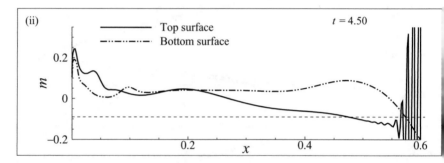

Fig. 5.20 Variation of azimuthal component of velocity (u) at indicated height of 1.242×10^{-6} from the top surface of the SHM1 airfoil at the indicated time instants

disturbances related to round-off error, and this along with other sources of numerical errors, can trigger disturbance growth and eventual transition, as seen in Fig. 5.19, near the trailing edge of the airfoil.

Figure 5.21 shows the sequence by which the top surface of the airfoil experiences bypass transition. This airfoil has concavity on the top and bottom surfaces near the trailing edge which manifests itself in flow instability, near the trailing edge on both surfaces resulting in the local unsteady separation. The region over which the disturbances are noted is seen to travel upstream. The upstream travel of disturbances generate new disturbance packets upstream, in exactly the same sequence bypass transition is noted to take place for a ZPG boundary layer in Sect. 5.4 by high amplitude localized harmonic excitation at moderate to high frequencies. By $t = 4.5$, the disturbed region on the top surface is noted to extend from $x/c = 0.6$ to the trailing edge. These observations also corroborate that bypass transition is caused by upstream propagating disturbances, as illustrated in Sect. 5.4 with respect to ZPG boundary layer.

However, the transition location is far aft of what is experimentally observed [17]. The reason for this difference is due to the fact that in computed flows, a perfectly smooth geometry placed in a uniform flow is considered and the transition is caused by the numerical disturbances acting, as the seed for the APG region over the airfoil surface. Figure 5.22 shows the case, when SBS harmonic excitation is applied

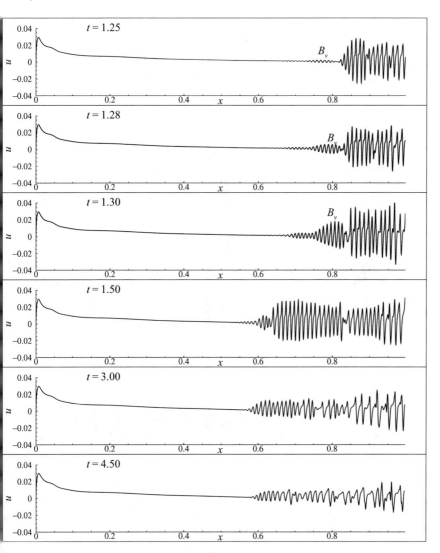

Fig. 5.21 Variation of azimuthal component of velocity (u) at indicated height from the top surface SHM1 airfoil at the indicated time instants

with a nondimensional frequency $F_f = 1.11441 \times 10^{-5}$, on the top surface of the airfoil. The location of the exciter is indicated by an arrowhead in Fig. 5.22. A little before $t = 1.75$, a STWF is noted to be created downstream of the harmonic exciter. With time, this grows and convects downstream. A little after $t = 1.95$, the STWF merges with main disturbance packet created due to the bypass transition without the imposed excitation. This enlarges the region over which transition is noted to be present. The frame at $t = 6.50$ shows that the transitional location point moves to $x/c = 0.4$, as also noted experimentally [17]. Therefore, to match the experimental

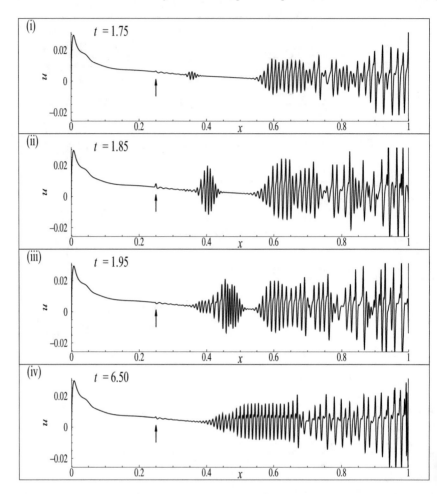

Fig. 5.22 Variation of azimuthal component of velocity (u) at a height of 1.242×10^{-6} from the top surface of the SHM1 airfoil at the indicated time instants. Wall excitation corresponds to an SBS frequency of $F = 1.11441 \times 10^{-5}$, with an amplitude of 0.001

drag coefficients, one must know the level of background disturbance present and simulate the flow accordingly [21, 22].

References

1. Batchelor, G. K. (1969). Computation of the energy spectrum in homogeneous two-dimensional decaying turbulence. *Physics of Fluids,12*, 233–239 [suppl. II].
2. Bhaumik, S. (2013). *Direct Numerical Simulation of Inhomogeneous Transitional and Turbulent Flows*. Ph. D. Thesis, I. I. T. Kanpur, INDIA.

3. Bhaumik, S., & Sengupta, T. K. (2017). Impulse response and spatio-temporal wave-packets: the common feature of rogue waves, tsunami and transition to turbulence. *Physics of Fluids, 29*, 124103.

4. Brillouin, L. (1960). *Wave Propagation and Group Velocity*. New York: Academic Press.

5. Chomaz, J.-M (2005). Global instabilities in spatially developing flows: non-normality and nonlinearity. *Annual Reviews Fluid Mechanics, 37*, 357–392.

6. Davidson, P. A. (2004). *Turbulence: An Introduction for Scientists and Engineers*. UK: Oxford University Press.

7. Drazin, P. G., & Reid, W. H. (1981). *Hydrodynamic Stability*. UK: Cambridge University Press.

8. Fasel, H., & Konzelmann, U. (1990). Non-parallel stability of a flat plate boundary layer using the complete Navier–Stokes equation. *Journal of Fluid Mechanics, 221*, 331–347.

9. Fasel, H. F., Rist, U., & Konzelmann, U. (1990). Numerical investigation of the three-dimensional development in boundary- layer transition. *AIAA Journal, 28*(1), 29–37.

10. Gaster, M. & Grant, I. (1975). An experimental investigation of the formation and development of a wave packet in a laminar boundary layer. *Proceedings of the Royal Society of London A: Mathematical, Physical and Engineering Sciences, 347*(1649), 253–269.

11. Huerre, P., & Monkewitz, P. A. (1985). Absolute and convective instabilities in free shear layers. *Journal of Fluid Mechanics, 159*, 151.

12. Kraichnan, R. H. (1967). Inertial ranges in two-dimensional turbulence. *Physics of Fluids, 67*, 1417–1423.

13. Kraichnan, R., & Montgomery, D. (1980). Two-dimensional turbulence. *Reports in Progress in Physics, 43*, 547–619.

14. Medeiros, M. A. F., & Gaster, M. (1999). The production of subharmonic waves in the nonlinear evolution of wavepackets in boundary layers. *Journal of Fluid Mechanics, 399*, 301–318.

15. Rajpoot, M. K., Sengupta, T. K., & Dutt, P. K. (2010). Optimal time advancing dispersion relation preserving schemes. *Journal of Computational Physics, 229*(10), 3623–3651.

16. Schubauer, G. B., & Skramstad, H. K. (1947). Laminar boundary layer oscillations and the stability of laminar flow. *Journal of the Aeronautical Science, 14*(2), 69–78.

17. Sengupta, T. K. (2012). *Instabilities of Flows and Transition to Turbulence*. Taylor & Francis Group, Florida, USA: CRC Press.

18. Sengupta, T. K., & Bhaumik, S. (2011). Onset of turbulence from the receptivity stage of fluid flows. *Physical Review Letters, 154501*, 1–5.

19. Sengupta, T. K., Bhaumik, S., & Bhumkar, Y. (2012). Direct numerical simulation of two-dimensional wall-bounded turbulent flows from receptivity stage. *Physical Review E, 85*(2), 026308.

20. Sengupta, T. K. & Bhumkar, Y. G. (2013). Direct numerical simulation of transition over a NLF aerofoil: Methods and validation. *Frontiers in Aerospace Engineering* (FAE) 2(1).

21. Sengupta, T. K., Das, D., Mohanamuraly, P., Suman, V. K., & Biswas, A. (2009). Modelling free stream turbulence based on wind tunnel and flight data for instability studies. *International Journal of Emerging Multidisciplinary Fluid Science, 1*(3), 181–201.

22. Sengupta, T. K., De, S., & Gupta, K. (2001). Effect of free-stream turbulence on flow over airfoil at high incidences. *Journal of Fluids Structures, 15*(5), 671–690.

23. Sengupta, T. K., Rao, A. K., & Venkatasubbaiah, K. (2006). Spatiotemporal growing wave fronts in spatially stable boundary layers. *Physical Review Letters, 96*(22), 224504.

24. Sengupta, T. K., Rao, A. K., & Venkatasubbaiah, K. (2006). Spatiotemporal growth of disturbances in a boundary layer and energy based receptivity analysis. *Physics of Fluids, 18*, 094101.

25. Sengupta, T. K., Singh, N., & Suman, V. K. (2010). Dynamical system approach to instability of flow past a circular cylinder. *Journal of Fluid Mechanics, 656*, 82–115.

26. Sengupta, T. K., Singh, N., & Vijay, V. V. S. N. (2011). Universal instability modes in internal and external flows. *Computers & Fluids, 40*, 221–235.

27. Tollmien, W. (1931). The Production of Turbulence. *NACA Report-TM 609*.

Chapter 6
3D Routes of Transition to Turbulence by STWF

6.1 Introduction

In Chap. 5, we have discussed the dynamics of the STWF for 2D transition. We have also shown the inadequacy of the linear spatial instability studies in determining the evolution of disturbances. For monochromatic wall-excitation, the spatio-temporal evolution of disturbance was noted [2] to depend on various factors like (a) excitation frequency, (b) amplitude, (c) exciter location and its width and (d) nature of excitation onset. In the present chapter, we would discuss about the 3D evolution of disturbances and the associated process of transition to turbulence. We first start with the governing equations, followed by numerical methods, problem definition and a brief description of boundary conditions. We have chosen the velocity-vorticity formulation [4] of the incompressible NSE for its inherent accuracy to compute the 3D excitation of a nominally 2D ZPG boundary layer. Growth and evolution of disturbances, nature of vortical structures in the transitional and turbulent zones, and integral properties of the turbulent boundary layer (in terms of displacement and momentum thickness, shape factor and skin friction coefficient) are described subsequently.

6.1.1 Governing Equations and Numerical Methods

Schematic diagram of the 3D receptivity problem for a 2D ZPG boundary layer and the computational domain is shown in Fig. 6.1. A 2D ZPG boundary layer is excited from the wall either through a Gaussian circular patch (GCP) exciter or a spanwise modulated (SM) exciter. While simulating the receptivity problem, the leading edge of the plate has been retained inside the computational domain to capture instability arising from the leading edge, which is a source of disturbance creation [37]. The origin of the reference co-ordinate system is located at the mid-point of the leading edge of the plate. The computational domain is given as $x_{in} \leq x \leq x_{out}$ along the streamwise direction, with $x_{in} < 0$; $0 \leq y \leq y_{max}$ along the wall-normal direction

© Springer Nature Singapore Pte Ltd. 2019
T. K. Sengupta and S. Bhaumik, *DNS of Wall-Bounded Turbulent Flows*,
https://doi.org/10.1007/978-981-13-0038-7_6

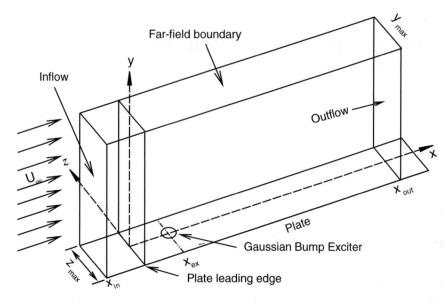

Fig. 6.1 Schematic diagram of the 3D receptivity problem for a 2D ZPG boundary layer

and $-z_{max}/2 \leq z \leq z_{max}/2$, along the spanwise direction. For the GCP exciter, the wall-excitation is provided through a circular patch centered around $(x_{ex}, 0, 0)$ and of radius r_0. For the SM exciter, wall-excitation is provided by an exciter strip which is from x_1 to x_2 along streamwise direction and along the whole spanwise extent of the computational domain.

The simulations are performed in the transformed (ξ, η, ζ)-plane, such that $x = x(\xi)$, $y = y(\eta)$ and $z = z(\zeta)$. The rotational variant of the $(\vec{V}, \vec{\Omega})$-formulation of the incompressible NSE is given by

$$\frac{\partial \vec{\Omega}}{\partial t} + \nabla \times \vec{H} = 0 \tag{6.1}$$

where $\vec{H} = \vec{\Omega} \times \vec{V} + (1/Re_L)\nabla \times \vec{\Omega}$. Equation (6.1) in transformed coordinate can be expressed as

$$\frac{\partial \Omega_\xi}{\partial t} + \left(\frac{1}{h_2} \frac{\partial H_\zeta}{\partial \eta} - \frac{1}{h_3} \frac{\partial H_\eta}{\partial \zeta} \right) = 0 \tag{6.2}$$

$$\frac{\partial \Omega_\eta}{\partial t} + \left(\frac{1}{h_3} \frac{\partial H_\xi}{\partial \zeta} - \frac{1}{h_1} \frac{\partial H_\zeta}{\partial \xi} \right) = 0 \tag{6.3}$$

$$\frac{\partial \Omega_\zeta}{\partial t} + \left(\frac{1}{h_1} \frac{\partial H_\eta}{\partial \xi} - \frac{1}{h_2} \frac{\partial H_\xi}{\partial \eta} \right) = 0 \tag{6.4}$$

where, $\vec{\Omega} = (\Omega_\xi, \Omega_\eta, \Omega_\zeta)$ and $\vec{V} = (u, v, w)$. The scale factors h_1, h_2 and h_3 are given as $h_1 = \frac{\partial x}{\partial \xi}$, $h_2 = \frac{\partial y}{\partial \eta}$ and $h_3 = \frac{\partial z}{\partial \zeta}$. In Eqs. (6.2) to (6.4), the terms H_ξ, H_η and H_ζ are given as

$$H_\xi = (w\Omega_\eta - v\Omega_\zeta) + \frac{1}{Re_L}\left(\frac{1}{h_2}\frac{\partial \Omega_\zeta}{\partial \eta} - \frac{1}{h_3}\frac{\partial \Omega_\eta}{\partial \zeta}\right) \tag{6.5}$$

$$H_\eta = (u\Omega_\zeta - w\Omega_\xi) + \frac{1}{Re_L}\left(\frac{1}{h_3}\frac{\partial \Omega_\xi}{\partial \zeta} - \frac{1}{h_1}\frac{\partial \Omega_\zeta}{\partial \xi}\right) \tag{6.6}$$

$$H_\zeta = (v\Omega_\xi - u\Omega_\eta) + \frac{1}{Re_L}\left(\frac{1}{h_1}\frac{\partial \Omega_\eta}{\partial \xi} - \frac{1}{h_2}\frac{\partial \Omega_\xi}{\partial \eta}\right) \tag{6.7}$$

For the $(\vec{V}, \vec{\Omega})$-formulation, the attendant velocity vectors are obtained from the velocity Poisson equation $\nabla^2 \vec{V} = -\nabla \times \vec{\Omega}$, and in the transformed (ξ, η, ζ)-plane these are given as,

$$\nabla^2_{\xi\eta\zeta} u = \left(h_1 h_2 \frac{\partial \Omega_\eta}{\partial \zeta} - h_3 h_1 \frac{\partial \Omega_\zeta}{\partial \eta}\right) \tag{6.8}$$

$$\nabla^2_{\xi\eta\zeta} v = \left(h_2 h_3 \frac{\partial \Omega_\zeta}{\partial \xi} - h_1 h_2 \frac{\partial \Omega_\xi}{\partial \zeta}\right) \tag{6.9}$$

$$\nabla^2_{\xi\eta\zeta} w = \left(h_3 h_1 \frac{\partial \Omega_\xi}{\partial \eta} - h_2 h_3 \frac{\partial \Omega_\eta}{\partial \xi}\right) \tag{6.10}$$

where the operator $\nabla^2_{\xi\eta\zeta}$ is given as

$$h_1 h_2 h_3 \nabla^2_{\xi\eta\zeta} = \frac{\partial}{\partial \xi}\left(\frac{h_2 h_3}{h_1}\frac{\partial}{\partial \xi}\right) + \frac{\partial}{\partial \eta}\left(\frac{h_3 h_1}{h_2}\frac{\partial}{\partial \eta}\right) + \frac{\partial}{\partial \zeta}\left(\frac{h_1 h_2}{h_3}\frac{\partial}{\partial \zeta}\right)$$

Note that the velocity field should also satisfy the divergence-free condition $D_v = \nabla \cdot \vec{V} = 0$ and in the transformed (ξ, η, ζ)-plane this is given as,

$$D_v = \frac{1}{h_1}\frac{\partial u}{\partial \xi} + \frac{1}{h_2}\frac{\partial v}{\partial \eta} + \frac{1}{h_3}\frac{\partial w}{\partial \zeta} = 0 \tag{6.11}$$

In deriving these equations, the free-stream velocity U_∞ and L are used as the velocity and length scales. The Reynolds number based on L is $Re_L = U_\infty L/\nu = 10^5$, for all the simulations reported here. While solving the receptivity problem, only the Poisson equations for u- and w-components of velocity given by Eqs. (6.8) and (6.10) are solved. The v-component of the velocity is calculated by integrating Eq. (6.11) from the wall as

$$v(\xi, \eta, \zeta) = v(\xi, 0, \zeta) - \int_0^\eta \left(\frac{h_2}{h_1}\frac{\partial u}{\partial \xi} + \frac{h_2}{h_3}\frac{\partial w}{\partial \zeta}\right) d\eta \tag{6.12}$$

Application of Eq. (6.12) not only identically satisfies the solenoidality condition on velocity, but also nullifies the requirement of imposition of any boundary condition on the v-component of the velocity, at the far-field boundary, as shown in Fig. 6.1.

6.1.2   Boundary Conditions

The boundary condition at the inflow of the computational domain for the components of velocity $\vec{V}$, and vorticity $\vec{\Omega}$, are given as

$$u = 1 \ \text{ and } \ \frac{\partial v}{\partial \xi} = \frac{\partial w}{\partial \xi} = \frac{\partial \Omega_\xi}{\partial \xi} = \Omega_\eta = \Omega_\zeta = 0 \tag{6.13}$$

At the far-field of the computational domain, one does not require any boundary condition on the v-component of velocity, as Eq. (6.12) is used to compute it. For the other five variables, the boundary conditions used in the far-field are given as

$$u = 1 \ \text{ and } \ w = \Omega_\xi = \frac{\partial \Omega_\eta}{\partial \eta} = \Omega_\zeta = 0 \tag{6.14}$$

Periodic condition on all the six variables (three components of velocity and three components of vorticity) are used in both the spanwise boundaries. At the wall, time-dependent wall-normal velocity, corresponding to the type of excitation is prescribed, along with the no-slip boundary conditions on u- and w-components of velocity. The boundary conditions on the six variables on the plate surface are given as

$$u = w = 0, \quad v = v_w(x, z, t), \quad \Omega_\eta = 0,$$
$$\Omega_\xi = \left(\frac{\partial w}{\partial y} - \frac{\partial v_w}{\partial z} \right) \ \text{ and } \ \Omega_\zeta = \left(\frac{\partial v_w}{\partial x} - \frac{\partial u}{\partial y} \right) \tag{6.15}$$

The sharp leading edge of the flat plate is assumed as the locus of stagnation points for this flow and hence, at the bottom plane ahead of the leading edge, as shown in Fig. 6.1, symmetry conditions are used on all the six variables. This prescribes the boundary conditions on the six variables given as

$$\frac{\partial u}{\partial \eta} = v = \frac{\partial w}{\partial \eta} = \Omega_\xi = \frac{\partial \Omega_\eta}{\partial \eta} = \Omega_\zeta = 0 \tag{6.16}$$

At the outflow boundary, the convective Sommerfeld boundary conditions are applied on the variables u, Ω_η and Ω_ζ as

$$\frac{\partial u}{\partial t} + U_c \frac{\partial u}{\partial x} = 0 \tag{6.17}$$

$$\frac{\partial \Omega_\eta}{\partial t} + U_c \frac{\partial \Omega_\eta}{\partial x} = 0 \tag{6.18}$$

$$\frac{\partial \Omega_\zeta}{\partial t} + U_c \frac{\partial \Omega_\zeta}{\partial x} = 0 \tag{6.19}$$

he boundary condition on Ω_ξ at the outflow boundary is derived from the solenoidal-ty condition of vorticity as

$$\frac{\partial \Omega_\xi}{\partial x} = -\left(\frac{\partial \Omega_\eta}{\partial y} + \frac{\partial \Omega_\zeta}{\partial z}\right) \tag{6.20}$$

he boundary conditions of small amplitude disturbance on the v- and w-components f velocity at the outflow boundary are derived from the definition of the vorticity omponent Ω_η and Ω_ζ as

$$\frac{\partial v}{\partial x} = \frac{\partial u}{\partial y} + \Omega_\zeta \tag{6.21}$$

$$\frac{\partial w}{\partial x} = \frac{\partial u}{\partial y} - \Omega_\eta. \tag{6.22}$$

5.1.3 Initial Condition

'or the 2D equilibrium flow, Ω_ξ, Ω_η and w are identically zero. Thus, one has to olve only the transport equation for Ω_ζ, while u- and v-components of velocity are •btained by solving the simplified Poisson equation and integrating the continuity quation for 2D flows, respectively. The equilibrium flow is simulated with the initial ondition of impulsive start, i.e.,

$$u = 1, \quad v = 0 \text{ and } \Omega_\zeta = 0 \tag{6.23}$$

Once the 2D equilibrium flow is established, the 3D equilibrium flow is obtained •y specifying u, v and Ω_ζ variables at all the discrete spanwise stations, while •rescribing other variables, i.e. w, Ω_ξ and Ω_η to be zero at all locations. With these nitial conditions, the 3D solver is run for approximately 1000 iterations (when all he unsteady terms falls below machine zero), so that the flow field adjusts itself o the 3D domain and boundary conditions. Subsequently, 3D periodic excitation is nitiated, whose receptivity are simulated by solving Eqs. (6.2) to (6.12), subject to he boundary conditions described in Sect. 6.1.2.

6.1.4   Grid Generation

The grid is generated such that grid-points are clustered near the leading edge of the plate and which becomes uniform after $x = 5$. The grid points are also clustered near the wall of the plate to accurately resolve the boundary layer. Both the clustering in the streamwise and wall-normal directions are performed using tangent hyperbolic function, as it produces minimum aliasing error during computation [10]. Along the spanwise direction uniform grid-points are used. The grid transformation function along the x-direction are given as for, $x_{in} \leq x \leq x_s$ $(0 \leq \xi \leq \xi_1)$

$$x(\xi) = x_{in} + (x_s - x_{in})\left[1 - \frac{\tanh[\beta_x(1 - \xi)]}{\tanh \beta_x}\right] \qquad (6.24)$$

while for $x_s \leq x(\xi) \leq x_{out}$ $(\xi_1 \leq \xi \leq 1)$,

$$x(\xi) = x_s + (x_s - x_{in})\left[\left(\frac{\beta_x}{\tanh \beta_x}\right)\left(\frac{\xi - \xi_1}{\xi_1}\right)\right] \qquad (6.25)$$

where $\xi_1 = \frac{1}{1+A_1}$ and $A_1 = \left(\frac{x_{out}-x_s}{x_s-x_{in}}\right)\left(\frac{\tanh \beta_x}{\beta_x}\right)$. The grid-transformation function along the wall-normal direction is given as

$$y(\eta) = y_{max}\left[1 - \frac{\tanh[\beta_y(1 - \eta)]}{\tanh \beta_y}\right] \qquad (6.26)$$

where $0 \leq \eta \leq 1$. Here, β_x and β_y are parameters that control the grid clustering in the streamwise and wall-normal direction, respectively. Here, for all the cases $\beta_x = 1$ and $\beta_y = 2$, are used.

For some of the simulations described here, $x_{in} = -0.05$, $x_{out} = 20$ and $y_{max} = 0.75$ have been used, while $z_{max} = 2$ for all the cases, except two cases, where $z_{max} = 4$ is taken. The reference length scale L is taken such that $L = 41\delta_{out}^*$, where δ_{out}^* is the displacement thickness at the outflow for the equilibrium flow of the computational domain. Note the streamwise length of the domain for these 3D simulations is much longer than the 2D simulation in [11].

For a typical simulation performed with $x_{in} = -0.05$, $x_{out} = 20$, $y_{max} = 0.75$ and 1001 and 301 points along x- and y-directions, the minimum and maximum resolutions along the streamwise direction are $\Delta x_{min} = 9.1 \times 10^{-3}$ and $\Delta x_{max} = 2.1 \times 10^{-2}$, respectively, while the wall-normal resolution is given as $\Delta y_{min} = 3.68 \times 10^{-4}$, which stretches up to $\Delta y_{max} = 5.18 \times 10^{-3}$ at the far-field boundary.

5.1.5 Numerical Method and Solution Technique

While simulating the flow, optimized staggered compact schemes (OSCS) are used or the purpose of both interpolation of the function and evaluation of first derivative of the function [35]. The OSCS scheme has been described in Chap. 2. The second or mixed derivative terms are evaluated by repeated application of the OSCS scheme for the evaluation of first derivative. This way of evaluating second derivative is distinctly different from that is use in discretizing diffusion term in self adjoint form [36], which as been used for 2D transitional flow simulations [34, 35]. The ORK_3 scheme is used to integrate the VTEs (Eqs. (6.2) to (6.4)) with a time-step of $\Delta t = 8 \times 10^{-5}$. To suppress numerical spanwise spurious oscillations, periodic sixth-order filter with the filter coefficient $\alpha_f = 0.45$ is used in the spanwise direction. For the purpose of de-aliasing, numerical fourth-order diffusion term is used with coefficient $\varepsilon = 0.06$, in both the streamwise and wall-normal directions. Proper way of adding fourth-order numerical diffusion in the vorticity transport equation is described next. Consider Eq. (6.2), which can be further expressed as

$$\frac{\partial \Omega_\xi}{\partial t} + \frac{1}{h_2 h_3}\left[\frac{\partial}{\partial \xi}(h_1 u \Omega_\xi) + \frac{\partial}{\partial \eta}(h_2 v \Omega_\xi) + \frac{\partial}{\partial \zeta}(h_3 w \Omega_\xi)\right]$$

$$= \frac{1}{h_2 h_3}\left[\frac{\partial}{\partial \xi}(h_1 u \Omega_\xi) + \frac{\partial}{\partial \eta}(h_2 u \Omega_\eta) + \frac{\partial}{\partial \zeta}(h_3 u \Omega_\zeta)\right] - \frac{1}{Re_L}[\nabla \times \vec{\Omega}]_\xi \quad (6.27)$$

Terms included within the first square braces indicate nonlinear convection of Ω_ξ, whereas the first set of terms on the right hand side of Eq. (6.27) indicate the generation of Ω_ξ due to stretching of the other components of $\vec{\Omega}$. Therefore, one has to add numerical diffusion in the respective convection terms for effective upwinding, whereas the vortex stretching terms are to be discretized with central difference schemes. Here, fourth-order numerical diffusion for Ω_ξ is added by combining the following terms with the convection terms,

$$\frac{\varepsilon}{h_2 h_3}\left[\frac{h_1|u|\Delta_i^3 \Omega_\xi}{\Delta \xi} + \frac{h_2|v|\Delta_j^3 \Omega_\xi}{\Delta \eta} + \frac{h_3|w|\Delta_k^3 \Omega_\xi}{\Delta \zeta}\right] \quad (6.28)$$

where $|\cdot|$ indicates the absolute value and $\Delta_p^3 f$ is given as

$$\Delta_p^3 f = -3(f_{p+1} - f_p) + (f_{p+2} - f_{p-1})$$

The value of ε fixes the amount of diffusion added, and here $\varepsilon = 0.06$ is taken. Equations (6.2) to (6.4) are solved in the computational domain shown in Fig. 6.1, by domain decomposition technique and using MPI parallelization framework. For solving the resulting tridiagonal matrix equations in ξ- and η-directions, domain-decomposition technique of [38] is used, where an overlap of six points are taken. In the spanwise direction, one has to retain periodicity of the problem, by solving the periodic tridiagonal equations. As the technique of [38] is essentially for non-

periodic problem, it is inherently incapable of retaining symmetry of the problem
a modified periodic version of the algorithm proposed in [27] to solve the periodic
tridiagonal equations in the spanwise directions is used. Most of the simulations are
carried out using 128 computing units (CU) with 16 CU in the streamwise, 4 CU in
the wall-normal and 2 CU in the spanwise direction. For some simulations a total of
512 CU are used, with 32 CU along streamwise, 8 CU along wall-normal and 2 CU
along spanwise direction.

6.2   Gaussian Circular Patch (GCP) Excitation

For the simulation results reported here, time-harmonic Gaussian type excitation
are provided in a circular patch. This is different from the experiments performed
for impulse response [12]. Present receptivity study is also different from other
theoretical effort [15]. The imposed wall-normal velocity component $v_w(x, z)$ on the
patch is given as

$$v_w(x, z) = \alpha_1 A_m(x, z) \sin(\bar{\omega}_0 t) \tag{6.29}$$

where α_1 is the amplitude control parameter, $A_m(x, z)$ is the amplitude function
with absolute value varying from zero to one, and $\bar{\omega}_0$ is the non-dimensional circular
frequency given by $\bar{\omega}_0 = F_f \times Re_L$. The amplitude function $A_m(x, z)$ for the circular
Gaussian type excitation is given as

$$A_m(x, z) = \frac{1}{2} \left(1 + \cos\left(\frac{\pi r}{r_{max}} \right) \right) \tag{6.30}$$

for $r \le r_{max}$ and $A_m(x, z) = 0$ for $r > r_{max}$. Here, $r = \sqrt{(x - x_0)^2 + (z - z_0)^2}$ with
x_0 and z_0 denoting the center of the circular patch. For the results reported here
$r_{max} = 0.09$, $x_0 = 1.5$ ($Re_{\delta^*} = 666$) and $z_0 = 0$ are used. In Fig. 6.2a, the amplitude
function $A_m(r)$ is plotted as a function of r/r_{max}. In Fig. 6.2b, a typical snapshot of
the receptivity result at $t = 15$ is shown, by plotting the streamwise component of
the disturbance velocity u_d, in the (x, z)-plane, for $F_f = 5 \times 10^{-5}$ and $\alpha_1 = 0.01$
at $y = 0.00189$. The spatio-temporal front is marked, with the oblique TS wave
packets shown clearly, along with the local solution. One notes that the amplitude of
the STWF is several orders of magnitude higher than the TS wave-packet, at this time.
In the left frames of Fig. 6.3, evolution of u_d at $z = 0$, $y = 0.00189$ and indicated
times are shown, as a function of x. The Fourier transform of the signals, displayed
in the left frame, are shown in the right frames of Fig. 6.3. One observes that the
STWF exhibits significant growth in time, in comparison to the TS wave-packet, as
it propagates downstream. This is also evident from the spectrum of u_d shown in the
right frames of Fig. 6.3. Induction of similar high velocity, very close to the wall
gives rise to the evolution of streamwise elongated puffs, similar to what is described
in [9, 44], whose amplification gives rise to transition, via formation of turbulent
spots. The wave-front is also shown to induce an additional disturbance wave-packet

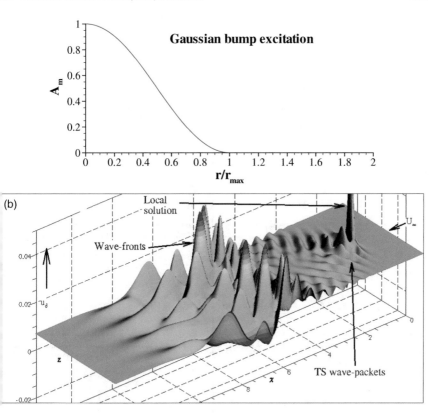

ig. 6.2 a The amplitude function $A_m(r)$ plotted as a function of r/r_{max}. **b** u_d plotted in (x, z)-
lane at $y = 0.00189$ and $t = 15$ for 2D ZPG boundary layer excited by a circular wall bump exciter
with non-dimensional frequency $F_f = 5 \times 10^{-5}$

narked as A, just upstream of it. This induced wave-packet is also seen to suffer
narginal growth in time.

In Figs. 6.4, 6.5, 6.6, 6.7, 6.8 and 6.9 evolution of disturbances for the Gaussian
vall bump exciter are shown, for $F_f = 0.5 \times 10^{-4}, 0.75 \times 10^{-4}$ and 1.0×10^{-4}, by
lotting the contours of u_d in the (x, z)-plane, at $y = 0.00189$ and indicated times.
Dne notes that with progress in time, a STWF evolves, whose shape resembles an
arrowhead. The width of these structures increase, as these propagate downstream.
Due to the enforcement of periodic conditions in the spanwise directions, soon neigh-
ooring wave-fronts starts interacting with each other, resulting in the creation of high
vavenumber oscillations, at the spanwise boundaries of the computational domain.
Df the three frequencies considered here, the STWF for the lowest frequency grows
astest, and hence it interacts most vigorously with the neighboring wave-fronts,
creating a turbulent spot at $t = 27.30$, as marked by P in Fig. 6.9a. At $t = 30$, one
also observes the wave-front for $F_f = 0.75 \times 10^{-4}$ to induce a turbulent spot at
$\simeq 13.5$ (Fig. 6.9b). However, its intensity is less than the spot shown in Fig. 6.9a.

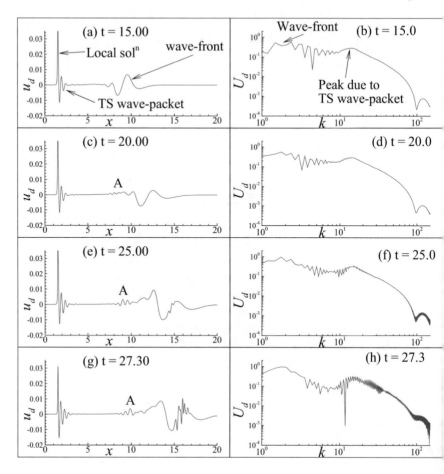

Fig. 6.3 u_d plotted as a function of x at $z = 0$, $y = 0.00189$ and indicated time-instants in the righ column for 2D ZPG boundary layer circular wall bump exciter with non-dimensional frequency $F_f = 5 \times 10^{-5}$. The spectrum for the plotted u_d in the left column is shown in the right

The wave-front created for $F_f = 1.0 \times 10^{-4}$, though interacts with the neighbor ing wave-fronts, but does not induce any turbulent spot by $t = 30$. Figures 6.4–6.9 clearly reveals that, it is only the STWF, which not only exhibits growth, but also induces additional perturbations upstream of it, similar to the 2D receptivity problem discussed in the previous chapter. One also notices in Figs. 6.8 and 6.9, that at $t = 25$ and 27.3, the magnitude of u_d is of the order of the free-stream velocity. Considering the fact that the height $y = 0.00189$, is only 3% of the boundary layer thickness a $x = 12$, one concludes that such high streamwise perturbation velocity inside the shear layer would not only give rise to elongated quasi-streamwise streaks, as found in [9, 44], but also accelerate the transition process violently.

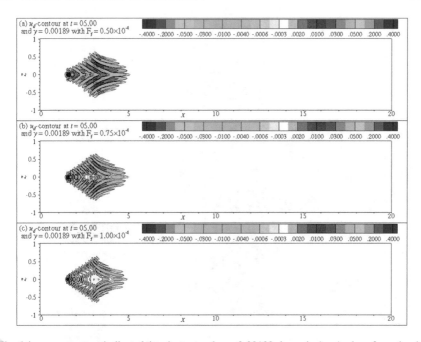

Fig. 6.4 u_d-contours at indicated time instant and $y = 0.00189$ shown in (x, z)-plane for a circular wall bump exciter for indicated frequencies of excitation

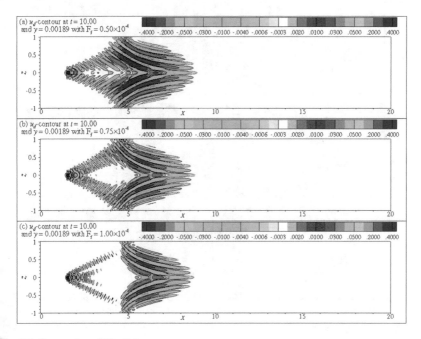

Fig. 6.5 See caption of Fig. 6.4

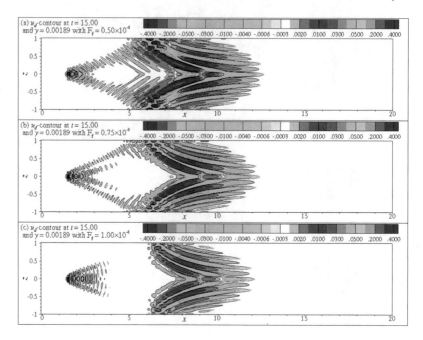

Fig. 6.6   See caption of Fig. 6.4

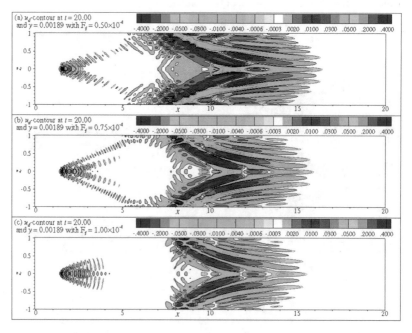

Fig. 6.7   See caption of Fig. 6.4

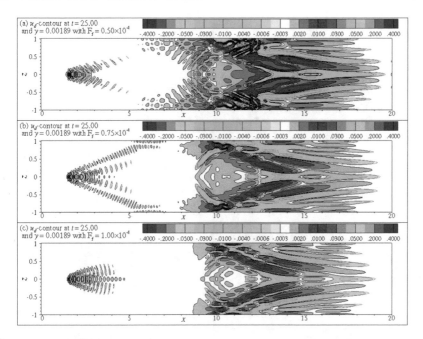

Fig. 6.8 See caption of Fig. 6.4

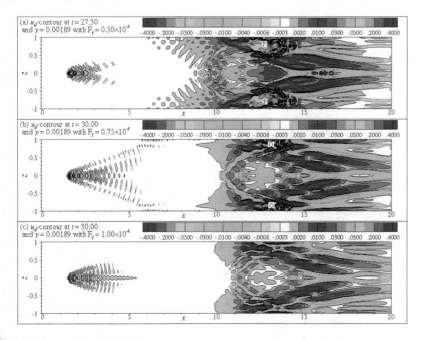

Fig. 6.9 See caption of Fig. 6.4

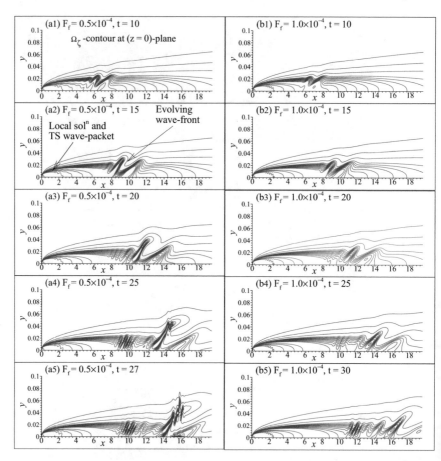

Fig. 6.10 Ω_ζ-contours plotted in (x, y)-plane for $z = 0$ at indicated times for a circular wall bump exciter with $F_f = 0.5 \times 10^{-4}$ (left frames) and $F_f = 1.0 \times 10^{-4}$ (right frames)

In Fig. 6.10, Ω_ζ-contours are plotted in the (x, y)-plane, and indicated times at $z = 0$ plane, for $F_f = 0.5 \times 10^{-4}$ (left frames) and $F_f = 1.0 \times 10^{-4}$ (right frames). This figure brings out the essential differences between 2D and 3D transition. For the 3D transition, one notes that the dominant vortices underneath the wave-front makes an angle with the streamwise direction. This is due to the 3D orientation of the vortical structures, as also noted in [6, 8, 26, 40]. For 2D transition, the created vortices are almost vertical on the plate surface, as noted in the previous chapter. One also notes from Fig. 6.10a4, a5, that the high wavenumber fluctuations initiate above the plate surface, close to the edge of the shear layer. In contrast, for all the 2D cases the breakdown is seen to be initiated very close to the plate surface.

In Fig. 6.11, u_d is plotted as a function of y/δ^*, at (a) $z = 0$ and (b) $z = 0.6625$. The time instants and the streamwise locations indicated in frames (a) and (b) of Fig. 6.11 correspond to the maximum amplitude of the wave-front at $z = 0$ and

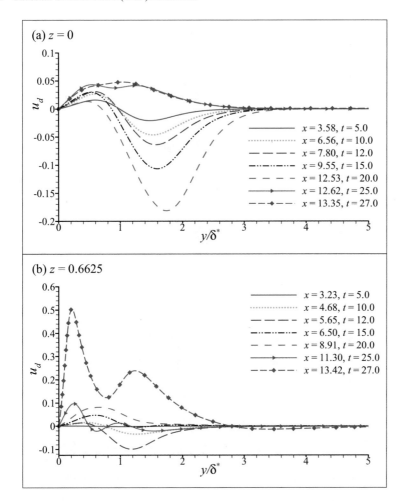

Fig. 6.11 u_d plotted as a function of y/δ^* at **a** $z = 0$ and **b** $z = 0.6625$. The time instants and the streamwise locations correspond to the maximum amplitude of the wave-front at $z = 0$ and $z = 0.6625$, respectively for $y = 0.00189$

$z = 0.6625$, respectively. One notes that at the initial phases of evolution, the wall-normal variation of u_d resembles the eigenfunction $\phi'(y)$ of the OSE noted in Chap. 3. At $z = 0$, as the wave-front moves downstream, significant amplification of u_d is noted, near the outer maxima of its wall-normal profile. This observation is in contrast to the frame (b) where u_d close to the inner maxima displays maximum growth. At later times of the evolution of the wave-front, the wall-normal profile of u_d displays marked variation with the Orr-Sommerfeld eigenfunction, with u_d attaining a value, which is of the order of the free-stream speed at $y/\delta^* \simeq 0.2$. One also notes the disturbances to rapidly penetrate higher wall-normal distances, as one approaches the turbulent spot.

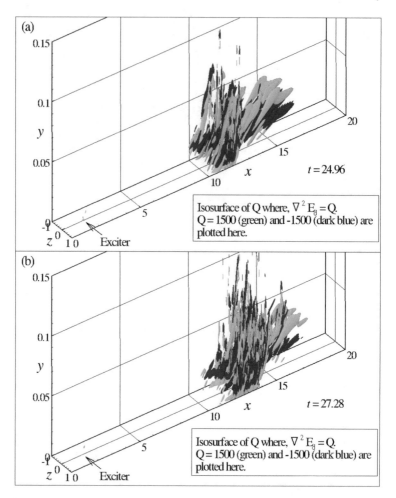

Fig. 6.12 Iso-surface of Q shown plotted at **a** $t = 24.96$ and **b** $t = 27.28$ for a circular wall bump exciter for $F_f = 0.5 \times 10^{-4}$

In Sengupta et al. [37], as well as in the previous chapter, the bypass transition on a flat plate caused by a slowly convecting vortex of anti-clockwise circulation is studied, which creates transition/unsteady separation ahead of it. The theoretical explanation in [37], is based upon the time evolution of the disturbance mechanical energy (E_d), defined as the difference between the total mechanical energy (E_t) and its equilibrium value (E_m), where $E_t = p/\rho_\infty + |\vec{V}|^2/2$ and $E_m = p_m/\rho_\infty + |\vec{V}_m|^2/2$. The governing equation for E_d is given as

$$\nabla^2 E_d = Q \tag{6.31}$$

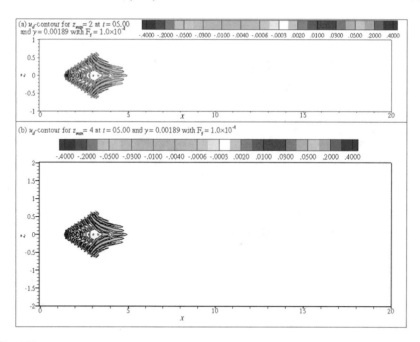

Fig. 6.13 u_d-contours plotted in the (x, z)-plane at indicated time for **a** $z_{max} = 2$ and **b** $z_{max} = 4$ when excited by a circular wall bump exciter for $F_f = 1.0 \times 10^{-4}$

where

$$Q = 2\vec{\omega}_m \cdot \vec{\omega}_d + |\vec{\omega}_d|^2 - \vec{V}_m \cdot (\nabla \times \vec{\omega}_d) - \vec{V}_d \cdot (\nabla \times \vec{\omega}_m) - \vec{V}_d \cdot (\nabla \times \vec{\omega}_d)$$

Hence E_d grows when $Q < 0$, and decays when $Q > 0$. This generic mechanism s based on NSE, without any simplifying assumptions. In Fig. 6.12, the iso-surface of $Q = 1500$ and $Q = -1500$ are plotted at $t = 24.96$ and 27.28, in frames (a) and b), respectively, for $F_f = 0.5 \times 10^{-4}$ case. The maximum of the absolute value of Q at both the time instants are of the order of 10^4. The $Q = 1500$ and -1500 iso-urfaces are depicted by the green and dark blue colors, respectively. One notes the imultaneous presence of the positive and negative values of Q in a very confined one, gives rise to enhanced instability of the flow. With advance in time from $t = 24.96$ to 27.28, the fluctuation in the value of Q is seen not only to intensify, but also o cover larger extent of the computational domain. At $t = 27.28$, the wall-normal xtent of these fluctuations are seen to pierce through the boundary layer at $x \simeq 12$.

The effects of the spanwise width of the computational domain is studied in Figs. 6.13, 6.14, 6.15, 6.16, 6.17 and 6.18, where u_d-contours are plotted in the x, z)-plane from $t = 5$ to 30, at an interval of five nondimensional time unit for he computational domain with (a) $z_{max} = 2$ and (b) $z_{max} = 4$, when ZPG flow past flat-plate is excited by a GCP exciter with $F_f = 1.0 \times 10^{-4}$. One finds that in

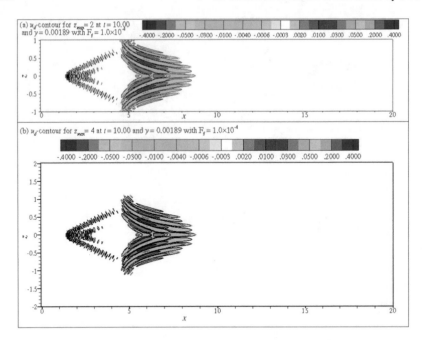

Fig. 6.14 See the caption for Fig. 6.13

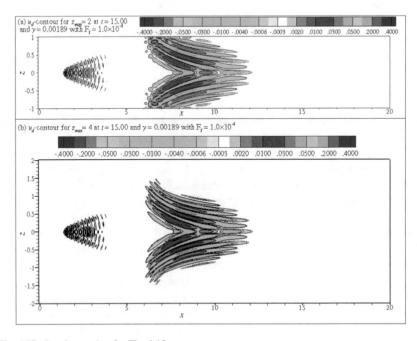

Fig. 6.15 See the caption for Fig. 6.13

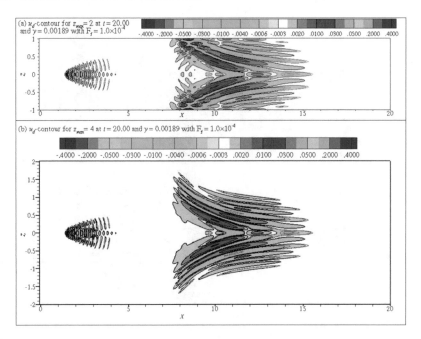

Fig. 6.16 See the caption for Fig. 6.13

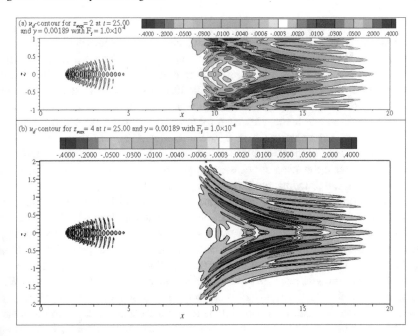

Fig. 6.17 See the caption for Fig. 6.13

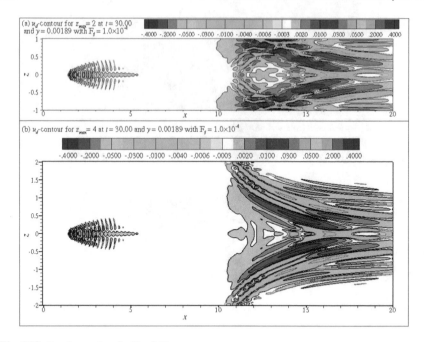

Fig. 6.18 See the caption for Fig. 6.13

the initial times up to $t = 10$, the evolution of disturbance is identical for both the cases. At $t = 10$, the disturbance corresponding to $z_{max} = 2$ case is seen to strike the spanwise boundaries of the corresponding computational domain. Thereafter the neighboring zones' disturbances, corresponding to the STWF, start to interact causing development of high wavenumber fluctuations at the spanwise boundaries as noted in Figs. 6.15a and 6.16a. However, for the $z_{max} = 4$ case, the spanwise extent for the u_d growth corresponding to the STWF increases. At $t = 25$, the STWF hits the spanwise boundaries of the computational domain in this case. However for both the cases, one finds that the structure of the wedge-shaped TS wave-packet is identical in nature, for all the time instants shown in Figs. 6.13–6.18.

To further understand the nature of transition, in Fig. 6.19 amplitude of u_d corresponding to STWF is plotted as a function of time, at four spanwise locations for $y = 0.00189$, and excitation frequency, $F_f = 0.5 \times 10^{-4}$. One notes that the time variation of the STWF amplitude is different from its 2D counterpart. For $z = 0.65$ and 0.9 (Fig. 6.19a) one notes that there exists a period of constant amplitude, after the initial linear growth, because of streamwise and spanwise dispersion of STWF. However, no such time duration exists for the spanwise location of $z = 0.325$. For $z = 0.325$, dispersion causes the amplitude to slightly decay, after the initial linear growth stage. This is followed by a period of continuous nonlinear growth, followed by eventual nonlinear saturation of amplitude. At the mid-spanwise location ($z = 0$)

Fig. 6.19 Amplitude of u_d
corresponding to the
wave-front plotted as a
function of time at indicated
spanwise locations of $z = 0$,
0.65 and 0.9 for
$\gamma = 0.00189$. The frequency
of excitation is
$\omega_f = 0.5 \times 10^{-4}$

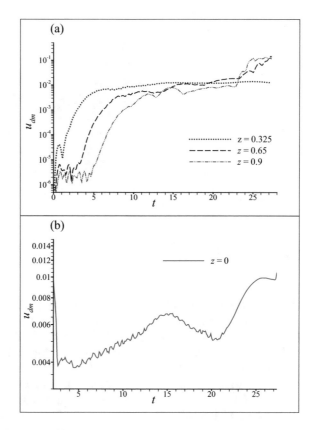

The growth of STWF is significantly different from the other three spanwise locations, because of the enhanced spanwise domain.

6.3   Spanwise Modulated (SM) Excitation

In this section, effects of spanwise modulated excitation on 2D ZPG boundary layer are studied. Here the amplitude function $A_m(x, z)$, as defined in Eq. (6.29), for the imposed wall-normal velocity component $v_w(x, z)$, is given as

$$A_m(x, z) = \frac{1}{2}\left(1 + \cos\left[\pi\frac{x - x_m}{x_2 - x_1}\right]\right)\sin\left(2\pi n\frac{z}{z_{max}}\right)$$
$$\text{for } x_1 \leq x \leq x_2 \qquad (6.32)$$

and $A_m(x, z) = 0$ for $x < x_1$ or $x > x_2$. Here $x_m = (x_1 + x_2)/2$, $n = 4$ and $z_{max} = 2$. The amplitude control parameter α_1, and the non-dimensional excitation frequency

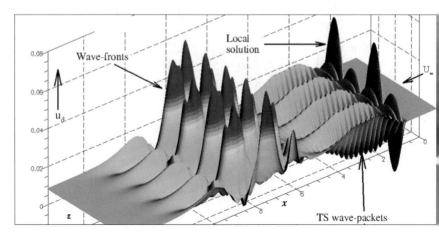

Fig. 6.20 Perspective plot of u_d at a fixed height shown for SM exciter. The frequency of excitation is $F = 0.5 \times 10^{-4}$

F_f, are taken as 0.01 and 1.0×10^{-4}, respectively. One notes that at the spanwise locations, $z = m z_{max}/8$: $A_m(x, z) = 0$ where $m = 0$, ± 1, ± 2, ± 3 and ± 4. As for this case $z_{max} = 2$, $A_m(x, z)$ is identically zero at $z = 0$, ± 0.25, ± 0.5, ± 0.75 and ± 1. These spanwise locations are the nodes. Similarly, maximum excitation is imposed at $z = \pm 0.125$, ± 0.375, ± 0.625 and ± 0.875, and which are the anti nodes or peaks. Out of these peak locations, excitation at $z = -0.875$, -0.375, 0.125 and 0.625 are in phase, and which have 180° phase difference with the imposed excitation at other peaks at $z = -0.625$, -0.125, 0.375 and 0.875. In the following discussion, the first set of spanwise locations are termed as "type-1 peak-locations" whereas the second set of spanwise locations are termed as "type-2 peak-locations" for the ease of explanation. It should be noted that such an excitation closely mim ics the experimental excitation of [21], where a spanwise modulated excitation is obtained by applying spacers in periodic succession. The present type of excitation also creates streamwise fluctuating vortices, with the center at the nodes. Perspective view of a typical disturbance evolution, corresponding to SM exciter, is shown in Fig. 6.20. Here also, one notes the existence of three-component solution structure. The local solution is seen very close to the exciter, and is followed by the TS wave-packet corresponding to the frequency of excitation and its super-harmonics. For the frequencies investigated, one also notices the presence of STWF for the displayed cases at all spanwise stations. The amplitude and phase of STWF show significant variation, as STWF propagates downstream. One notes that TS wave-packet does not grow and cause transition; instead STWF is the main precursor of flow transition. Detailed dynamics of the STWF is illustrated next.

In Fig. 6.21, the u_d-contours are plotted in the (x, z)-plane for $y = 0.00189$ and indicated time instants, for this case. One notes the evolution of a STWF, induc ing perturbations at $t = 10$, whose magnitude are higher than the corresponding disturbances induced by the TS waves. At $t = 10$, one also notes the prominent

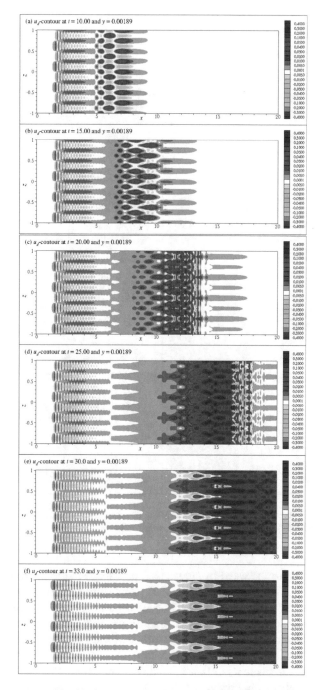

Fig. 6.21 u_d-contours plotted in the (x, z)-plane at $y = 0.00189$ and indicated time, when 2D ZPG boundary layer excited by spanwise modulated exciter strip for $F_f = 1.0 \times 10^{-4}$

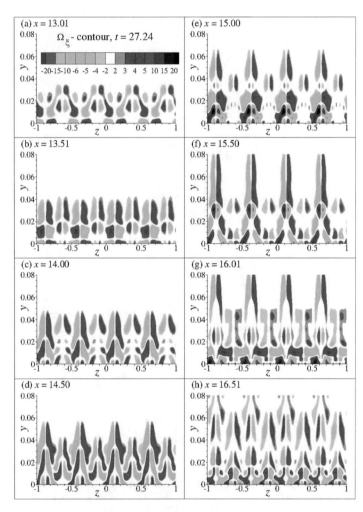

Fig. 6.22 The Ω_ξ-contours at indicated streamwise locations and $t = 27.24$ plotted in the (y, z) plane for spanwise modulated excitation for $F_f = 1.0 \times 10^{-4}$

disturbances, due to the wave-front arranged in a staggered formation. Such staggered formation is obtained due to the 180° phase difference between two successive peak locations. At $t = 15$, these staggered perturbations tend to merge with each other, inducing high wavenumber fluctuations, in the range $10.5 \le x \le 13.5$ at $t = 20$. These fluctuations induce four turbulent spots, as noted at $t = 25$. These turbulent spots are located at "type-2 peak-locations", and are denoted by dark blue spots in Fig. 6.21d, where $u_d \simeq 0.4$. In contrast, a relatively elongated, but slightly reduced perturbations are observed at the "type-1 peak-locations" at $t = 25$, trailing the "type-2 peak-location" structures. Gradually, these disturbances penetrate upstream to give rise to periodic spanwise-modulated, and streamwise-elongated streaks at $t = 30$, while the above turbulent spots are seen to reach the outflow of the computational domain. These streaks at $t = 30$ are located at the nodes, and are possibly intensified, due to lift-up effect of the streamwise counter-rotating vortices, which while lifting fluid with low velocity from the wall, and forcing high-speed fluids towards the wall, are most effective in creating streamwise-oriented streaks, as postulated in [5, 23, 24]. In [1, 25], a pair of counter-rotating streamwise vortices have been reported, as being most effective for the generation and evolution of such streamwise streak, whose stability analysis is performed in [32]. Similar streamwise streaks are also shown in [44], to be the precursor to the formation of turbulent spots.

The spanwise modulation of the perturbation across the shear layer is shown in Fig. 6.22, where Ω_ξ-contours are plotted in the (y, z)-plane, at $t = 27.24$, for streamwise locations varying from 13.01 to 16.01. From Fig. 6.22, one notes the presence of positive-negative vortex pairs, close to the wall, and negative-positive vortex pairs, away from the wall. The positive-negative vortex pair directs fluids of higher momentum towards the wall, whereas the negative-positive vortex pair pushes fluids of lower momentum, away from the wall. One notes that for the peak locations of "type-1", the first pair is stronger, whereas for the "type-2 peak locations", the second pair is stronger. This causes the perturbations corresponding to the "type-1 locations" to amplify more vigorously, than the perturbations at the "type-2 peak locations". This circulation of fluid also causes maximum u_d close to the wall to occur at the positions of the nodes, as explained in [3]. This establishes the crucial role streamwise vortices play, in determining transition for 3D flows, which has been pointed out experimentally earlier in [21]. However, continuous fully developed turbulence is not shown in the present simulations, due to short streamwise length of the computational domain, which occurs at downstream first, and then induces events upstream.

In Fig. 6.23, u_d for $y = 0.00189$ is plotted as a function of x, for the spanwise locations corresponding to the "type-1" and "type-2" peaks. One finds that the amplitude variation of the TS wave-packet is identical for both types of peak locations, with $180°$ phase shift between these. However the evolution of the wave-front is almost identical for both the types of the peak locations, at initial times up to $t = 10$. It is also interesting to note that contrary to the corresponding TS wave-packet, the wave-fronts generated at both types of the peak locations, are completely in phase at initial times. The wave-fronts at "type-2 peak locations" develop higher wavenumber fluctuations, than the "type-1" counterpart, as noted from Fig. 6.23c at $t = 15$. As a result of this, the wave-front at the "type-2 location" nonlinearly saturates and breaks down earlier, than the wave-front at "type-1" locations, as noted at $t = 20$ and 25.

Results also show that the wave-front at nodal spanwise locations have a phase difference of approximately $180°$, with the wave-fronts at both "type-1" and "type-2 peak locations", which mutually are in phase at initial times. This feature is illustrated in Fig. 6.24a, b, where u_d is plotted as a function of x for nodal, "type-1" and "type-2 peak locations". This along with the already noted feature that the TS waves are $180°$ phase apart for "type-1" and "type-2 peak locations", indicate that the spanwise wavelength of the wave-front, at initial times, are half the spanwise wave-

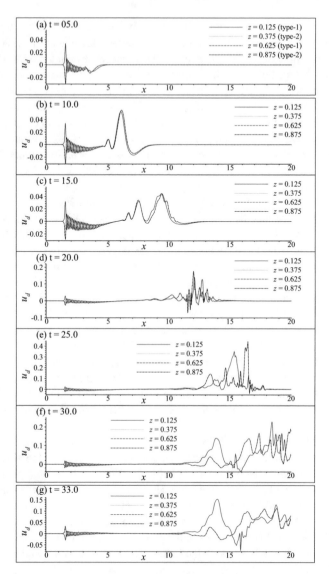

Fig. 6.23 u_d plotted at $y = 0.00189$, indicated times and spanwise locations for spanwise modu lated excitation for $F_f = 1.0 \times 10^{-4}$. All the indicated z-locations correspond to the peak position of the excitation

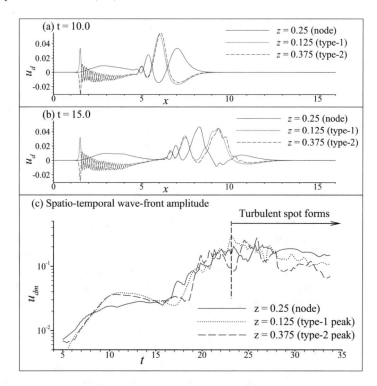

ig. 6.24 u_d at $y = 0.00189$ and spanwise locations plotted for **a** $t = 10$ and **b** $t = 15$. Maximum mplitude of the STWF u_{dm} at $y = 0.00189$ plotted as a function of time for $z = 0.25, 0.125$ and .375. Here, spanwise modulated excitation is provided for $F_f = 1.0 \times 10^{-4}$

:ngth of the TS wave-packet. In other words, the fundamental spanwise wavenumber f the wave-front, for this case, is twice the fundamental spanwise wavenumber of ne TS wave-packet. In Fig. 6.24c the maximum amplitude of the STWF for stream- vise disturbance velocity, u_{dm}, is plotted as a function of time, for $z = 0.25$ (nodal)cation), 0.125 (type-1 peak location) and 0.375 (type-2 peak location). One notes nat for the wave-front at "type-1" and "type-2 peak location", the amplitude of ne wave-front decays for $10 \leq t \leq 17$, after initial phase of exponential growth. 'his is because of the nonlinear self-and/ or multi-modal interactions, coupled with .ispersion. However, this stage is followed by a stage of rapid growth, because of sec- ·ndary and higher order instabilities [14], which leads to the formation of turbulent pots at $t \simeq 24$. However, for the wave-front at the nodal locations, the growth rate f the wave-front is moderated during the time duration of $10 \leq t \leq 17$, instead of 'ecaying, as noted for STWF at both "type-1" and "type-2" peak locations. For these panwise locations also, rapid growth of the wave-front takes place, after $t = 15$, ue to secondary and higher order instabilities.

6.4   Routes of Flow Transition: K- and H-Type Routes

Significant progress have been made for different aspects of early stages of 3D transition by experiments and computations in recent times, yet the actual routes traversed by flows from laminar to eventual turbulent stage are not completely understood till recent times. To explain the unit processes and routes of 3D flow transition mechanisms, experimental efforts have been undertaken by researchers for monochromatic deterministic excitation of wall bounded shear layers, as reported in [13, 21, 22, 33]. In [21], a rectangular ribbon with spanwise spacers was vibrated monochromatically at 1489Hz (corresponding non-dimensional frequency $F_f = 2\pi \nu f / U_\infty^2 = 6.03 \times 10^{-4}$), near the surface of a flat plate. It has been shown in [21], that longitudinal vortices are associated with the nonlinear 3D wave motions. The transition was shown to be characterized by the downstream growth of spanwise modulation of the disturbance amplitude, with the formation of peaks and valley along the spanwise direction [17]. At downstream locations, spikes appear suddenly in the peak spanwise locations, where the disturbance amplitudes reach the local maxima. Flow transition has been caused by rapid amplification and multiplication of these spikes, which has been attributed to high-frequency secondary instability due to inflectional instantaneous velocity profiles and high-shear [20–22].

Excitations, such as used in the experiments of [21], caused the evolution of an aligned pattern of Λ vortices in the transitional zone, termed subsequently as K type transition [17, 30]. The Λ-vortices have the appearance of horse-shoe shape whose central portion is lifted up with respect to the legs of the hairpin vortex. An alternate route of 3D transition by monochromatic excitation at a much lower frequency of 120 Hz (corresponding non-dimensional frequency $F_f = 2\pi \nu f / U_\infty^2 = 1.37 \times 10^{-5}$) was reported in [19], where Λ vortices are found to be in staggered arrangement; classified later as H- or N-type transition [17, 30].

Theoretical approaches attempted to explain the K- and H-type breakdown to a resonant mechanism [7, 17, 18]. K-type transition route is described to happen where 2D disturbance wave interacts with two oblique 3D waves of identical frequency [7, 17]. In contrast, H-type transition is noted, when a 2D disturbance wave interacts with two oblique 3D waves, corresponding to half the frequency of the 2D wave, as theoretically predicted in [7, 45]. The computational efforts to induce K- and H-type breakdown is attempted [31, 43] by simultaneously exciting waves at fundamental frequency and its spanwise modulated counterpart. None of these efforts register the role of STWF, as described here as the precursor of flow transition. This is because of the use of significantly shorter computational domain [31, 43]. In the previous chapters it has been noted that STWF are the spatio-temporal eigenmodes of NSE, as observed from the solution of the OSE in spatio-temporal framework [41]. It originates due to the onset of excitation, as a combination of spatio-temporal eigenmodes with weightage, which depends on the exciter location and excitation frequency. Therefore, properties of STWF are different from the pure spatial eigenmodes of the OSE [39]. Here, we show by the solution of NSE that transition of wall-bounded flows caused by the the growth of STWF follows both

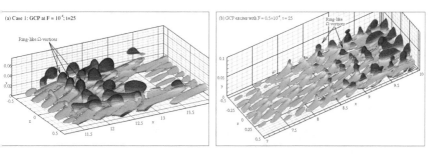

ig. 6.25 Perspective view of $\lambda_2 = -0.0025$ iso-surface is shown in (x, z)-plane at $t = 25$ after he onset of excitation corresponding to GCP exciter case for **a** $F_f = 1.0 \times 10^{-4}$ and **b** $F_f = 1.5 \times 10^{-4}$. Flow is from left to right

K- and H- or N-types of routes for low amplitude, monochromatic, deterministic vall-excitation, due to the growth of STWF. While the H-type transition is noted or lower frequency excitation cases, K-type is seen to occur for monochromatic excitations with higher frequency.

Transition process in 3D flows are dominated by highly unsteady vortical structures, with all three components important. Accurately representing these vortical structures unambiguously is quite difficult. Here, we capture the unsteady vortical structures by negative λ_2 iso-surfaces, following the method proposed in [16], o capture the unsteady vortical structures in late stages of 3D transition process. Here, λ_2 is the second eigenvalue of the symmetric matrix $S_{ik}S_{kj} + \Omega_{ik}\Omega_{kj}$ representing rate of strain tensor, where $S_{ij} = (1/2)(\partial u_i/\partial x_j + \partial u_j/\partial x_i)$ and $\Omega_{ij} = 1/2)(\partial u_i/\partial x_j - \partial u_j/\partial x_i)$ are symmetric and anti-symmetric part of the velocity gradient tensor, respectively.

In Fig. 6.25a, b, we have plotted the perspective of $\lambda_2 = -0.0025$ iso-surface colored by streamwise velocity), for the GCP exciter cases with $F_f = 10^{-4}$ and $F_f = 0.5 \times 10^{-4}$, respectively. In Fig. 6.25, trailing part of the STWF is focused to how later stages of transition in the perspective plot. One notes the formation of Λ-type vortices, in all cases with lifted Ω-like element of the hairpin vortices at the enter. In Fig. 6.25a, the darker spots denote the Ω-like element at the top of the Λ-vortices.

In Fig. 6.25a, we note an aligned pattern of hairpin-vortices, while these are arranged as Λ vortices in Fig. 6.25b, in a staggered arrangement. Therefore, the ase shown in Fig. 6.25b, has been conjectured to display subharmonic route of transition [17, 19, 30]. Unlike in [31, 43], here both the cases are due to monochromatic excitation, with the only difference in the excitation frequency. Thus, the present esults clearly indicate that for moderate frequencies, one notices K-type transition, as in Fig. 6.25a, while H-type transition occurs at significantly lower frequencies.

Similar differences in the arrangement of the vortices are also noted for the SM exciter cases in Fig. 6.26a–e, for $F_f = 10^{-4}$ and 0.5×10^{-4}, respectively. While definitive staggered pattern of Λ-vortices are noted for $F_f = 0.5 \times 10^{-4}$ case, almost aligned arrangement of these vortices are seen for $F = 10^{-4}$ case. In both the

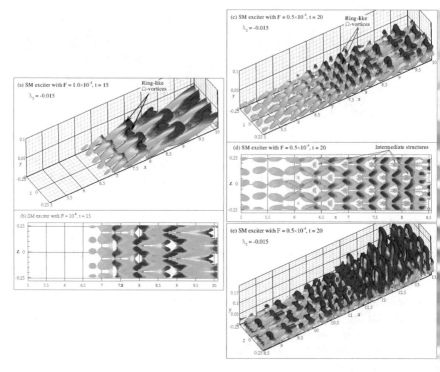

Fig. 6.26 Perspective and top view (in (x, z)-plane) of $\lambda_2 = -0.015$ iso-surface are shown for the SM exciter case for **(a, b)** $F_f = 10^{-4}$ and **(c, d, e)** $F_f = 0.5 \times 10^{-4}$. Flow is from left to right

cases, existence of ring-like Ω-vortices connecting the legs of the vortices are noted. These are also known as hairpin vortices [40]. One notes from Fig. 6.26a, c that, while for higher frequency, the hairpin vortices are aligned along the x-direction, these make an oblique angle for the lower frequency case. For $F_f = 10^{-4}$ case, Ω-vortices connect the two spanwise neighboring hairpin vortices (see Figs. 6.25a and 6.26a). For lower frequency case, Ω-vortices only connect the hairpin vortices, whose lifted front parts are close to each other. In Fig. 6.26d, between the dominant Λ-vortices, one notes the existence of some intermediate structures, which appear due to the induction of pressure variation by the adjoining Λ-vortices [40]. At further downstream locations, these structures become streamwise elongated, giving rise to U-shaped vortices, noted in Fig. 6.26e, which penetrates the shear-layer.

It is to be pointed out that here both K- and H-type transition is noted for deterministic monochromatic excitation via the growth of STWF. So far, all the H-type transition routes demonstrated in experiments [19] and simulations [31, 43] are stated to be via the amplification of the spatial TS wave-packet, where a fundamental frequency along with its sub-harmonic is excited. This follows the theoretical view-point of triad resonant interaction of a fundamental frequency f with its sub-harmonic component $f/2$ [7, 45]. In contrast the presented results show flow transition to occur via both K and H/N-type routes for monochromatic deterministic excitation, due to the growth of STWF, without requiring any subharmonic frequency to be explicitly excited.

6.5   Formation of Turbulent Spots and Fully Developed Turbulent Flow

The formation of turbulent spots, and consequent fully developed turbulent flow for the case of GCP exciter with $F = 0.5 \times 10^{-4}$, is illustrated here. In Fig. 6.27a–d, we show contours of Ω_ζ at the indicated time instants, along $z = 0$. At $t = 30$, an intermittent turbulent zone spans from $x \simeq 13$ to 19. With time, one notes the front of this highly perturbed zone to move downstream, at a speed comparable to the free stream velocity. The trailing edge of STWF shows very minor movement downstream up to $t = 35$, and beyond that time, this almost remains frozen at $x \simeq 13$. The transitional flow, which spans from $x \simeq 12$ to 15 in frame (b), keeps elongating, while causing vortical eruptions, which grow and merge with the turbulent part ahead of it. One such set of vortical eruptions is marked as A in frames (b) and (c). One also notes, thickening of the boundary layer gradually, from the laminar value at $x = 12$ to 15 (which can be construed as the point of transition) and beyond for later times. The intermittent zone is characterized by highly unsteady vortical eruptions. Presence of unsteady vortical eruptions, and constant regeneration mechanism was also noted for deterministically created 2D turbulent flow in Chap. 5. Here, the flow is seen to be turbulent beyond $x = 15$, however, the intermittency is lower beyond $x \simeq 25$.

To characterize the turbulent zone, we plot the wall-normal variation of u^+ (mean streamwise velocity $< U >$, non-dimensionalized by wall-friction velocity $u_\tau = \sqrt{< \tau_w > /\rho}$, where $< \tau_w >$ is the mean wall-shear and ρ is fluid density) as a function of $y^+ = u_\tau y/\nu$, in Fig. 6.28a, for indicated streamwise stations. According to descriptions of Reynolds-averaged fully developed turbulent boundary layer in [28, 42],

$$u^+ = y^+$$

in the viscous sub-layer ($0 \leq y^+ \leq 10$) and

$$u^+ = \frac{1}{\kappa_1} \ln(y^+) + a$$

in the inertial layer - above the viscous sub-layer and buffer layer (for $30 \leq y^+ \leq 2000$), where $\kappa_1 = 0.41$, is the von Karman coefficient. This is the 'logarithmic law of the wall' in Fig. 6.28a, marked by dashed lines. To determine the mean stream-wise velocity for all the streamwise stations, we have time-averaged the data from $t = 40$ to 50. At $x = 20$ to 23, the match of u^+ at the outer part is not good, because these stations are not in fully developed turbulent flow region. One obtains very

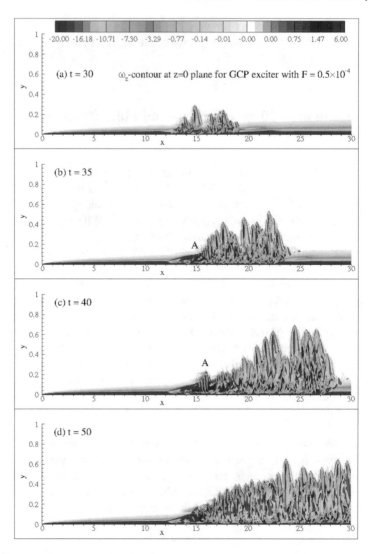

Fig. 6.27 ω_z-contours at $z = 0$ station plotted in (x, y)-plane at **a** $t = 30$, **b** $t = 35$, **c** $t = 40$ and **d** $t = 50$ for GCP exciter case for $F = 0.5 \times 10^{-4}$

good match with the above expressions for u^+, at $x = 25$ and beyond. In Fig. 6.28b, Reynolds stress variation with y^+ is shown at the indicated stations, which demonstrates how this stress changes in the transitional flow. One notes that the Reynolds stress is predominantly negative between $10 \le y^+ \le 2000$, for $x \ge 25$. The range, $10 \le y^+ \le 2000$, corresponds to that part of the boundary layer, where predominant negative values signify enhanced production of turbulent kinetic energy, as reported in the literature. In Fig. 6.28c, the non-dimensional root mean square streamwise

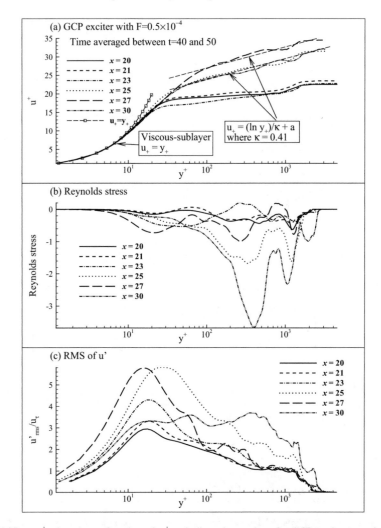

ig. 6.28 a u^+ plotted as a function of y^+ at indicated x-stations for GCP exciter case for $F =$ .5 × 10^{-4}. **b** Non-dimensional Reynolds stress plotted as a function of y^+ at indicated x-stations nd **c** nondimensional streamwise component *rms* velocity shown at the indicated streamwise ocations

elocity component is shown as function of y^+. While in all the profiles one notices n inner maximum, at downstream locations, one can also note the effects of con-ecting vortices, via the presence of another maximum in the outer part.

In Fig. 6.29a–c, the skin friction coefficient C_f variation for the section along = 0 are shown with the streamwise co-ordinate x, for the indicated times. For aminar flows, skin friction coefficient varies as $C_f = 0.664 \times Re_x^{-1/2}$, where Re_x s Reynolds number based on streamwise co-ordinate x, and shown in the frames

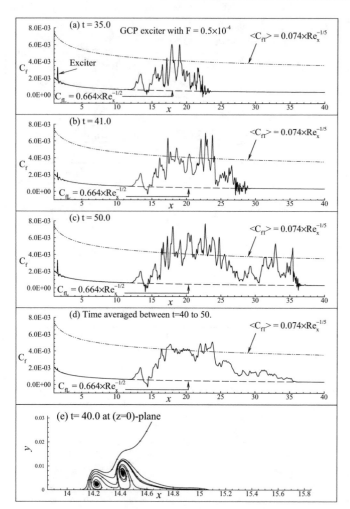

Fig. 6.29 **a–c** Instantaneous and **d** time-averaged skin-friction coefficient C_f along $z = 0$ spanwise station plotted as a function of x. In **d** Time-averaged C_f during the interval $40 \leq t \leq 50$ are plotted as function of x. For comparison, corresponding correlations for laminar and turbulent flows are shown in all these frames. **e** Stream-trace is shown at $t = 40$ to indicate the recirculating region at the onset of transition

as dashed line. For fully developed turbulent boundary layer, mean skin friction coefficient varies as $< C_f > = 0.74 \times Re_x^{-1/5}$, given in [28, 42]. This line is shown in all the frames by dash-dotted line. The passage of the STWF is clearly evident in all the frames, as highly intermittent flow, with the value of C_f fluctuating significantly. When we plot the time-averaged skin-friction coefficient (averaged between $t = 40$ and 50) as a function of x, we note that in the central part of the turbulent spot, the time-averaged C_f displays very good match with the turbulent boundary layer value

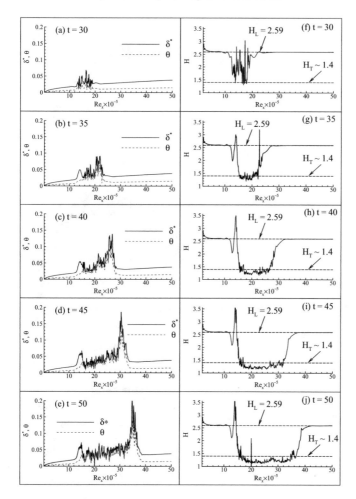

Fig. 6.30   a–e Displacement thickness δ^*, and momentum thickness θ, at $z = 0$ section plotted at indicated time instants. **f–j** Shape factor H at $z = 0$ section plotted at indicated time instants

One notes from Fig. 6.29, that near the location $x \simeq 14$, the averaged C_f shows a dip, and this is due to the onset of unsteady separation associated with the transition. This is shown in the stream trace plot shown in Fig. 6.29c.

In Fig. 6.30, corresponding influence on other integrated quantities are shown in terms of displacement and momentum thickness (δ^* and θ), at the indicated times on the left column. The right column of this figure shows the variation of the shape factor ($H = \delta^*/\theta$), as a function of Reynolds number based on current length. In these frames, the dashed lines indicate the typical values of laminar (top) and turbulent flows (bottom). Once again, one can notice an excellent match with the known experimental and theoretical trends for these quantities.

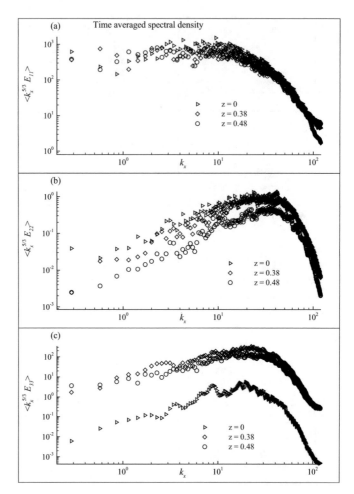

Fig. 6.31 a–c Time averaged compensated streamwise spectral density for streamwise, wall-normal and spanwise velocity components plotted as a function of streamwise wavenumber k_x for $z = 0$, 0.38 and 0.48

In Fig. 6.31, compensated energy spectra, as defined in Saddoughi and Veeravalli [29] by E_{11}, E_{22} and E_{33}, are shown along three spanwise locations given by $z = 0.0$, 0.38 and 0.48, plotted in different frames. These are obtained by squaring the Fourier–Laplace transform of u_d, v_d and w_d, respectively. For example, E_{11} is obtained as,

$$E_{11} = |U_d(k_x)|^2 \qquad (6.33)$$

where $U_d(k_x)$ is the Fourier–Laplace transform of u_d, with respect to the streamwise wavenumber, k_x. Similarly E_{22} and E_{33} can also be defined. Variation of the compensated energy spectra shows deviation from the homogeneous isotropic turbulence values given due to Kolmogorov [28]. From this figure, one notes the existence of an intermediate wavenumber region, where spectral densities vary as $k_1^{-5/3}$ – a variation in the inertial sub-range, predicted for 3D isotropic homogeneous turbulence. More importantly, displayed computed spectra show similarity with the experimental data for inhomogeneous flows in [29]. In this latter reference, local isotropy of the boundary layer about the streamwise wavenumber (k_x) is explored experimentally. Thus, the STWF is noted to take the flow all the way from receptivity to an equilibrium flow, obtained for local isotropic turbulence, explored experimentally for inhomogeneous flows by Saddoughi and Veeravalli [29].

References

1. Andersson, P., Berggren, M., & Henningson, D. S. (1999). Optimal disturbances and bypass transition in boundary layers. *Physics of Fluids, 11*, 134–150.
2. Ashpis, D. E., & Reshotko, E. (1990). The vibrating ribbon problem revisited. *Journal of Fluid Mechanics, 213*, 531–547.
3. Bhaumik, S. (2013). *Direct Numerical Simulation of Inhomogeneous Transitional and Turbulent Flows*. Ph. D. Thesis, I. I. T. Kanpur, INDIA.
4. Bhaumik, S., & Sengupta, T. K. (2015). A new velocity-vorticity formulation for direct numerical simulation of 3D transitional and turbulent flows. *Journal of Computational Physics, 284*, 230–260.
5. Brandt, L., & Henningson, D. S. (2002). Transition of streamwise streaks in zero-pressure-gradient boundary layers. *Journal of Fluid Mechanics, 472*, 229–261.
6. Chen, L., & Liu, C. (2011). Numerical study on mechanisms of second sweep and positive spikes in transitional flow on a flat plate. *Computers and Fluids, 40*, 28–41.
7. Craik, A. D. D. (1971). Non-linear resonant instability in boundary layers. *Journal of Fluid Mechanics, 50*, 393–413.
8. Duguet, Y., Schlatter, P., Henningson, D. S., & Eckhardt, B. (2012). Self-sustained localized structures in a boundary-layer flow. *Physical Review Letters, 108*, 044501.
9. Durbin, P. A., & Wu, X. (2007). Transition beneath vortical disturbances. *Journal of Fluid Mechanics, 39*, 107–128.
10. Eiseman, P. R. (1985). Grid generation for fluid mechanics computation. *Annual Review of Fluid Mechanics, 17*, 487–522.
11. Fasel, H., & Konzelmann, U. (1990). Non-parallel stability of a flat-plate boundary layer using the complete Navier-Stokes equations. *Journal of Fluid Mechanics, 221*, 311–347.
12. Gaster, M., & Grant, I. (1975). An experimental investigation of the formation and development of a wave packet in a laminar boundary layer. *Proceedings of the Royal Society of London Series A. Mathematical and Physical Sciences, 347*(1649), 253–269.
13. Hama, F. R., & Nutant, J. (1963). Detailed flow-field observations in the transition process in a thick boundary layer. In *Proceedings of the 1963 Heat Transfer and Fluid Mechanics Institute*, (pp. 77–93). Stanford: Stanford University Press.
14. Herbert, Th. (1988). Secondary instability of boundary layers. *Annual Review of Fluid Mechanics, 20*, 487–526.
15. Hill, D. C. (1995). Adjoint systems and their role in the receptivity problem for boundary layers. *Journal of Fluid Mechanics, 292*, 183–204.
16. Jeong, J., & Hussain, F. (1995). On the identification of a vortex. *Journal of Fluid Mechanics, 285*, 69–94.
17. Kachanov, Y. S. (1994). Physical mechanisms of laminar-boundary-layer transition. *Annual Review of Fluid Mechanics, 26*, 411–482.

18. Kachanov, Y. S. (1987). On the resonant nature of the breakdown of a laminar boundary layer. *Journal of Fluid Mechanics, 184*, 43–74.

19. Kachanov, Y. S., & Levchenko, V. Y. (1984). The resonant interaction of disturbances at laminar turbulent transition in a boundary layer. *Journal of Fluid Mechanics, 138*, 209–247.

20. Klebanoff, P. S., & Tidstrom, K. D. (1959). Evolution of amplified waves leading to transition in a boundary layer with zero pressure gradient. In *N.A.S.A. Technical Note*, D–195.

21. Klebanoff, P. S., Tidstrom, K. D., & Sargent, L. M. (1962). The three-dimensional nature o boundary-layer instability. *Journal of Fluid Mechanics, 12*, 1–34.

22. Kovasznay, L. S. G., Komoda, H., & Vasudeva, B. R. (1962). Detailed flow-field in transition. In *Proceedings of the Heat Transfer and Fluid Mechanics Institute*, Palo Alto, California Stanford University Press.

23. Landahl, M. T. (1975). Wave breakdown and turbulence. *SIAM Journal on Applied Mathemat ics, 28*, 735.

24. Landahl, M. T. (1980). A note on an algebraic instability of inviscid parallel shear flows. *Journa of Fluid Mechanics, 98*, 243–251.

25. Luchini, P. (2000). Reynolds-number independent instability of the boundary layer over a fla surface. Part 2: Optimal perturbations. *Journal of Fluid Mechanics, 404*, 289–309.

26. Lu, P., Wang, Z., Chen, L., & Liu, C. (2012). Numerical study on U-shaped vortex formation in late boundary layer transition. *Computers and Fluids, 55*, 36–47.

27. Mattor, N., Williams, T. J., & Hewett, D. W. (1995). Algorithm for solving tridiagonal matri problems in parallel. *Parallel Computing, 21*, 1769–1782.

28. Pope, S. B. (2000). *Turbulent flows*. UK: Cambridge University Press.

29. Saddoughi, S. G., & Veeravalli, S. V. (1994). Local isotropy in turbulent boundary layers a high Reynolds number. *Journal of Fluid Mechanics, 268*, 333–372.

30. Saric, W. S., & Thomas, A. S. W. (1984). Experiments on the subharmonic route to turbulenc in boundary layers. In T. Tatsumi (Ed.), *Turbulence and Chaotic Phenomena in Fluids*. Elsevier USA: North Holland.

31. Sayadi, T., Hamman, C. W., & Moin, P. (2013). Direct numerical simulation of complete H type and K-type transitions with implications for the dynamics of turbulent boundary layers *Journal of Fluid Mechanics, 724*, 480–509.

32. Schoppa, W., & Hussain, F. (2002). Coherent structure generation in near-wall turbulence *Journal of Fluid Mechanics, 453*, 57–108.

33. Schubauer, G. B., & Skramstad, H. K. (1947). Laminar boundary layer oscillations and th stability of laminar flow. *Journal of the Aeronautical Sciences, 14*, 69–78.

34. Sengupta, T. K., Bhaumik, S., & Bose, R. (2013). Direct numerical simulation of transitiona mixed convection flows: Viscous and inviscid instability mechanisms. *Physics of Fluids, 25(9)* 094102. (1994-present).

35. Sengupta, T. K., Bhaumik, S., & Bhumkar, Y. G. (2012). Direct numerical simulation o two-dimensional wall-bounded turbulent flows from receptivity stage. *Physical Review E, 85* 026308.

36. Sengupta, T. K., Bhaumik, S., & Usman, S. (2011). A new compact difference scheme fo second derivative in non-uniform grid expressed in self-adjoint form. *Journal of Computationa Physics, 230*(5), 1822–1848.

37. Sengupta, T. K., De, S., & Sarkar, S. (2003). Vortex-induced instability of incompressible wall-bounded shear layer. *Journal of Fluid Mechanics, 493*, 277–286.

38. Sengupta, T. K., Dipankar, A., & Rao, A. K. (2007). A new compact scheme for paralle computing using domain decomposition. *Journal of Computational Physics, 220*, 654–677.

39. Sengupta, T. K., Lele, S. K., Sreenivasan, K. R., & Davidson, P. A. (2015). Advances i Computation, Modeling and Control of Transitional and Turbulent Flows. In *IUTAM Symposic Proceedings*, Singapore: World Scientific Publishing Company.

40. Singer, B. A., & Joslin, R. D. (1994). Metamorphosis of a hairpin vortex into a young turbulen spot. *Physics of Fluids, 6*, 3724–3736.

41. Sengupta, T. K., Rao, A. K., & Venkatasubbaiah, K. (2006). Spatio-temporal growing wav fronts in spatially stable boundary layers. *Physical Review Letters, 96*(22), 224504.

42. Tennekes, H., & Lumley, J. L. (1972). *A first course in turbulence*. Cambridge, MA: MIT Press.
43. Würz, W., Sartorius, D., Kloker, M., Borodulin, V. I., & Kachanov, Y. S. (2012). Detuned resonances of Tollmien-Schlichting waves in an airfoil boundary layer: Experiment, theory, and direct numerical simulation. *Physics of Fluids, 24*, 094103.
44. Wu, X., Jacobs, R. G., Hunt, J. C. R., & Durbin, P. A. (1999). Simulation of boundary layer transition induced by periodically passing wakes. *Journal of Fluid Mechanics, 399*, 109–153.
45. Zelman, M. B., & Maslennikova, I. I. (1993). Tollmien-Schlichting-wave resonant mechanism for subharmonic-type transition. *Journal of Fluid Mechanics, 252*, 449–478.

Appendix A

A.1 Boundary Layer Equation for Mixed Convention Problem

The physical problem considered is a 2D flow of a fluid over a semi-infinite inclined flat plate with the leading edge as the stagnation point. The free-stream velocity and temperature is U_∞ and T_{infty}, respectively. The plate is inclined at an angle β with the direction of the flow as shown schematically in Fig. A.1. This implies that the velocity at the edge of the boundary layer is different from U_{infty} and therefore, is a function of the streamwise coordinate x^*. Let it be denoted as $U_e^*(x^*)$. The surface of the plate is maintained at a temperature $T_w^*(x^*)$. The plate temperature is greater or less than T_∞ depending upon the plate under consideration is hot or cold. Here, all the dimensional quantities are represented with an asterisk as superscript, whereas those without the superscript asterisk symbol denote the non-dimensional quantities. The governing equations for this case is given by the incompressible Navier–Stokes equation subjected to boundary-layer approximation [1, 3], where Boussinesq approximation is used to account for the buoyancy effects due to heat transfer. One also has to consider an additional energy equation as the governing equation for the temperature field. These equations in the dimensional form are given as,

$$\frac{\partial u^*}{\partial x^*} + \frac{\partial v^*}{\partial y^*} = 0 \tag{A.1}$$

$$u^*\frac{\partial u^*}{\partial x^*} + v^*\frac{\partial u^*}{\partial y^*} = -\frac{\partial p^*}{\partial x^*} + \nu\frac{\partial^2 u^*}{\partial y^{*2}} \tag{A.2}$$

$$\frac{\partial p^*}{\partial y^*} - g\beta_T(T - T_\infty) = 0 \tag{A.3}$$

$$u^*\frac{\partial T^*}{\partial x^*} + v^*\frac{\partial T^*}{\partial y^*} = \alpha_T\frac{\partial^2 T^*}{\partial y^{*2}} \tag{A.4}$$

© Springer Nature Singapore Pte Ltd. 2019
T. K. Sengupta and S. Bhaumik, *DNS of Wall-Bounded Turbulent Flows*,
https://doi.org/10.1007/978-981-13-0038-7

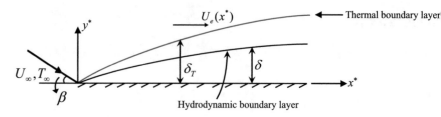

Fig. A.1 Schematic diagram for the mixed-convection flow over a semi-infinite inclined flat plate. The plate is inclined with the free-stream velocity direction at an angle β. Here, δ and δ_T are the hydrodynamic and thermal boundary layer thickness, respectively

where ν, g, β_T and α_T denote the kinematic viscosity, gravitational acceleration, volumetric thermal expansion coefficient and the thermal diffusivity, respectively. Note that while deriving Eqs. (A.1)–(A.4), we have not considered viscous dissipation of kinetic energy and also neglected any possible presence of source for internal heat generation. The relevant boundary conditions are given as

$$u^*(x^*, y^*) = v^*(x^*, y^*) = 0 \text{ and } T^*(x^*, y^*) = T_w^*(x^*) \, at \, y^* = 0 \qquad (A.5)$$

$$u^*(x^*, y^*) \to U_e^*(x^*) \text{ and } T^*(x^*, y^*) \to T_\infty \, as \, y^* \to \infty \qquad (A.6)$$

To nondimensionalize Eqs. (A.1)–(A.4), a length scale (L), a velocity scale (U_∞), a temperature scale ($\Delta T_L = T_w^*(L) - T_\infty$) and a pressure scale ($\rho U_\infty^2$) are adopted. The nondimensionalized form of Eqs. (A.1)–(A.4) are given as,

$$\frac{\partial u}{\partial x} + \frac{\partial v}{\partial y} = 0 \qquad (A.7)$$

$$u\frac{\partial u}{\partial x} + v\frac{\partial u}{\partial y} = -\frac{\partial p}{\partial x} + \frac{1}{Re}\frac{\partial^2 u}{\partial y^2} \qquad (A.8)$$

$$\frac{\partial p}{\partial y} - \frac{Gr}{Re^2}\theta = 0 \qquad (A.9)$$

$$u\frac{\partial \theta}{\partial x} + v\frac{\partial \theta}{\partial y} = \frac{1}{Re}\frac{1}{Pr}\frac{\partial^2 \theta}{\partial y^2} \qquad (A.10)$$

where, $\theta = (T^* - T_\infty)/\Delta T_L$, $Gr = g\beta_T \Delta T_L L^3/\nu^2$ is the Grashof number, $Re = U_\infty L/\nu$ is the Reynolds number and $Pr = \nu/\alpha_T$ is the Prandtl number. The Grashof number (Gr) gives the ratio of buoyancy and viscous forces present in the fluid and the Richardson number (Ri), given by $Ri = \frac{Gr}{Re^2}$, indicates the relative dominance of natural to forced convection. For such cases, $Ri \geq 0$ or $Ri \leq 0$ refer to assisting and opposing flows. In the mixed convection regime Ri is of order one. In general $Pr = 0.71$ is used, which is the Prandtl number for air as the working medium. Following Eqs. (A.5)–(A.6), corresponding boundary conditions for the nondimensional Eqs. (A.7)–(A.10) are given as

$$u(x, y) = v(x, y) = 0 \text{ and } \theta(x, 0) = \Theta_w(x) \text{ at } y = 0 \tag{A.11}$$

$$u(x, y) \to U_e(x) \text{ and } \theta(x, y) \to 0 \text{ as } y \to \infty \tag{A.12}$$

where $\Theta_w(x) = (T_w^*(x^*) - T_{infty})/\Delta T_L$. For flows with high Reynolds number, the boundary layer thickness is orders of magnitude smaller than the dimensions of the plate. Therefore, it is customary to scale up the boundary layer thickness, so that both x- and y-scales are of similar orders of magnitude. This can be accomplished by following coordinate transformation from (x, y) to (X, Y). Let, $X = x$, $Y = y\sqrt{Re}$, $U = u$; $V = v\sqrt{Re}$ and $P = p$. Following these coordinate and variable transformations, Eqs. (A.7)–(A.10) changes to

$$\frac{\partial U}{\partial X} + \frac{\partial V}{\partial Y} = 0 \tag{A.13}$$

$$U\frac{\partial U}{\partial X} + V\frac{\partial V}{\partial Y} = -\frac{\partial P}{\partial X} + \frac{\partial^2 U}{\partial Y^2} \tag{A.14}$$

$$\frac{\partial P}{\partial Y} - K\theta = 0 \tag{A.15}$$

$$U\frac{\partial \theta}{\partial X} + V\frac{\partial \theta}{\partial Y} = \frac{1}{Pr}\frac{\partial^2 \theta}{\partial Y^2} \tag{A.16}$$

where $K = Gr/Re^{5/2}$. The corresponding boundary conditions are

$$U(X, 0) = V(X, 0) = 0, \text{ and } \theta(X, 0) = \Theta_w(X) \text{ as}$$
$$Y \to \infty, U(X, Y) \to U_e(X) \text{and } \theta(X, Y) \to 0 \tag{A.17}$$

One notes from Eqs. (A.14) and (A.15) that at the free-stream the quantity $P + \frac{1}{2}U_e^2$ is function of the wall-normal coordinate Y only. This prompts one to define a modified pressure as

$$\tilde{P} = P + \frac{1}{2}U_e(X)^2 \tag{A.18}$$

Now, one can also define the streamfunction ψ such that

$$U = \frac{\partial \psi}{\partial Y} \tag{A.19}$$

$$V = -\frac{\partial \psi}{\partial X} \tag{A.20}$$

Equations (A.19) and (A.20) automatically satisfies the continuity equation (A.13). Substituting Eqs. (A.18)–(A.20) into (A.13)–(A.16) one gets,

$$\frac{\partial \psi}{\partial Y}\frac{\partial^2 \psi}{\partial X \partial Y} - \frac{\partial \psi}{\partial X}\frac{\partial^2 \psi}{\partial Y^2} = -\frac{\partial \tilde{P}}{\partial X} + U_e \frac{dU_e}{dX} + \frac{\partial^3 \psi}{\partial Y^3} \qquad (A.21)$$

$$\frac{\partial \tilde{P}}{\partial Y} - K\theta = 0 \qquad (A.22)$$

$$\left(\frac{\partial \psi}{\partial Y}\frac{\partial \theta}{\partial X} - \frac{\partial \psi}{\partial X}\frac{\partial \theta}{\partial Y} \right) = \frac{1}{Pr}\frac{\partial^2 \theta}{\partial Y^2} \qquad (A.23)$$

The appropriate boundary conditions for Eqs. (A.21)–(A.23) are given as

$$\text{at } Y = 0 \ \ \psi(X, Y) = 0, \ \ \frac{\partial \psi}{\partial Y} = 0 \ \text{ and } \ \theta(X, 0) = \Theta_w$$

$$\frac{\partial \psi}{\partial Y} \to U_e(X) \ \text{ and } \ \theta \to 0 \ \text{ as } Y \to \infty$$

A.2 Similarity Transformation

The similarity transformation [2, 4] is performed following the coordinate transformation from (X, Y)- to (ξ, η)-coordinate system where $\xi = X$ and $\eta = Y\sqrt{U_e/X}$. Let $\psi = \sqrt{U_e X} f(\xi, \eta)$, $\tilde{P} = U_e^2 q(\xi, \eta)$, $\theta = \Theta_w g(\xi, \eta)$. Let us further assume that the $U_e(X)$ and $\Theta_w(X)$ varies as X^n and X^r, respectively [4]. Therefore,

$$\frac{X}{U_e}\frac{dU_e}{dX} = n \ \text{ and } \ \frac{X}{\Theta_w}\frac{d\Theta_w}{dX} = r$$

These transformations lead to the following relationships

$$\frac{\partial \psi}{\partial X} = \frac{1}{2}\sqrt{\frac{U_e}{X}}\left((n+1)f + \eta(n-1)f' + 2Xf_\xi \right)$$

$$\frac{\partial^2 \psi}{\partial X \partial Y} = \left(\frac{U_e}{X} \right)\left(Xf_\xi' + nf' + \frac{1}{2}\eta(n-1)f'' \right)$$

$$\frac{\partial \psi}{\partial Y} = \sqrt{\frac{U_e}{X}}\sqrt{U_e X} f' = U_e f'$$

$$\frac{\partial^2 \psi}{\partial Y^2} = \left(\frac{U_e}{X} \right)\sqrt{U_e X} f'' = \frac{U_e^{3/2}}{\sqrt{X}} f''$$

$$\frac{\partial^3 \psi}{\partial Y^3} = \left(\frac{U_e}{X} \right)^{3/2}\sqrt{U_e X} f''' = \frac{U_e^2}{X} f'''$$

$$\frac{\partial \tilde{P}}{\partial X} = \frac{X}{U_e^2}\left(Xq_\xi + 2nq + \frac{1}{2}\eta(n-1)q'\right)$$

$$\frac{\partial \tilde{P}}{\partial Y} = \frac{U_e^{5/2}}{X^{1/2}}q'$$

$$\frac{\partial \theta}{\partial X} = \frac{\Theta_w}{X}\left(rg + \frac{1}{2}\eta(n-1)g' + Xg_\xi\right)$$

$$\frac{\partial \theta}{\partial Y} = \sqrt{\frac{U_e}{X}}\Theta_w g'$$

$$\frac{\partial^2 \theta}{\partial Y^2} = \frac{U_e}{X}\Theta_w g'' \qquad (A.24)$$

n Eq. (A.24), the subscript ξ and the prime (') denote the partial derivatives with espect to ξ and η, respectively. Substituting Eq. (A.24) in Eqs. (A.21)–(A.23), one gets

$$f''' + \left(\frac{n+1}{2}\right)ff'' + n(1 - f'^2) - \left(2nq + \frac{1}{2}\eta(n-1)q'\right)$$
$$= \xi(f'f'_\xi - f_\xi f'' + q_\xi) \qquad (A.25)$$

$$\frac{U_e^{5/2}}{\xi^{1/2}\Theta_w}q' - Kg = 0 \qquad (A.26)$$

$$\frac{1}{Pr}g'' - rf'g + \left(\frac{n+1}{2}\right)fg' = \xi(f'g_\xi - f_\xi g') \qquad (A.27)$$

?or self-similarity, Eqs. (A.25)–(A.27) would be independent of ξ. Therefore all the partial derivatives with respect to ξ would be zero. Following Eq. (A.26), one also equires that

$$\frac{U_e^{5/2}}{\xi^{1/2}\Theta_w} = Const. \qquad (A.28)$$

As U_e X^n and Θ_w X^r, Eq. (A.28) indicates that

$$r = \frac{5n - 1}{2}$$

Following these restrictions, one obtains the self-similar equations for the boundary layer profiles for mixed convection flow past a flat-plate as

$$\left[f''' + \left(\frac{n+1}{2} \right) ff'' + n(1 - f'^2) \right] - \left(2nq + \frac{1}{2}\eta(n-1)q' \right) = 0 \quad \text{(A.29)}$$

$$q' - Kg = 0 \quad \text{(A.30)}$$

$$\frac{1}{Pr}g'' - \left(\frac{5n-1}{2} \right) f'g + \left(\frac{n+1}{2} \right) fg' = 0 \quad \text{(A.31)}$$

The appropriate boundary conditions for these nonlinear ordinary differential equations are given following Eqs. (A.25)–(A.27) as

$$f(0) = f'(0) = 0, \quad \text{and} \quad g = 1 \quad \text{(A.32)}$$

$$f' \to 1, \quad \text{and} \quad g \to 0 \quad \text{as} \quad \eta \to \infty \quad \text{(A.33)}$$

If the edge velocity is constant i.e., $U_e = Const. = U_\infty$ then $n = 0 \, (r = -1/2)$, and Eqs. (A.29)–(A.31) simplifies to [2]

$$f''' + \frac{1}{2}ff'' + \frac{1}{2}K\eta g = 0 \quad \text{(A.34)}$$

$$\frac{1}{Pr}g'' + \frac{1}{2}(f'g + fg') = 0 \quad \text{(A.35)}$$

Equations (A.34) and (A.35) and the boundary condition $f(0) = 0$ show that at $\eta = 0$, $g' = 0$ i.e., the flat plate surface has an adiabatic wall condition. This is true for all streamwise locations $x > 0$ except at the leading edge $(x = 0)$, where all heat transfer occurs singularly while the wall temperature varies as $x^{-1/2}$ for both hot and cold plate cases.

References

1. Schlichting, H. (1933). Zur entstehung der turbulenz bei der plattenströmung *Nach. Gesell. d. Wiss. z. Gött., MPK, 42*, 181–208.
2. Schneider, W. (1979). A similarity solution for combined forced and free convection flow over a horizontal plate. *International Journal of Heat and Mass Transfer, 22*, 1401–1406.
3. White, F. M. (2008). *Fluid mechanics* (6th ed.) New York: The McGraw Hill Companies.
4. Mureithi, E. W., & Denier, J. P. (2010). Absolute-convective instability of mixed forced-free convection boundary layers. *Fluid Dynamic Research, 42*(5) 055506.

Index

© Springer Nature Singapore Pte Ltd. 2019
T. K. Sengupta and S. Bhaumik, *DNS of Wall-Bounded Turbulent Flows*,
https://doi.org/10.1007/978-981-13-0038-7

Printed in the United States
By Bookmasters